Progress in Motor Control

VOLUME ONE

Bernstein's Traditions in Movement Studies

Mark L. Latash, PhD
Pennsylvania State University

Editor

Human Kinetics

QP
301
.P767
1998
v.1

Library of Congress Cataloging-in-Publication Data

Progress in motor control / Mark L. Latash, editor.
     p.   cm.
     Includes bibliographical references and index.
     Contents: v. 1. Bernstein's traditions in movement studies.
     ISBN 0-88011-674-9
     1. Human locomotion.   2. Motor ability.   3. Bernshteîn, N. A.
(Nikolaî Aleksandrovich), 1896-1966.   I. Latash, Mark. L., 1953   .
QP301.P767   1998
612.7'6--DC21   97-44950
                                           CIP

ISBN: 0-88011-674-9

Permission notices for material reprinted in this book from other sources can be found on page vii.

Acquisitions Editors: Richard Frey, PhD, and Judy Patterson Wright, PhD; **Managing Editors:** Joanna Hatzopoulos and Alesha G. Thompson; **Assistant Editors:** Jennifer Miller and Erin Sprague; **Copyeditor:** Michelle Sandifer; **Proofreader:** Pam Johnson; **Graphic Designers:** Judy Henderson and Robert Reuther; **Graphic Artist:** Francine Hamerski; **Cover Designer:** Jack Davis; **Cover photo:** Courtesy of *Fizkultura i Sport*: **Printer:** Edwards Brothers

Printed in the United States of America   10 9 8 7 6 5 4 3 2 1

Human Kinetics
Web site: http://www.humankinetics.com/

United States: Human Kinetics, P.O. Box 5076, Champaign, IL 61825-5076
1-800-747-4457
e-mail: humank@hkusa.com

Canada: Human Kinetics, 475 Devonshire Road Unit 100, Windsor, ON N8Y 2L5
1-800-465-7301 (in Canada only)
e-mail: humank@hkcanada.com

Europe: Human Kinetics, P.O. Box IW14, Leeds LS16 6TR, United Kingdom
(44) 1132 781708
e-mail: humank@hkeurope.com

Australia: Human Kinetics, 57A Price Avenue, Lower Mitcham, South Australia 5062
(088) 277 1555
e-mail: humank@hkaustralia.com

New Zealand: Human Kinetics, P.O. Box 105-231, Auckland 1
(09) 523 3462
e-mail: humank@hknewz.com

Contents

Preface

Nikolai A. Bernstein, one of the premier minds of our century, was the first scientist to work in an area now defined as motor control. Research within this area is directed toward understanding the basic principles of controlling natural voluntary movements. This area, motor control, differs from traditional biomechanical, behavioral, or neurophysiological studies of movements by its clear emphasis on the processes of control. It may also be considered as a particular window providing insights into the basic principles of brain function. The study of motor control has both basic and applied aspects. In particular, understanding how humans control voluntary movements is important in the development of treatment and rehabilitation strategies for patients with movement disorders.

Motor control has already developed into a well-defined research area. Its most prominent, distinctive feature is the integration of information from many seemingly independent areas of movement studies. However, until recently, it lacked a forum (a journal, a book series, or a conference) where different aspects of movement studies were all represented and analyzed. In other words, a forum for a Bernsteinian integration of information from different fields did not exist.

Recently, the situation has started to improve. In particular, a new journal, *Motor Control,* has been launched by the Human Kinetics publishing house. Another step was an international conference entitled *Bernstein's Traditions in Motor Control,* which took place at Pennsylvania State University (University Park) in August 1996. The conference was dedicated to Bernstein's centenary. Conference speakers wrote fifteen chapters contained in this volume. Their names are well known to scholars and students working in movement science. These contributors are united by a common feature. They all acknowledge the outstanding contribution of Nikolai Bernstein to the field in general and to their scientific views in particular. Contributors from the former USSR knew Nikolai A. Bernstein personally and worked with him: Profs. Josef M. Feigenberg, Anatol G. Feldman, Victor S. Gurfinkel, and Lev P. Latash.

The chapter authors were asked to write specific reviews addressing urgent problems of motor control, rather than focusing on their personal recent research activities. Thus, the book has been designed as a reference volume that will also be useful to graduate students in movement studies.

Another major purpose of the present volume has been to start a tradition of integrating information relevant to motor control. Despite the relatively small number of chapters, the volume covers a spectrum of topics including biomechanics, motor behavior (kinesiology), motor disorders and rehabilitation, neurophysiology of movements, motor development, psychophysiology of movements, and models and theories in motor control. I hope that subsequent volumes using a similar format will appear every few years and summarize recent progress in motor control using a similar format, i.e., a relatively small number of chapters written by prominent scientists who summarize their understanding of the area and formulate problems in urgent need of investigation.

Publication of this book would have been impossible without the generous financial support of the Whitaker Foundation; Continuing and Distance Education, and the Department of Kinesiology of Pennsylvania State University; and Human Kinetics. I would also like to thank personally those who shared with me the load of organizing and running the conference as well as putting together this volume: Alexander Aruin, Karl Newell, David Rosenbaum, Takako Shiratori, and Vladimir Zatsiorsky.

Mark L. Latash

Credits

Figure 2.1 Reprinted from F.V. Severin, G.N. Orlovsky, and M.L. Shik. *Biophysics 12,* ©1967, 576, with kind permission from Elsevier Science Ltd., The Boulevard, Langford Lane, Kidlington OX5 1GB, UK.

Figure 2.2 Reprinted from M.C. Wetzel and D.G. Stuart. *Progress in Neurobiology 7,* ©1976, 26, with kind permission from Elsevier Science Ltd., The Boulevard, Langford Lane, Kidlington OX5 1GB, UK.

Figure 2.3 Reprinted from M.C. Wetzel and D.G. Stuart. *Progress in Neurobiology 7,* ©1976, 84, with kind permission from Elsevier Science Ltd., The Boulevard, Langford Lane, Kidlington OX5 1GB, UK.

Figure 2.4 Reprinted, by permission, from A. Prochazka, 1993, *IEEE transactions on rehabilitation engineering 1* (New York: IEEE), 7–17.

Figure 5.1 Reprinted, by permission, from the cover of N.A. Bernstein, 1935, *The Problem of Interrelation of Co-ordination and Localization* (Leipzig: J.A. Barth).

Figure 5.2 Reprinted, by permission, from N.A. Bernstein, 1988, *Bewegungsphysiologie,* 2d ed., (Leipzig: J.A. Barth), 1–272.

Figure 5.5 Reprinted, by permission, from A. Bethe, 1917, "Beiträge zum Problem der willkürlich beweglichen Armprothesen. III. Die Konstruktionsprinzipien willkürlich beweglicher Armprothesen," *Medizinische Wochenschrift* 64: 1625–1629.

Figure 5.6B Reprinted, by permission, from N.A. Bernstein, 1988, *Bewegungsphysiologie,* 2d ed., (Leipzig: J.A. Barth), 1–272.

Figure 5.8 Reprinted, by permission, from M. Wiesendanger, 1997, *Perspective of Motor Behavior and Its Neural Basis,* edited by M.C. Hepp-Reymond, G. Marini, (Basal, Switzerland: Karger), 103–134.

Figure 5.10 Reprinted, by permission, from R. Jung and R. Hassler, 1960, The extrapyramidal motor system. In *Handbook of physiology neurophysiology,* Volume II, edited by J. Field, H.W. Magoun, and V.W. Hall (Washington, D.C.: American Physiological Society), 904.

Figure 7.5, 7.6 Reprinted, by permission, from Cordo et al., 1995, *Journal of Neurophysiology,* Vol. 74 (Washington, D.C.: American Physiological Society).

Figure 7.7, 7.8 Reprinted, by permission, from Nature, 1996, "Stochastic resonance in human muscle spindle afferents," Macmillan Magazine 383: 269.

Figure 11.1a,b, 11.3a,b Reprinted, by permission, from N.G. Kim and M.T. Turvey, 1996, "Wayfinding and the Sampling of Optical Flow by Eye Movements," Journal of Experimental Psychology: Human Perception and Performance 22: 1314–1319.

Figure 12.2, 12.3 Adapted, by permission, from E. Thelen, D. Corbetta, K. Kamm, J. Spencer, K. Schneider, and R.F. Zernicke, 1993, "The transition to reaching: Matching intention and intrinsic dynamics," *Child development* 64: 1058–1098.

Figures 12.7, 12.8 Reprinted, by permission, from E. Thelen, 1995, Time scale dynamics and the development of embodied cognition. In *Mind as motion: Explorations in the dynamics of cognition,* edited by Port and van Gelder (Cambridge, MA: MIT Press), 69–99.

Figure 13.1, 13.2, 13.3 Adapted, by permission, from A.G. Feldman and M.L. Levin, 1995, "The Origin and Use of Positional Frames of Reference in Motor Control," *Behavioral and Brain Sciences* 18: 723–806.

Figure 15.2 Reprinted, by permission, from D.M. Schneider and R.A. Schmidt, 1995, "Units of Action in Motor Control: Role of Response Complexity and Target Speed," *Human Performance* 8: 2.

Figure 15.3, 15.4 Reprinted, by permission, from D.E. Young and R.A. Schmidt, 1990, Attention and Performance XIII,(NJ: Erlbaum), 763–795.

Figure 15.5, 15.6 Reprinted, by permission, from H. Heuer, R.A. Scmidt, and D. Ghodsian, 1995, "Generalized Motor Programs for Rapid Bimanual Tasks: A Two-Level Multiplicative-Rate Model," *Biological Cybernetics* 73: 343–356.

Figure 16.1 Reprinted, by permission, from M.F. Bobber and A.J. Van Soest, 1994, "Effects of Muscle Strengthening on Vertical Jump Height: a simulation study," *Medicine and Science in Sports and Exercise* 26: 1012–1020.

Figure 16.2 Reprinted from *Journal of Biomechanics* 26, "The influence of the biarticularity of the gastroenemius muscle on vertical-jumping achievement," 5-6, 1993, with kind permission from Elsevier Science Ltd, The Boulevard, Langford Lane, Kidlington OX5 1GB, UK.

Figure 16.3, 16.4 Reprinted, by permission, from M.F. Bobber and A.J. Van Soest, 1993, "The Contribution of muscle of properties in the control of explosive movement," *Biological Cybernetics* 69: 195–204.

Nikolai A. Bernstein
1896–1996

1

CHAPTER

The Scientific Legacy of Nikolai Bernstein

Victor S. Gurfinkel

Institute for Information Transmission Problems,
Russian Academy of Sciences, Moscow, Russia

Paul J. Cordo

Robert S. Dow Neurological Sciences Institute,
Portland, OR, U.S.A.

Nikolai Alexandrowitsch Bernstein was born in 1896. He dedicated most of his professional life to the study of how the brain controls motion. His scientific legacy is broad and comprehensive. It includes monographs (Bernstein 1926a, b, 1966, 1975), chapters in manuals and textbooks (Bernstein 1934a), papers, and instructive materials about different research methods. Much of his published work concerns different aspects of motor control—one of the most integrative branches of physiology—which includes the biomechanics of the skeletomotor system, muscular physiology, and neurophysiology. However, Bernstein's main achievement was that he conceived a new scope of studies—ahead of his time—whose meaning and great potential are still being discovered by modern research. Today, Bernstein's name is known to a great number of biomechanicians, specialists in motor control, and neuropsychologists.

Thirty years have passed since Bernstein's death, giving us the opportunity to look at his legacy through the impartial filter of time and progress. We are reminded not only of his scientific heritage but of his extraordinary personality as well. Bernstein was a man of science and intellectual discipline, distinguished by independence of opinion, personal dignity, firmness, and moral staunchness. He was highly dedicated to his work. He surrounded himself with other investigators whose scientific interests included problems of the physiology and pathology of movement: physiologists of labor and sport, neurologists, prosthetists, and specialists in musical pedagogy. During his life, Bernstein had great prestige, which gave him the opportunity to expand the areas of his experimental work. At different times, he created laboratories of motor control at the Labor Protection Institute, All-Union Institute of Sports, Institute of Musical Science, All-Union Institute of Experimental Medicine, Institute of Neurology (Academy of Medical Sciences), and Moscow Institute of Prosthetic Appliances.

His scientific contacts with foreign colleagues were rather limited. In 1924 and 1928, Bernstein went on business trips to Germany, France, and Czechoslovakia. Around the same time, he published several methodological papers in German (Bernstein 1928, 1930, 1933). However, like most Soviet physiologists, he did not publish in Western journals.

As with all original researchers, Bernstein had supporters as well as detractors, perhaps due to the precocious nature of his research. His views were often criticized, sometimes because they were not understood, but at other times because they did not agree with views dominating Soviet biology. Despite such criticism, he was recognized by leading Soviet physiologists, including A.A. Ukhtomsky, L.A. Orbely, and I.S. Beritashvily. In 1935, he was nominated Doctor of Sciences without submitting a thesis. He was also included in the initial membership of the Academy of Medical Sciences of U.S.S.R., created in 1946. In 1948, Bernstein was awarded the Stalin prize for science.

Bernstein did not belong to any scientific school and often described himself as a self-made man. The development of his scientific ideas was probably influenced by the Victorian physician and philosopher J. Hughlings Jackson. Jackson pointed out that most lesions of the nervous system produce both negative and positive symptoms. Not only does a loss of function occur, expressing destructive consequences, but positive symptoms also occur due to the release of lower centers from higher levels of control. Based on clinical observations, Jackson concluded that the brain had a hierarchical organization, and that its structure and integration were driven largely by evolutionary principles. He distinguished younger levels of the nervous system from more ancient levels. He graded movements on a scale ranging from most automatic to least automatic.

Bernstein was highly familiar with the physiological and neurological literature, which is clearly evident in his unpublished book *Current Research in the Physiology of Nervous Processes*—a critical review of the main problems of brain function (Bernstein 1936). Nevertheless, the development of his ideas was influenced primarily by his own experimental results.

Early Work—Manual Labor and Cyclogrammetry

Bernstein's first scientific work was carried out at the Central Institute of Labor, in Moscow, where he was invited in 1922 to study the biomechanics of human movements during manual labor. Here he joined and worked with a group of physiologists including K. Kektchejev, N. Tychonov, and A. Bruzhes. The ultimate purpose of their research was the optimization of human movement, tool placement, and workplace design. This topic of research was of value at the time because of a low level of automation and the predominance of manual labor. In his first experimental work, Bernstein studied the movements involved in cutting metal with a chisel. During this operation, the worker holds the chisel in one hand, orienting its cutting edge relative to a sheet of metal; using his other hand, the worker strikes the upper area of the chisel with a hammer. This movement requires bimanual coordination. One hand sets the position of the chisel, while the other moves rapidly and precisely to the same position, accelerating an inertial load.

While working in the Central Institute of Labor, Bernstein realized that movements could indicate processes occurring in the brain. However, adequate methods of registration and analysis were necessary. He then set out to improve the technique of movement registration and methods for the evaluation of experimental results. The cyclographic method permitted the registration of movements of part or all of the body, providing a picture of changes in any number of points on the body over space and time. Cyclography allowed researchers to pursue kinematic descriptions over very short time intervals by using shutter frequencies of 150–200 frames/s. More importantly, it allowed them to obtain accurate quantitative data from those pictures.

Under Bernstein's leadership, the cyclographic technique was refined. Cyclographic mirror methods were developed that allowed registration of movements. For cyclical movements whose trajectories could be superimposed upon each other, methods for cyclogram evaluation were further developed and nomograms were constructed, rendering the calculations simpler. The combination of these methodological and technical procedures made possible a microscopic analysis of movement, which was termed cyclogrammetry. A detailed presentation of this material is contained in the book *Techniques for the Investigation of Movement*, which appeared in 1934 and was edited by Bernstein (1934b). It is now believed that this first work greatly predetermined the methodology of Bernstein's later studies as well as the essence of his theoretical notions.

Movements as Morphological Objects

In 1928, Bernstein proposed the concept that the motor skills of a living organism could be regarded as morphological objects (1929). He wrote,

The basic vital properties that exist in the movements of living beings clearly confirm their close analogy to anatomical organs or tissues. Firstly, a live movement reacts and secondly it regularly evolves and involutes. I noted and described the former of these properties as early as 1924 [Bernstein 1924]. Studying the biodynamics of movements involved in cutting with a chisel I was able to show that it is impossible to alter selectively any one given detail in this movement without affecting others. If, for example, the trajectory of the elbow is slightly altered, the form of the trajectory of the hammer is also unavoidably changed; as are the relationships between the velocity of swing and the impact; between the velocities of the wrist and the hammer head; and a whole series of other nuances of the movement.

The analysis of dynamics during striking movements showed that inertial forces are the dominant factors the CNS must control to produce accurate movement. Probably the most important conclusions from the chisel-cutting study were that even highly complex movements are organized as integral units. Changes in one part of the motion lead to corresponding changes in the whole movement.

Using cyclogrammetry, Bernstein and his coworkers conducted a large number of investigations about movement kinematics. Since the movements of animals and humans are highly diverse, it is impossible to make judgments about general principles of motor control based on the study of a single movement, no matter how complex or detailed that study might be. This is why Bernstein and his colleagues studied exercises such as parallel bars and rings, discus throwing, walking, running, jumping, skiing, and swimming. These data were published in two books in 1935 and 1940 (Bernstein 1935a, 1940).

These conclusions were further strengthened by experiments about locomotion that Bernstein carried out in 1926. He had to obtain data for engineering designs of pedestrian bridges (Bernstein 1927). Locomotor movements attracted Bernstein because they are phylogenetically old, highly automated, and unusually stable. They display an extremely widespread synergy incorporating most of the body's muscles. The first task was to study the dynamics of walking. Bernstein examined the development of locomotion in children and its involution in the elderly. He compared locomotion of healthy people with patients who had different forms of brain pathology. From these preliminary studies, he concluded, "We regard the locomotor process as a living morphological object of inexhaustible complexity."

Bernstein's studies of locomotion have inspired an entire generation of work at the Institute for Information Transmission Problems in Moscow and at Moscow State University, which has made available a large body of information about the neurophysiology of locomotion. Researchers at those institutions have conducted hundreds of experiments using the mesencephalic cat on a treadmill while stimulating and recording from the brain and spinal cord with microelectrodes during locomotion (Orlovsky 1970; Severin et al. 1967; Shik et al. 1966a, b; Shik and Orlovsky 1976).

In subsequent years, Bernstein's locomotion studies revealed that the reactivity of movements can simultaneously be extremely selective and abstract. He observed that the nervous system reacts to changes in a single movement detail by changing a whole series of other details that are sometimes spatially and temporally remote from the causative agent. At the same time, elements that are closely adjacent to the first detail are left unchanged. He concluded that movements cannot be readily described as chains of elementary components but, rather, as structures that are differentiated into details. He viewed movements as structurally whole. They exhibit a distinct differentiation of elements within the movement while, at the same time, having different types of relationships among these elements. It was his view, therefore, that it is inappropriate to study movement coordination by constitutive reductionism—the reduction of the movement into its most basic constituents. Bernstein's account of changes in locomotion during development from infancy and of the involution of locomotion in senescence further supports this understanding.

This point of view is not restricted to the physiology of movement, as illustrated by the following statement of the physicist Philip Anderson,

> The ability to reduce everything to simple fundamental laws does not imply the ability to start from those laws and reconstruct the universe. . . . The behavior of large and complex aggregates of elementary particles, it turns out, is not to be understood in terms of a simple extrapolation of the properties of a few particles. Instead, at each level of complexity entirely new properties appear, and the understanding of the new behaviors requires research which I think is as fundamental in its nature as any other. (Anderson 1972)

Bernstein's strategy for studying motor control was described by the well-known English physiologist Charles Phillips in the XVII Sherrington Lecture *Movements of the Hand*. Phillips contrasted Bernstein's approach with that of Sir Charles Sherrington,

> Sherrington's strategy was to start from "most automatic" movements and work toward their cellular basis. A different strategy was adopted by N. Bernstein and his pupils in Moscow between and after the two World Wars. They started from skilled "least automatic" performances such as hammering a nail, using a file, and striking piano keys, and resolved their four dimensional structures by ingenious methods of the greatest accuracy and refinement. They were working toward complex neuronal networks rather than short chains of cells. (Phillips 1986, page 11)

Hierarchical Theory of Motor Coordination

By studying a diverse range of movements, Bernstein was able to formulate general concepts of the structure and function of the motor-control system. One of the more prominent of these is the hierarchical theory of motor coordination. It was presented in complete form in a book entitled *On the Construction of Movements,* published in 1947 in Russian (Bernstein 1947). This book remains to be translated into English. For this book, Bernstein was awarded the highest scientific prize of the U.S.S.R. in 1948.

According to Bernstein's theory of movement, the motor-control system is comprised of a number of structural/functional levels:

1. The level of *paleokinetic regulation*, also called the *rubrospinal* level

2. The level of *synergies*, also called the *thalamo-pallidar* level

3. The level of *spatial fields*, also called the *pyramido-striatal* level (subdivided into two sublevels: the striatal sublevel, belonging to the extrapyramidal system, and the pyramidal level)

4. The level of *actions* (actions with objects, semantic chains, and so forth), also the called *parietal-premotor* level

5. A group of *higher cortical* levels

Rubrospinal Level

The first three of these levels—the rubrospinal, thalamo-pallidar, and pyramido-striatal levels—are of particular interest. The rubrospinal level is very ancient and highly developed in animals that locomote without legs. This level controls mainly the neck and trunk muscles. The movements controlled by this level are smooth and quasi-static. The rubrospinal level supplies tone—termed background tension by Bernstein—to the skeletal muscles. It controls the excitability of spinal structures, and ensures reciprocal innervation and coactivation of antagonistic muscles. Actions at this level are completely involuntary. Afferent signals to this level include the direction and magnitude of muscle forces, otolith signals, and contact perception in the form of protopathic sensitivity. This afferent information is synthesized in a rather simple way to signal the position and orientation of the body in the gravity field as well as the length and tension of muscles.

Thalamo-Pallidar Level

The types of control exerted by the thalamo-pallidar level on movement include the organization of muscle synergies, the coordination of cyclical movements, and the execution of highly practiced, highly learned movements. Examples of movements controlled by the thalamo-pallidar level include facial expressions and expressive gestures of the body, as in pantomime. The effector system of this level is the pallidum. Afferent signals to this level include proprioception, deep sensation, and the epicritical skin sensations of touch, friction, pain, pricking, and temperature, all characterized by local signs. In general, these are the afferent signals that give rise to self-perception of the body.

Today, the concepts and mechanisms of self-perception bring to mind feedback. However, Bernstein developed his theories at a time before feedback theories were widespread. Nevertheless, he was acutely aware of the relationship between sensory signals, body perception, and movement, which must have seemed circular and almost paradoxical to his contemporaries.

Without a coordinate system upon which to interpret sensory signals, it is impossible to organize these signals into a coherent whole. Thus, afferent signals are combined and integrated into a perceptual frame of reference for the body. At the same time, this frame of reference serves as the starting point for sensory data acquisition and feedback control.

The representation of the body at the thalamo-pallidar level is therefore both a starting coordinate for sensory reception and the final target of perception, making movement possible. Merging and overlapping local signs are generalized according to the reference frame of the body, which is common to all signals. These diverse tactile and proprioceptive inputs combine to give exceptionally full and thorough information about the motor apparatus.

Pyramido-Striatal Level

The pyramido-striatal level includes the synthetic spatial field, which is the perception and possession of the space external to the body. This field is extensive and stretches out over large distances around the body. It is uniform and, importantly, immovable. Bernstein stressed important features of the spatial field such as metrics and geometry. These manifest in a subject observing geometrical forms or geometrical likenesses. The internal construct of external space at the pyramido-striatal level is filled by objects having shapes, sizes, and masses and by the forces of their interaction. This level is associated with visual fields 17 and 18, tactile-proprioceptive centers (postcentral convolution), partial auditory and vestibular cortical areas, and the cerebellar cortex.

Internal and External Representations

Modern literature contains descriptions of other versions of multilevel systems of motor control. However, the Bernstein scheme even now seems to be quite modern, and perhaps it remains the most complete. As seen from previous summaries, the thalamo-pallidar level contains the representation of the body. At the same time, this representation serves as the body reference frame, without which the analysis of sensory signals would be impossible. As for the pyramido-striatal representation of space, Bernstein wrote,

> There is considerable reason to suppose that in the higher motor centers of the brain (very probably in the cortical hemispheres) the localizational pattern is nothing other than some form of projection of external space in the form present for the subject in his motor field. This projection . . . must be congruent with external space, but only topologically, and not metrically. (Bernstein 1935b)

Bernstein clearly realized that purposeful movements are impossible without taking into account the current state of the body and its relation to external space.

Study of the internal representation of the body and the external space has only recently become a major subject of physiological experiments (Gurfinkel et al. 1995b).

Parallel Structure of Movement

Bernstein's name is usually associated with hierarchical theories of motor coordination. The most important characteristic of multilevel systems of motor control is not hierarchical subordination of different levels of control but a complex distribution of tasks. The necessity of task distribution is, on one hand, imposed by the anatomical organization of the brain. It contains evolutionarily different structures that have preserved, to some degree, the specificity of their functioning. On the other hand, the extremely complex structures of the skeletomotor system impose tremendous dimensionality. Another specificity of function within the motor-control system consists of dividing these levels into foreground and background levels, depending on the current motor task and on conditions of its realization.

In a 1939 publication (Bernstein 1939), Bernstein regarded the movement control system as one in which parallel work by numerous efferent channels takes place. Modern data corroborate these notions. In one recent study of parallel output processing by Marsden and colleagues (Jahanshahi et al. 1995), the electrical activity and local blood circulation in different brain areas were recorded during a very simple triggered movement—extension of the right index finger while the hand rested on a table. This study showed that in normal human subjects relative to the resting state, self-initiated finger movements activated the left primary sensorimotor cortex, bilateral supplementary motor areas, bilateral anterior cingulate and lateral premotor cortices, bilateral insular cortex, left thalamus, left putamen, bilateral parietal areas 40, and right dorsolateral prefrontal cortex. The activation of such a large number of diverse anatomical structures to execute such a simple movement suggests the presence of massive parallel processing.

Motor Rehabilitation

Bernstein is primarily known for work in which he developed the concept of motor control. For the most part, this aspect of his work is known to English-speaking readers from two books, *The Coordination and Regulation of Movements* (Bernstein 1967) and *Human Motor Action: Bernstein Reassessed* (Whiting 1984). However, Bernstein also made major contributions to the field of motor rehabilitation, which are little known to Russian and English readers alike.

Clinical observations have often served as stimulus for elaboration of different fields within physiology. The history of brain function studies provides many such examples. Disorders of the brain evoked by trauma, tumor growth, or vascular abnormality have provided extensive data for the formation of initial notions about localization of brain functions. At the same time, neurology is rich in observations that underscore the extraordinary adaptability of the brain to compensate for the disease or destruction of nervous tissue (Bernstein and Buravtzeva 1954). Particularly illustrative examples of plasticity in the motor-control system can be seen in amputees with and without prostheses. The literature contains descriptions of remarkable feats of adaptation, such as artists painting pictures by holding a brush in their teeth and armless women embroidering using their feet.

Over a long period of his career, Bernstein engaged in prosthetic design. He established the laboratory of Physiology and Pathology of Movements at the Moscow Research Institute of Prosthetic Appliances (see figure 1.1). Bernstein's interest in prosthetic problems was quite natural. He believed that prostheses should be systematically designed based on a scientific background—knowledge of the skeletomotor system and its exact mechanisms of function. This aspect of Bernstein's studies is little known. Even in the most complete bibliography of his works (Jansons

Figure 1.1 Photograph of N.A. Bernstein with the staff of his laboratory at the Moscow Research Institute of Prosthetic Appliances. Upper row, from left to right: Y. Yakobson, PhD—engineer, V. Zudina—technician, N. Smolyansky, PhD—Doz., G. Dosina—technician, and V. Gurfinkel—postgraduate. Lower row: T. Vinogradova—senior researcher, Professor N. Bernstein, and E. Izbushkina—technician who worked with N. Bernstein for 20 years. The picture was taken in 1947.

1992), his publications about prostheses are not presented. Bernstein's works in the field of prosthetics reveal another side to his talent—translating his theories into functional engineering designs.

The prosthesis is a mechanical device that attempts to compensate for lost functions. Unpredictable performance in prostheses, orthoses, and other prosthetic appliances most often arises from their interaction with the human body. Not surprisingly, therefore, one of the most complex problems in prosthetic design is to adapt the prosthesis to the human body optimally. This problem is solved differently in the case of arm or leg amputation.

Knowledge of biomechanics and movement physiology is required at different stages of prosthetic management. For instance, this knowledge is needed while determining the prerequisites and functional demands when constructing prostheses. It is also necessary for the objective evaluation of the prosthetic appliances' quality and for the development of training methods.

A number of Bernstein's published reports (Bernstein 1948, 1992) are devoted to lower limb prostheses for patients with above-knee amputation. The most important requirements for prostheses of this kind are stability and mobility. According to Bernstein, the leg works in two different modes in the stance and swing of locomotion. During the stance phase, the leg absorbs longitudinal compression of the body's weight. During the swing phase, it produces a pendulum-like distention. In normal gait by the beginning of the stance phase, the leg is fully extended at the knee joint, touching the supporting surface only with the heel. In the course of loading the leg, plantar flexion in the ankle joint and flexion in the knee takes place, placing the foot on the ground. Subsequently, the leg straightens under the load. By the middle of the stance phase, it is straightened in the knee joint. In the second half of the stance phase, the leg flexes again at the knee joint under load. This flexion lasts until the end of the stance phase and continues into the swing phase. In this way during the period in which the leg is loaded, it provides support for the body, as well as mobility. To compensate for amputation, above-knee prostheses must provide reliable support for the body and, at the same time, retain mobility when loaded.

In the leg of a healthy person, the contradictory demands of providing reliable support while retaining mobility are met through muscle activity. To meet these demands with a prosthetic leg, Bernstein conceived a principle, which he called the stability hole (Bernstein 1992), as illustrated in figure 1.2. He proposed the construction of a knee junction utilizing this principle. Figure 1.3 (Bernstein 1992) illustrates a draft of polycentric articulations incorporating the principle of the stability hole.

Another example of Bernstein's work in prosthetic applications is the arm prosthesis. The arm prosthesis is intended to assist a disabled person in the restoration of motor function, particularly to restore some of the functions needed for self-care. One of the most important is the ability to eat independently. Bernstein designed an arm prosthesis that would enable an amputee to eat with a spoon (Bernstein and Salzgever 1948). The design of this arm prosthesis serves as a good example of Bernstein's style of work.

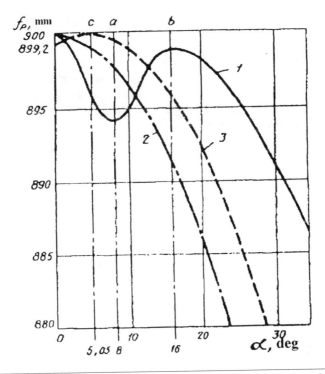

Figure 1.2   The curves of the length changes of the f_p segment (the distance from the hip articulation to the center of ankle hinge as a function of knee joint angle). Curve 1 shows a stability hole; curves 2 and 3 correspond to a usual prosthetic knee joint.

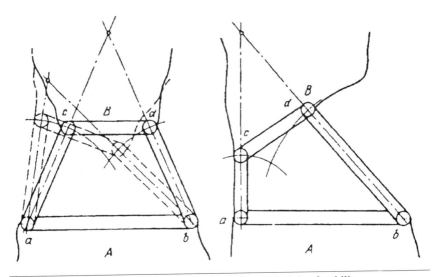

Figure 1.3   Drafts of polycentric articulations with a principle of stability.

Using his mirror-recording method, Bernstein recorded three-dimensional arm movements while a subject with an intact arm ate with a spoon. Figure 1.4 illustrates the changes of arm joint angles over time. The movement of the spoon to the mouth requires a complex synergy that includes simultaneous movements in all joints of the arm. During this movement, elbow flexion is accompanied by pronation-supination movement of the lower arm. After analyzing these synergies, Bernstein proposed a prosthesis in which pronation and supination movements of the artificial wrist were combined with elbow flexion (see figure 1.5). This prosthesis was designed and constructed in the Moscow Research Institute of Prosthetic Appliances.

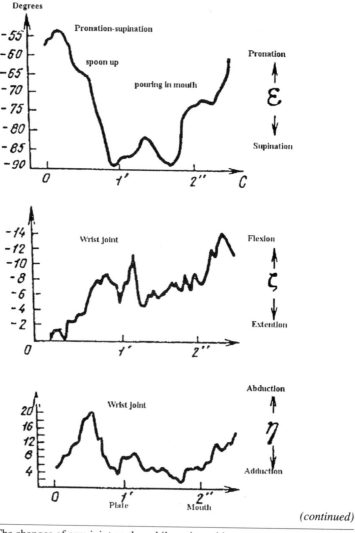

(continued)

Figure 1.4   The changes of arm joint angles while eating with a spoon.

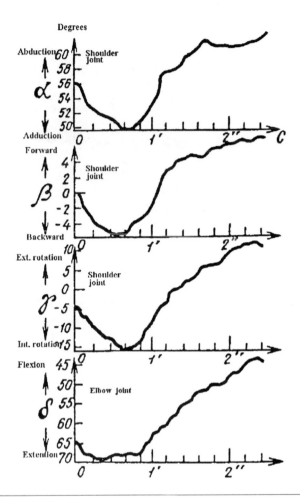

Figure 1.4   *(continued)*

One other example of Bernstein's work in the field of prosthetics (see figure 1.6) was a motorized leg prosthesis, which he developed in 1947 (Bernstein 1948). Pneumatic actuators in the hip and knee joints drew power from an external power source, which was worn on a belt about the waist. The actuators were controlled by a fixed program stored in a controlling device, which was also attached to the belt. Even now, the design of powered lower limb prostheses presents difficult technical problems. These problems occur not only because the prosthesis is difficult to control but also because the actuators and the power source must be compact, light, and powerful. Researchers who trained under Bernstein continued his work into the 1950s and developed an EMG-controlled arm prosthesis, which was particularly useful for bilateral amputees (Gurfinkel et al. 1972).

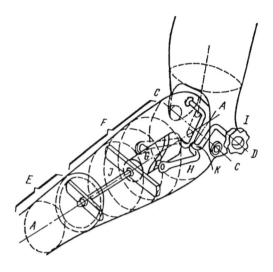

Figure 1.5   The draft of an arm prosthesis that realizes a synergy of elbow flexion-extension with wrist pronation-supination. (AA) longitudinal forearm axis, (CC) elbow joint axis, and (E) lower part of forearm is connected by rod J to pronation-supination mechanism G and H. This mechanism is connected to the upper arm by hinge K.

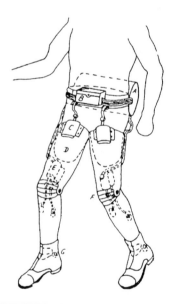

Figure 1.6   The draft of a bilateral, motorized prosthesis for high-level, above-knee amputation: (A) a bag with a motor and a tank for compressed gas, (B) controller, (C) hip actuator, (D) stump, (E) knee joint actuator, (F) cosmetic covers for knee joints, and (G) ankle joints.

Tonus

The notion of motor programs is attributed to Bernstein (1935b). According to Jeannerod (1988), the motor program concept greatly influenced the development of ideas about central programming. In the paper *The Problem of Interrelation Between Coordination and Localization*, Bernstein (1935b) wrote, "At the moment when movement begins there is already in existence in the central nervous system a whole collection of engrams which are necessary for the movement to be carried on to its conclusion" (see also Gurfinkel et al. 1971).

The motor program is realized through several parallel channels acting simultaneously. Those channels might be used to transfer commands of different types. Some commands implement the main part of the motor program and are addressed directly to specific muscles. Other channels carry tuning commands and also participate in the coordination of movements. Bernstein wrote, "The role of coordination at this level must therefore consist in the preparatory organization of the motor periphery. . . . Coordination . . . organizes and prepares the periphery for reception of the right impulse at the right moment" (Bernstein 1940). The term he used for tuning was tonus. Bernstein emphasized that the role of tonus extended considerably beyond the originally proposed scope, which incorporated only the idea of a condition of elasticity of the muscle fibers. In the vocabulary of physiologists, tonus gradually began to be applied to a wide range of phenomena, beginning with decerebrate rigidity and extending to Magnus and de Klein's tonus—tensing of the musculature of the neck and body prior to movements.

In part due to the early work of Bernstein on motor preparation, tonus is now viewed as an ongoing physiological adaptation and organization of the periphery that is not a condition of elasticity but of readiness. Tonus is not merely a condition of the muscles. It is a condition of the entire neuromuscular apparatus, minimally including the final spinal synapse and the final common pathway. Therefore, tonus is related to coordination as a state is related to an action or as a precondition to an effect (Bernstein 1940). The relation between tuning and motor commands is perhaps best illustrated by the example, given by Bernstein, of the violinist compressing a string with his finger. That compression does not create sound by itself but specifies which sound will be elicited by the next stroke of the bow.

Bernstein's ideas about tonus inspired many experimental studies of the pretuning of movement. Even in a relatively simple motion such as a voluntary knee extension triggered by an auditory click, EMG activity was subsequently shown to be preceded by an enhancement in the excitability of the spinal cord 60 ms earlier. These changes can be revealed by monosynaptic testing using the T- or H-reflex (Alexeief and Naidel 1972; Gurfinkel et al. 1971; Ioffe 1973; Kots 1977), or by transcranial electrical stimulation (Starr et al. 1988). Pretuning effects take a more complex form in motor tasks where postural mechanisms are involved. For example, the beginning of a phasic arm motion is preceded by state changes in postural muscles of the spine, neck, and legs (Belenkii et al. 1967).

Recently, new data have been published demonstrating that, in healthy persons under a level of tonus, tuning modifications like those described by Magnus and de Klein may be observed (Gurfinkel et al. 1995a). Bernstein's idea that two concurrent control processes exist—direct commands, which drive movements, and indirect commands, which provide proper tuning in the effector system—seems credible. However, this hypothesis has not yet been assessed by modern researchers.

Legacy of Bernstein

Leibnitz emphasized that the way important results were achieved may be as informative as the results by themselves. Table 1.1 provides a chronology of Bernstein's main publications. Bernstein's road to the understanding of motor-control problems starts at biomechanics, techniques of motion study, and gathering of sufficient experimental data. It is followed by data analysis, physiological interpretation, and generalization.

Bernstein's extensive background in general biology, medicine, and mathematics gave clarity to his experimental studies and general theoretical ideas. His close friend, the well-known psychologist A.R. Luria, wrote in his foreword to the English

Table 1.1   Chronological List of Important Publications of N.A. Bernstein

1975	Bewegungsphysiologie
1967	The Co-ordination and Regulation of Movements
1966	Current Problems in the Theoretical Physiology of Activity
1966	Physiology of Movement and Physiology of Activity
1966	Some Emergent Problems of the Regulation of Motor Acts
1947	On the Construction of Movements
1940	Studies of the Biodynamics of Walking, Running and Jumping
1939	Modern data on the structure of the neuro-motor process. In: To the Musician-Pedagogue
1936	Current Research in the Physiology of Nervous Process
1935	Biodynamics of Locomotion
1934	Techniques for the Investigation of Movement
1933	Ein Zeit-Okular zu der Zeitlupe
1930	Analyse der Korperbewegungen und stellungen in Raum mittels Spiegel-Stereoaufnahmen
1928	Die kymocyclographische Methode der Bewegungsuntersuchungen
1926	General Biomechanics
1926	Biomechanics for Instructors
1924	Studies of Biomechanics of the Stroke by Means of Photo-registration

edition of Bernstein's 1967 book that "Bernstein was a man who had the remarkable gift to penetrate into the future." Bernstein's work is distinguished by a remarkably deep scientific grasp of the world, defining an era in the physiology of motion.

In Russia, there is a saying—half serious and half joking—that the prominence of a scientist can be measured by the number of years over which his own ideas suppressed new progress in his field. This principle does not apply to Bernstein since his work stimulated rather than impeded research. In a preface to the book *On the Construction of Movements* (Bernstein 1947), Bernstein wrote, "This book is possibly more a program of impending vital research than a dogmatic report about ultimately stated results." Thus, it is more relevant when speaking about the scientific heritage of Bernstein to consider what influence his ideas had on subsequent progress in motor-control studies. It would be hard to do this in an exhaustive manner under any conditions, and it would be impossible to do within the limits of this paper. For additional descriptions of Bernstein's influence on current motor-control research, the reader is referred to the introductory article "Bernstein's Significance Today" by C. Boylls and P. Green in the book *Human Motor Actions: Bernstein Reassessed* (Boylls et al. 1984) and a series of other publications (Fitch et al. 1982; Tuller et al. 1982; Whiting 1984).

Acknowledgments

The authors gratefully acknowledge assistance from Scott Buckley in preparing this manuscript.

References

Alexeief, M.A., Naidel, A.V. (1972) The mechanisms of interrelationship between human muscle activity in complex motor tasks. *Physiol. Zh. U.S.S.R.*, 58: 1721–1730.

Anderson, P.W. (1972) More is different. *Science*, 177: 393–396.

Belenkii, V.E., Gurfinkel, V.S., Palítsev, E.I. (1967) Elements of control of voluntary movements. *Biophysics*, 12: 154–161.

Bernstein, N.A. (1924) Studies of the biomechanics of hitting by means of optical recording, *Res. Central Institute of Labour, Moscow*, N1: 19–79 (In Russian).

Bernstein, N.A. (1926a) *Biomechanics for Instructors*. New Moscow Publ. Moscow: meditsina (in Russian).

Bernstein, N.A. (1926b) *General Biomechanics. Basic Theory on Human Movements* Moscow: All-Union Central Council of Trade Unions (in Russian).

Bernstein, N.A. (1927) Studies on the biodynamics of walking and running. Question of the dynamics of bridges. In *Research of Scientific and Engineering Committee of the People's Commissariat of Transport*. Vol. 63. Moscow: Scientific and Engineering Committee of the People's Commissariat of Transport, 51–76 (In Russian).

Bernstein, N.A. (1928) Die Kymozyclographische Methode der Bewegungsuntersuchungen. Handbuch der biologischen Arbeitsmethoden. Ed. E. Abderhalden, 5: 629–680.

Bernstein, N.A. (1929) Clinical ways of modern biomechanics. In *Collection of papers of the Institute for Medical Improvement*. Vol. 1. Kazan,U.S.S.R.: Glavnauka, 249–270 (In Russian).

Bernstein, N.A. (1930) Untersuchung der Korperbewegungen und Stellungen in Raum mittels Spiegel-Stereoaufnahmen. *Arbeitsphysiologie*, 3: 179–206.

Bernstein, N.A. (1934a) Physiology of movement. In *Fundamentals of Physiology of Work*. G.P. Conrady, A.D. Slonim, V.S. Farfel. (Eds.), Moscow-Leningrad: Biomedgiz, 366–450 (In Russian).

Bernstein, N.A. (Ed.). (1934b) *Techniques for the Investigation of Movement. Practical Handbook on Cyclogrammetry.* Moscow: Standartizatsia (In Russian).

Bernstein, N.A. (1935a) *Studies of the Biodynamics of Locomotion (Normal Gait, Load and Fatigue).* Moscow: Institute of Experimental Medicine, Moscow, (In Russian).

Bernstein, N.A. (1935b) The problem of interrelation between coordination and localization. *Arch. Biol. Sci., Moscow-Leningrad*, 38: 1–34 (In Russian).

Bernstein, N.A. (1936) *Current Research in the Physiology of Nervous Processes.* (Unpublished).

Bernstein, N.A. (1939) Modern data on the structure of the neuro-motor process. In *To the Musician-Pedagogue.* Moscow: Muzgiz, 207–239.

Bernstein, N.A. (1940) *Studies of the Biodynamics of Walking, Running and Jumping.* Moscow: Physical Culture and Sport.

Bernstein, N.A. (1947) *On the Construction of Movements.* Moscow: Medgiz (In Russian).

Bernstein, N.A. (1948) An account of the theory of the construction of prostheses for lower limbs. *Proc. Moscow Res. Inst. Prosthetics*, 1: 5–12 (In Russian).

Bernstein, N.A. (1966) *Physiology of Movement and Physiology of Activity.* Moscow: Meditsina (in Russian).

Bernstein, N.A. (1967) *The Coordination and Regulation of Movements.* Elmsford, NY: Pergamon Press.

Bernstein, N.A. (1975) *Bewegungsphysiologie (Movement Physiology).* Leipzig, Germany: Barth.

Bernstein, N.A. (1992) *Biomechanics and Prosthetics.* No. 1. Moscow: Central Institute for Prosthetic Appliance (In Russian).

Bernstein, N.A., Buravtzeva, G.R. (1954) Coordination disturbances and restitution of the biodynamics of gait after brain damage. *Abstr., 7th Session of the Inst. Neurology,* Moscow: Academy of Medical Sciences, 28–30 (In Russian).

Bernstein, N.A., Dementíev, E. (1933) Ein Zeit-Okular zu der Zeitlupe. *Arbeitsphysiologie*, 6: 4.

Bernstein, N.A., Salzgeber, O.A. (1948) Analysis of same spatial movements of an arm and proposals on construction of a working prosthesis. *Proc. Moscow Res. Inst. Prosthetics*, 1: 12–24 (In Russian).

Boylls, C.C., Green, P.H. (1984) Introduction: Bernstein's significance today. In *Human Motor Actions: Bernstein Reassessed.* H.T.A. Whiting (Ed.),. Amsterdam: North-Holland, xix–xxxv.

Fitch, H.L., Tuller, B., Turvey M.T. (1982) The Bernstein perspective: Tuning of coordinative structures with special reference to perception. In *Human Motor Behavior: An Introduction.* J.A.S. Kelso (Ed.), Hillsdale, NJ: Lawrence Erlbaum Associates, 271–281.

Gurfinkel, V.S., Ivanenko, Y.P., Levik, Y.S. (1995a) The influence of head rotation on human upright posture during balanced bilateral vibration. *Neuroreport*, 7: 137–140.

Gurfinkel, V.S., Ivanenko, Y.P., Levik, Y.S., Babakova, I.A. (1995b) Kinesthetic reference for human orthograde posture. *Neuroscience*, 68: 229–243.

Gurfinkel, V.S., Kots, Y.M., Krinskiy, V.I., Tsetlin, M.L., Shik, M.L. (1971) Concerning tuning before movement. In *Models of the Structural-Functional Organization of Certain Biological Systems.* Gelfand, I.M., Gurfinkel, V.S., Fomin, S.V., Tsetlin, M.L. (Eds.), Cambridge, MA: MIT Press, 361–372 (Translated by C.R. Beard).

Gurfinkel, V.S., Malkin V.B., Tsetlin, M.L., Schneider, A.Y. (1972) *Bioelectrical Control.* Moscow: Nauka (In Russian).

Ioffe, M.E. (1973) Supraspinal adjustment of the segmental apparatus prior to the performance of an instrumental movement in dogs. *Zh. Vyssh. Nerv. Deyat. im. I.P.Pavlova.* 23: 488–495.

Jahanshahi, M., Jenkins, I.H., Brown, R.H., Marsden, C.D., Passingham, R.E., Brooks, D.J. (1995) Self-initiated versus externally triggered movements. I. An investigation using measurement of regional cerebral blood flow with PET and movement-related potentials in normal and Parkinson's disease subjects. *Brain*, 118: 913–933.

Jansons, H. (1992) Bernstein: The microscopy of movement. In *Biolocomotion: A Century of Research Using Moving Pictures.* F.Cappozzo, M.Marchetti, V.Tosi (Eds.), Rome: Promograph, 137–174.

Jeannerod, M. (1988) *The Neural and Behavioural Organization of Goal-Directed Movement.* Oxford: Claredon Press.

Kots, Y.M. (1977) *The Organization of Voluntary Movement. Neurophysiological Mechanisms.* New York and London: Plenum Press.

Orlovsky, G.N. (1970) Connections of the reticulo-spinal locomotor sections of the brain stem. *Biophysics*, 15: 178–186.

Phillips, C.G. (1986) *Movements of the Hand. The Sherrington Lectures XVII.* Liverpool, England: Liverpool University Press.

Severin, F.V., Shik, M.L., Orlovsky, G.N. (1967) Work of the muscles and single motor neurones during controlled locomotion. *Biophysics*, 12: 762–772.

Shik, M.L., Orlovsky, G.N. (1976) Neurophisiology of locomotor automatism. *Physiological Reviews*, 56: 465–501.

Shik, M.L., Orlovsky, G.N., Severin, F.V. (1966) Organization of the locomotor synergism. *Biophysics*, 11: 1011–1019.

Shik, M.L., Severin, F.V., Orlovsky, G.N. (1966) Control of walking and running by means of electrical stimulation of the mid-brain. *Biophysics*, 11: 756–765.

Starr, A., Caramia, M., Zarola, F., Rossini, P.M. (1988) Enhancement of motor cortical excitability in humans by non-invasive electrical stimulation appears prior to voluntary movement. *EEG Clin. Neurophysiol.*, 70: 26–32.

Tuller, B., Turvey, M.T., Fitch, H.L. (1982) The Bernstein perspective: II. The concept of muscle linkage or coordinative structure. In *Human Motor Behavior: An Introduction.* J.A.S. Kelso (Ed.), Hillsdale, NJ: Lawrence Erlbaum Associates, 253–270.

Turvey, M.T., Fitch, H.L., Tuller, B. (1982) The Bernstein perspective: I. The problems of degrees of freedom and contex-conditioned variability. In *Human Motor Behavior: An Introduction.* J.A.S. Kelso (Ed.), Hillsdale, NJ: Lawrence Erlbaum Associates, 239–252.

Whiting, H.T.A. (Ed.). (1984) *Human Motor Action: Bernstein Reassessed,* Amsterdam: North-Holland.

2

CHAPTER

Reflections on a Bernsteinian Approach to Systems Neuroscience: The Controlled Locomotion of High-Decerebrate Cats

Douglas G. Stuart and Jennifer C. McDonagh

Department of Physiology, University of Arizona
College of Medicine, Tucson, AZ, U.S.A.

In a volume dedicated to the scientific contributions of Nikolai Bernstein, it is fitting to reflect (albeit idiosyncratically) on how one of his theoretical contributions to systems neuroscience was tested experimentally. A talented, innovative group of his countrymen performed this testing in the institute Bernstein had developed so painstakingly over the years and from which he was so unceremoniously deposed in the late 1940s.

In his 1947 monograph, *On the Construction of Movements,* Bernstein proposed that, "The higher sections of the nervous system determine the chains of motor activity, the lower level ties movements to spatial coordinates. Still lower levels solve the motor problem as such by organizing the necessary interaction of elements (muscles, joints, limbs) and by operatively controlling their work." This quotation is part of the introduction to a report by Shik et al. (1966b). Two companion articles (Shik et al. 1966a, 1967) and several further reports (Severin et al. 1967a, b;

Shik et al. 1968; Orlovsky 1969a, b; Budakova and Shik 1970; Kulagin and Shik 1970; Severin 1970) quickly followed. This body of work supported Bernstein's proposition in its *de novo* demonstration and quantification of the treadmill locomotion of high-decerebrate cats (non-goal-directed and in a straight line). These cats were activated (controlled) to locomote by the separate or combined stimulation (or both) of three specific sites in the brain stem that converged on a common reticulospinal pathway to spinal circuits. A schematic representation of the general experimental arrangement (Severin et al. 1967a) is shown in figure 2.1.

This chapter argues that the initial 1966–1970 sequence of reports on controlled locomotion and a follow-up 1970–74 sequence (Orlovsky 1970a–c, 1972a–e; Orlovsky and Feldman 1972a, b; Arshavsky et al. 1972 a–e, 1973, 1974; Orlovsky and Pavlova 1972a–c) provided new openings in three areas. These were the testing of precise hypotheses about the neural control of movement, studies about unitary neuronal activity during movement, and exploration of the interface between cellular neurophysiology and biomechanics. Interestingly, these articles were also a major force in the development of an interphyletic awareness (Stuart 1985) that was barely apparent in the early 1960s but firmly established by the late 1970s. This involved the progressive merger of concepts about motor-control neuroscience, derived from work on both vertebrates and invertebrates. It revealed an evolutionary conservation of motor-control mechanisms that extend from the molecular/cellular level of analysis to the behavioral level of investigation (Pearson 1993; Stuart and Callister 1993).

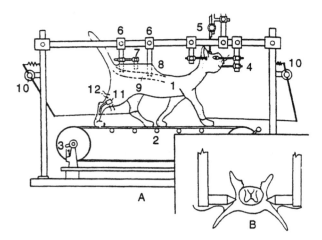

Figure 2.1 Experimental arrangement for inducing controlled locomotion in the high-decerebrate cat: *(A)* general view, *(B)* fixing of vertebra. Components are as follows: (1) high-decerebrate cat (plane of section *B* in figure 2.2), (2) treadmill, (3) belt tachometer, (4) sterotactic head holder, (5) electrode holder, (6) clamps to fix spine, (7) electrodes for recording unitary activity, (8) skin flap forming oil bath, (9) spinal cord, (10) detectors of longitudinal displacement of limbs, (11) joint angle detectors, (12) implanted electrodes in muscle.

Reprinted from Severin, Orlovsky, and Shik 1967.

This chapter first considers precedents to the Moscow sequence on controlled locomotion and then summarizes the key findings themselves. Next, it includes comments about the rapid dissemination and impact of the findings in Western countries. Finally, it relates them to subsequent advances in the understanding of the neural control of locomotion.

Precedents About the Use of Brain Stem Preparations to Study Locomotion

In their first report about controlled locomotion, Shik et al. (1966b) acknowledged previous work, part of which is summarized in figure 2.2.

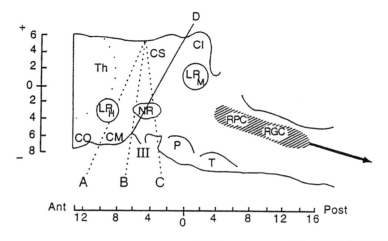

Figure 2.2   Neuroanatomical aspects of descending command signals for locomotion. Shown are brain stem structures and section levels important for determining the nature of spontaneous or controlled locomotion. Horsley-Clarke coordinate scales appear below and to the left of the drawing. The most rostral level of section is at Ant(erior) 12 to show a thalamic cat. Successively more caudal planes of sections are shown as dashed lines A, B, and C and as solid line D (the classical decerebrate preparation). The brain stem circle shows the dorsal portion of the effective locomotor region in the midbrain from which controlled locomotion can be induced (LR_M), while the diencephalic circle shows the locomotor region of the hypothalamus (LR_H). The hatched region shows the approximate position of neurons of the medial reticular formation of the pons and medulla that project to the spinal cord. Presumably, they are of major importance for switching on and sustaining the spinal stepping circuitry. Additional reticular substance, which is essential for controlled locomotion, should be visualized as extending even more rostrally, to at least the B plane. The labeled structures include: (CI) inferior colliculus, (CM) mammillary body, (CO) optic chiasm, (CS) superior colliculus, (NR) red nucleus, (P) pons, (RGC) giant cell reticular nucleus, (RPC) caudal reticular nucleus of pons, (T) trapezoid body, (Th) thalamus, and (III) oculomotor nerve.
Reprinted from Wetzel and Stuart 1976.

Previous, primarily Western work gave three results. First, spontaneous locomotion could occur in precollicular or premammillary (Wetzel and Stuart 1976)
cats (Hinsey et al. 1930), as shown in cut **A** in figure 2.2. Second, locomotor
movements could be elicited in intact cats by electrical stimulation of the subthalamus (Waller 1940; Grossman 1958), albeit somewhat imprecisely. Third, while
locomotion in premammillary cats was obviated by removal of the subthalamus
(Hinsey et al. 1930), this structure was not essential for locomotion to occur in
intact cats (Haertig and Masserman 1940). Shik et al. (1966a,b) contributed to this
area by exploiting the use of a cat preparation with a transection from the rostral
margin of the superior colliculus dorsally to the posterior margin of mammillary
bodies ventrally (the **B** cut in figure 2.2). This postmammillary preparation had
been previously shown to be incapable of spontaneous locomotion or that evokable
by exteroceptive stimuli (Hinsey et al. 1930). In the hands of Shik et al. (1966a,b),
however, this preparation was shown to locomote on a treadmill by sustained
electrical stimulation of a specific locomotion region of the midbrain. It would not
locomote if the brain stem transection was slightly more caudal, i.e., from the same
rostral margin of the superior colliculus to the posterior margin of the exit of cranial
nerve III (the **C** cut in figure 2.2). Subsequently, the Moscow group showed that
relatively identical controlled treadmill locomotion was evokable in postmammillary
preparations by similar ipsilateral stimulation of the pyramidal tract at the pontine
level (**P** in figure 2.2), with the pyramidal tract sectioned immediately below, at the
bulbar level (Shik et al. 1968). The Moscow group also obtained this result in
thalamic cats by similar ipsilateral stimulation of a locomotion region of the posterior hypothalamus during periods in which this latter preparation was not
locomoting spontaneously (Orlovsky, 1969b). The extent of the preserved brain
stem is shown in figure 2.2.

In these seminal Moscow studies (Shik et al. 1966a,b; Shik et al. 1968; Orlovsky
1969b), the locomotion analyzed included gait transitions from walking to trotting
and galloping, with due reference to the earlier work of Muybridge (1957) in this
field. In addition, the Moscow group had confidence in the relative normalcy of the
controlled treadmill locomotion under study by virtue of its similarity to the treadmill locomotion of unrestrained dogs that they had analyzed and described in previous reports (Arshavsky et al. 1965; Orlovsky and Shik 1965; Shik and Orlovsky
1965; Orlovsky et al. 1966a, b). These articles, too, supported Bernstein's proposition but not as directly as their subsequent work on high-decerebrate cats.

Clearly, a key advancement made by the Moscow group was their introduction
of a means of controlling the appearance, intensity, and pattern of locomotion in a
brain stem preparation. They immediately exploited this finding by demonstrating
that the neural control of locomotion is essentially tripartite. It involves continuous
interactions between descending command signals, active rhythm-generating spinal circuits, and sensory feedback.

Precedents Concerning the Large Spinal Contribution to the Control of Locomotion

In their second report about controlled locomotion, Shik et al. (1966a) concluded that the function of the descending command signals from the brain stem was to prime spinal circuits so they could elaborate the stepping movements of a single limb and contribute (presumably by interactions with the brain stem) to interlimb coordination. For this argument, the Moscow group recognized three precedents. First was their own immediately preceding work on intact dogs *(vide supra)*. Second, they included mathematical theories of the early 1960s, which were elaborated in their own institute (for summaries in English see Gelfand and Tsetlin 1971; Gelfand et al. 1971). Third, the Moscow group recognized the arguments put forward in the West concerning spinal locomotion in both nonmammalian (von Holst 1938; Gray 1950) and mammalian (Graham Brown 1914, 1935; Russian translation of Creed et al. 1932) vertebrates. Clearly, the Moscow group had a thorough grasp and appreciation of the Western literature about spinal locomotion. Wetzel and Stuart (1976) contains a detailed review on this issue.

In retrospect, it would have been valuable for Shik et al. (1966a) to have emphasized more strongly the remarkable early twentieth-century contributions of Graham Brown (see below) and to comment more fully concerning ganglionic and spinal pattern-generating circuits. By the mid-1960s, these included mechanisms of repetitive discharge by single neurons (Arvanitaki 1936), spontaneous neuronal activity in isolated earthworm (von Holst 1932, 1933) and crayfish (Prosser 1934a, b) nerve cords, limb movements in deafferented frogs (Weiss 1941), the concept of the command neuron (Wiersma 1947), the role of ganglionic pattern-generating circuits in locust flight (Wilson 1961) and crayfish swimming (Hughes and Wiersma 1960), and operation of similar spinal cord circuits in salamander walking (Székely 1965). In defense of the Moscow group, however, such a synthesis was not the style of the journal *(Biofizika)* in which their results were published, and a full synthesis was readily forthcoming from the authors subsequently (see below). Furthermore, political restraints, which were beginning to abate in the mid-1960s (Gelfand et al. 1971), may have discouraged too full an exposition on the merging of the new Moscow findings with Western results and thought (Feigenberg and Latash 1996). It should also be emphasized that no Western synthesis of this work was fully realized until the mid-1970s. As such, the more significant and realistic comment might be that in Shik et al. (1966a) there appears one of the first impactful, albeit in passing, references to the work of Graham Brown since the 1935 publication of Creed et al. (1932). Interestingly, Creed et al. (1932) included passing and somewhat obscure comments about the worthwhileness of Graham Brown's far earlier findings (see below).

Precedents to the Recording of Unitary Neuronal Activity During Locomotion

The third and fourth papers in the Moscow group's controlled locomotion sequence (Severin et al. 1967a, b) provided unitary extraaxonal recordings from muscle receptor (afferent) axons and motoneuron (efferent) axons, respectively. Selected key precedents to the results were duly noted (Björk and Kugelberg 1953; Eklund et al. 1964; Davey and Taylor 1966). The early readership considered the controlled locomotion approach to be quite novel for the 1960s and, indeed, a tour de force for recordings made during relatively unrestrained limb movements (Stuart et al. 1979).

Motoneuron Discharge

The strategy of relating the unitary discharge of motoneurons to the various phases of an ongoing movement was initiated earlier, in the 1920s. It advanced more rapidly (early 1950s) for respiratory movements than for movements of the eyes (Björk and Kugelberg 1953) and other body parts, the latter occurring in or after the 1960s. For selected reviews about respiratory movement motoneuronal recordings, see Adrian and Bronk (1928), Pitts (1943), Eccles et al. (1962), Sears (1962, 1963, 1964b), and Gill and Kuno (1963). Similarly, issues concerning the relative extent of coactivation of alpha versus gamma motoneurons were addressed first for respiratory movements (Eklund et al. 1963; Sears 1963, 1964a) and, a little later, were inferred for jaw-closing movements (Davey and Taylor 1966; Taylor and Davey 1968). Given the technical difficulties associated with such recordings during limb movement, it was a notable advance shortly thereafter for Severin et al. (1967b) to report about recordings from single, alpha motor axons during controlled walking, including an example of the discharge of a provisional gamma motoneuron.

Muscle Receptor Discharge

The relating of unitary muscle receptor discharge to movement was also first achieved for respiratory muscle afferents (Critchlow and von Euler 1963; Sears 1962). Again, this advance was next exploited for jaw-closing movements (Davey and Taylor 1966). Severin et al. (1967a) made a notable achievement by providing such recordings from limb-muscle afferents during controlled walking. They included in their sample the activity of single afferent axons supplying both muscle spindles and tendon organs. By far, the majority of most of the previous findings about the properties of these receptors had been made on hind limb muscles in deeply anesthetized cats (Matthews 1972).

Precedents to Recording From CNS Interneurons During Locomotion

In the follow-up sequence of 1970–1974 reports discussed previously, Arshavsky, Orlovsky, and their colleagues extended the strategy of relating unitary neuronal activity to the phase of a movement. They provided information about the behavior during controlled locomotion of a variety of interneurons in the spinal cord, brain stem, and cerebellum. Again, long-standing precedents to this approach existed in the study of respiratory (Gesell et al. 1940) and eye (Duensing and Schaefer 1957) movements in mammals. For rhythmic invertebrate movement, the analysis and modeling of the firing patterns of ganglionic interneurons were well under way by the early 1960s (Hughes and Wiersma 1960; Maynard 1965; Wilson 1965). Nonetheless, given technical difficulties associated with extracellular interneuronal records in the *in vivo* cat (particularly in the spinal cord), the technical perseverance and virtuosity of the 1970–1974 Arshavsky-Orlovsky sequence shines forth as brightly as the immediately preceding 1966–1970 sequence of Shik, Severin, and Orlovsky. In fact, the Arshavsky-Orlovsky sequence included, most remarkably, intracellular recording of largely spike discharge (Orlovsky 1970b).

Bernsteinian Features of Controlled Locomotion

The authors and others (see below) have previously provided detailed comments about the full array of controlled locomotion findings reported by the Moscow group in their two initial sequences of reports. The focus here is on those features of their 1966–1970 results that most obviously supported Bernstein's (1947) proposition.

• Controlled locomotion is brought on by a simple command signal. It is remarkable that such coordinated, global motor behavior was shown to be activated by sustained electrical stimulation of specific brain stem sites that converge on the reticulospinal pathway, using a relatively wide array of stimulus pulses (square to sinusoidal, 0.1 ms–1 ms duration), frequency (>10–1000 Hz) and duration (10–100 s at 3 min intervals). This finding supported Bernstein's (1947) concept of a relatively simple and facultative go signal emerging from "higher sections of the nervous system [that] determine the chains of motor activity" (Shik et al. 1966b).

• Controlled locomotion involves the coordinated stepping movements of the four limbs and assistive movements of the trunk. At times, the overall quality of controlled locomotion approaches that of the unrestrained, intact animal. This, too, is remarkable, considering that the step cycle of even a single limb involves the coordinated activation and deactivation of approximately 31 muscles. This behavior gave credence to Bernstein's idea of a "lower (CNS) level [that] ties movements to spatial coordinates" (Shik et al., 1966b). The lower level was presumably the brain stem's reticular formation, where the integration of postural and locomotor commands is achieved (Mori 1995; Grillner et al. 1997). From this lower level onward, the command for a specific gait, speed, and power of locomotion accomodates posutre, too.

• Controlled locomotion features gait conversions brought on by changes in either descending command signals or sensory feedback. The gait of the high-decerebrate cat can be converted from walking to trotting to galloping by either of two methods. First, the strength of brain stem stimulation can be progressively increased with the treadmill belt free running, thereby recruiting more descending axons into discharge. Alternatively, during constant (walking) strength stimulation, a progressively increasing treadmill belt speed can be imposed (Shik et al. 1966b). A key difference between the two forms of gait conversion is that an increased strength of descending command brings on not only an increased speed and gait change but also an increased power (force) of overall muscle contraction that is appropriate for the altered conditions. In contrast, while gait conversion by altered sensory input can certainly change the structure of individual step cycles and interlimb coordination, the power of the altered movements is not necessarily appropriate (Shik and Orlovsky 1976). Despite this caveat, the Moscow findings about gait conversion are particularly appealing conceptually because they illustrate the operation of spinal circuits whose timing can be altered by either descending command signals or sensory feedback. To the authors, this is the quintessence of Bernstein's idea of "still lower levels [of the CNS that] solve the motor problem as such by organizing the necessary interaction of the elements (muscles, joints and limbs) and by operatively controlling their work" (Shik et al. 1966b). In this quotation, the authors would argue that the necessary interaction of the elements includes not only the activity of the spinal pattern-generating circuits that control the structure of the step cycle of each limb (i.e., the actual movements of the muscles, joints, limbs), but also the efferent-afferent interactions between the spinal cord and the moving body parts.

• Controlled locomotion during a constant command signal can involve a feedback-initiated uncoupling of the step cycles of individual limbs. This finding extended the tripartite interaction theme in a particularly striking, indeed dramatic, way. Kulagin and Shik (1970) demonstrated that during constant strength brain stem stimulation with the right-side limbs on one treadmill belt and the left-side limbs on another, controlled walking could be changed profoundly by progressively increasing the speed of one of the belts. For up to a two to three times difference in belt speed, the overall duration of the step cycles remained the same for the control versus speeded limbs. However, the latter exhibited a markedly changed tempo, that is, a relatively longer swing phase of the step and a relatively shorter stance phase. With a further increase in treadmill speed, the speeded limbs produced two step cycles for each one produced by the two limbs on the control speed side. It seems that Kulagin and Shik interpreted this remarkable finding as a demonstration that afferentation of the homonymous limbs exerted a more powerful role on their own spinal stepping generators than did afferentation from the contralateral limbs. Presumably, the same result would ensue if separate treadmills could be used for the controlled locomotion of the forelimbs versus the hind limbs or, for that matter, the separate controlled locomotion of any combination of the limbs. The main point remains that the Kulagin and Shik (1970) findings give even further support to Bernstein's idea of the lowest level of CNS hierarchy (i.e., a single spinal stepping generator) "organizing the necessary interaction of the elements" (Shik et al. 1966b). These are the structure and speed of the step cycle of the generator's own limb. Based on their work on both the controlled locomotion of high-decerebrate cats and the unrestrained treadmill locomotion of intact dogs, Shik et al. (1966b) proposed that "there is an absence of a generator of the locomotor rhythm independent of the movements actually performed." This argument does not contradict the concept of central pattern generators. Rather, it emphasizes the importance of afferentation for the full expression of locomotion. For current opinions on this issue, see Pearson and Ramirez (1997) and Stein and Smith (1997).

• Controlled locomotion lends itself to the intracellular analysis of unitary neuronal activity during fictive movement. To advance understanding of CNS mechanisms that operate in locomotion, it is obviously of particular advantage to record intracellularly, as well as extracellularly, from neurons as movement ensues. For mammalian preparations, intracellular recording is far more robust in fictive (paralyzed muscle) than natural movement. Remarkably, however, Orlovsky (1970b) achieved stable intracellular recording from reticulospinal neurons of the pons and medulla oblongata relatively early, in the 1966–1974 Moscow sequence. For vertebrate movement, fictive preparations were first used to study the respiratory system (MacDonald and Reid 1898), with extracellular (Salmoiraghi and Burns 1960) preceding intracellular (Gill and Kuno 1963) recordings by only three years. For mammalian locomotion, the first exploitation of fictive movement appears to have been that of Viala and Buser (1965), followed shortly thereafter by Shik et al. (1966a) and Perret (1968). Almost hand in hand, Jankowska et al. (1967) recorded intracellularly from motoneurons during DOPA-driven fictive, rhythmic (locomotor-like) hind limb movement in spinal cats. This body of work forms the precedent to the first report of extracellular recording from CNS spinal interneurons during the fictive locomotion of high decerebrate cats (Feldman and Orlovsky 1975). Several years later, intracellular recording from spinal motoneurons was achieved during fictive locomotion of the decorticate (Perret and Cabelguen 1980) and high decerebrate (Jordan 1983) cat. Recall, also, the intracellularly recorded spike discharges reported by Orlovsky (1970b) in his report. The purpose here is not to expand on the fruitfulness of the intracellular analyses that subsequently followed (Jordan 1991; Jordan and McCrea 1997). Rather, the purpose is to reiterate the key role that one of the initial reports in the 1966–1974 Moscow sequence (Shik et al. 1966a) played in the subsequent advancement of locomotion neurobiology.

Impact of the Initial Recordings of Unitary Neuronal Activity During Controlled Locomotion

The focus here is not on Bernstein's (1947) proposition. Rather it is on the nearly immediate impact of a sequence of 1967–1974 Moscow findings. This work provided unitary recordings of the activity of several neuronal species during controlled locomotion. It stimulated subsequent attempts to extend the approach to making analogous recordings in freely moving, intact animals.

• Alpha motoneurons discharge at relatively fixed rates during controlled locomotion of variable speed. Severin et al. (1967b) found that extensor motoneurons to hind limb muscles fired off a steady train of approximately 5 action potentials at a fixed frequency (in the range of 15–45 Hz) during the appropriate phase of the step. The impulse train was remarkably consistent from step to step, even when the strength of brain stem stimulation was increased. This finding, which has been repeatedly confirmed subsequently for both controlled near-natural (Zajac and Young 1980) and fictive (Brownstone et al. 1992) locomotion, was initially thought to bring out a more prominent role for motoneuron recruitment than for rate coding during variable speed locomotion. Subsequently, it has been shown, however, that rate coding is a prominent feature of the unrestrained treadmill locomotion of intact cats (Hoffer et al. 1987). Thereby, this serves warning that not all segmental, and presumably suprasegmental, motor mechanisms that operate in controlled locomotion generalize to the normal condition. Interestingly, Brownstone et al. (1992) have recently shown that in response to intracellular-injected depolarizing current (I) in fictive stepping, motoneurons exhibit a disassociation between their action potentials, after-hyperpolarization (AHP) and their firing rate (f), and a profound qualitative alteration in their I-f relation. This latter finding cannot be interpreted on the basis of currently known motoneuron properties (Binder et al. 1996). Therefore, this invites a detailed examination of the AHP and I-f relations of motoneurons in experimental preparations like those pioneered by Severin et al. (1967b).

• Gamma motoneurons are coactivated with alpha motoneurons during controlled locomotion in a functionally relevant fashion. Severin et al. (1967a) and later Severin (1970) were the first to show a coactivation of alpha and gamma motoneurons during near-natural limb movements. Gamma discharge was shown to depend strongly on the phase of the step cycle. It was proposed to modulate spindle afferent discharge in a manner appropriate for the segmental reflex support of homonymous motoneuron discharge during the relevant phase of the step (Severin 1970). These findings created substantial interest immediately thereafter (see below), and they have stood the test of time. For example, they must surely have been a major impetus for the subsequent, similarly virtuosic development of techniques to record from single spindle afferents, both at the level of dorsal-root axons (Prochazka et al. 1976) and from the dorsal-root ganglia (Loeb et al. 1977), during the unrestrained locomotion and many other forms of voluntary movements of conscious cats. This development led to two major, present-day inferences about gamma motoneuron function. First, they may maintain spindle afferent discharge in the midrange of possible extremes when a movement proceeds as planned (Loeb 1984). In addition, they may activate muscle spindles to fire at particularly high rates when the conscious cat is confronted with a novel and/or difficult motor task (Prochazka 1996). Both of these roles may be considered a logical extension of the seminal findings of Severin et al. (1967b) and Severin (1970).

• During controlled locomotion, muscle spindles and tendon organs are coactivated when their parent muscle becomes active. This *de novo* finding of Severin et al. (1967a) has stood the test of time. It has also been extended to include not only other muscle receptors but cutaneous and joint afferents as well. It relates to the unrestrained voluntary movements of conscious animals, and the voluntary isometric contractions and limited, slow movements of humans (Loeb 1984; Prochazka 1996). The implication that sensory coactivation was a feature of natural movements was of high impact in 1967. It was a particular challenge to neuroscientists unraveling the spinal actions of muscle receptors in their subsequent attempts to show how length and tension information can work together rather than in opposition in the reflex support of locomotion. The importance of the Severin et al. (1967b) and Severin (1970) findings for this field of inquiry cannot be overstated.

• During controlled locomotion, the fundamental, spinally generated rhythm of the step cycle is reinforced by spino-cerebellar-spinal actions. Figure 2.3 shows a synthesis of the Moscow 1970–1974 findings on intralimb coordination that is still valuable conceptually, more than two decades after its initial presentation (Wetzel and Stuart 1976, their figure 21). It must be conceded, however, that the Moscow group's subsequent analogous analysis of fictive scratching (Arshavsky et al. 1986) provided a rigorous confirmation of ideas about cerebellar function that had emerged from their earlier work about controlled locomotion.

Figure 2.3 shows that neurons in spino-cerebellar-spinal pathways are exquisitely sensitive to phase-dependent signals that originate in both the movements themselves and the spinal circuits that generate the step cycle. In all, the 1970–1974 Moscow sequence provided information about the phase relations between locomotion movements (measured via limb kinematics and the EMG) and the firing patterns of 11 species of CNS interneurons. The ensemble rhythmic discharge of nine of these cell types is shown in figure 2.3. The figure shows cells of the dorsal spinocerebellar tract; the ventral spinocerebellar tract (rhythmic even with the hind limbs deafferented); the interpositus and fastigial nuclei; the cerebellar cortex (Purkinje cells); the reticulospinal, rubrospinal, and vestibulospinal tracts; and the ventral horn (many presumably last-order to motoneurons). Not shown in figure 2.3 but also exhibiting rhythmic discharge during controlled locomotion are cells of the cuneocerebellar tract (Arshavsky et al. 1973) and second-order interneurons in the spinoreticulocerebellar tract (Arshavsky et al. 1974). By 1975, the Moscow group had also reported about the rhythmic, functionally appropriate discharge of last-order spinal ventral horn interneurons in the segmental reciprocal Ia inhibitory pathway (Feldman and Orlovsky 1975), as recorded during controlled fictive locomotion. This firing pattern, too, can be incorporated into the figure 2.3 summary. Similarly,

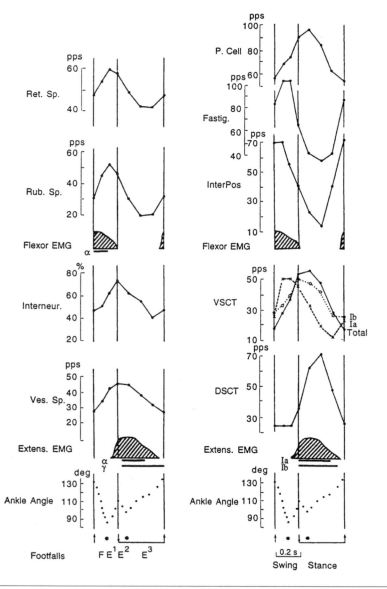

Figure 2.3 Rhythmic activity in spino-cerebellar-spinal pathways. Shown is the ensemble activity of nine types of neurons in spino-cerebellar-spinal pathways during a single walking step cycle normalized to a swing duration of 200 ms. All records are referred to events for the ankle joint. Efferent aspects appear in the left column and afferent aspects in the right column. Flexor EMG, extensor EMG, ankle angle, and footfalls are duplicated in both columns to simplify comparisons of timings. Most data were compiled during controlled locomotion experiments. However, the kinematic data were determined by cinematographic measurements for cats moving unrestrained overground.

Reprinted from Wetzel and Stuart 1976.

it was later shown that during fictive locomotion, Renshaw cells of the spinal recurrent inhibitory pathway discharged rhythmically, again in a manner that supported the fundamental locomotor rhythm (McCrea et al. 1980). Included in the 1970–1974 Moscow sequence were data and propositions about how neuronal activity throughout the entire CNS could be altered when monotonous, straight-line locomotion was perturbed in some fashion (Orlovsky 1970c, 1972c; Orlovsky and Pavlova 1972c) and how activity in the corticospinal tract, in addition to the above-mentioned descending ones, could be brought into play to make appropriate spinal corrections. Clearly, the 1970–1974 Moscow sequence invited consideration of how many CNS cell types could participate in the elaboration of controlled versus perturbed locomotion and by what mechanisms, for example cells of the motor cortex (Armstrong 1986; Grillner et al. 1997).

Figure 2.3 shows, for the elaboration of a single step cycle, how sensory information from different points in spinal and limb space and time can arrive at the cerebellum, be available for integration, and return via the descending tracts. This idea of a cerebellar reinforcement of the fundamental, spinally generated locomotor rhythm owes much to the original 1970–1974 Moscow findings. In time, they should lead to testable hypotheses: how the various neuronal (type) responses are truly represented by their ensemble discharge; how collective neuronal discharge profiles change with forward velocity; how signals from and to the four limbs are parceled and integrated within the cerebellum; and how corrective signals during perturbed and uneven-terrain locomotion are initiated and projected to the spinal cord, and the relative extent of their access to motoneurons versus the pattern-generating interneuronal network. For the short-range resolution of these issues, controlled near-natural and fictive locomotion preparations and experimental paradigms introduced by the Moscow group in their 1967–1974 reports are still of great usefulness, with their extension to unrestrained animals an ever-daunting challenge (Prochazka 1996).

• During controlled locomotion, the descending command signals from the brain stem and rhythmic reinforcement signals from the cerebellum are mediated by descending tracts in a facultative fashion. Figure 2.3 does not show that during controlled locomotion, reticulospinal, in particular, and possibly rubrospinal and vestibulospinal neurons are excited by a sustained stream of command signals from the brain stem site used to bring on the movement (see above). Rather, the figure emphasizes that the 1970–1974 Moscow sequence showed that about half the activated reticulospinal and almost all the activated vestibulospinal and rubrospinal neurons also receive phase-modulating signals from the cerebellum. This arrangement suggests a certain facultative (i.e., thriving on more than one overall condition) quality to the instigation and projection of the command signal (from the higher brain stem) and rhythm reinforcement (from the cerebellum) to the spinal centers. This was born out subsequently (Grillner et al. 1997). Again, the Moscow sequence led to the testing of hypotheses that took the study of locomotion to a new level of rigor and sophistication. Importantly, several open issues still exist concerning the parceling of command signals, rhythm reinforcement, and postural integration functions for the neurons of the descending tracts. For example, there is a particular need to advance understanding of the mechanistic details of the specific roles played by neurons of the reticulospinal pathways (Selionov and Shik 1990, 1991, 1992).

International Impact of the Initial Controlled Locomotion Findings

The initial 1966–1970 sequence of Moscow reports about controlled locomotion had a major impact in a relatively short time on the international community of motor-control neuroscientists. This would have pleased Bernstein. He would also have found it ironic, given the relative isolation in which he and his collaborators had labored throughout the years. For example, for motor-control neuroscientists, a pre–World War II monograph of particular significance was Creed et al. (1932), and it contained no reference to the Bernstein school. Similarly, several impactive symposia and handbooks that were featured in the training of post–World War II scientists in the 1945–1970 era contained no reference to their work. Examples include Hess (1957), Field et al. (1959, 1960a, b), Barker (1962), Eccles and Schadé (1964), and Granit (1966). The abrupt change that subsequently occurred seems attributable to several converging factors.

• The publication of *Biofizika* in 1955 and its English translation, *Biophysics,* gave access to Bernsteinian ideas. It has never been the custom for Western neuroscientists to read Russian literature, even though the work of Sechenov in the late nineteenth century and Pavlov in the late nineteenth through early twentieth centuries was well known throughout their active years (Brazier, 1959). This problem became exacerbated after World War II, with the emphasis solely on English as the international language of commerce and science rather than French and German, too. Hence, while regretful, it may not be surprising that the availability of *Biophysics* had such an impact on Western motor-control neuroscientists.

• Interdisciplinary motor-control bioengineering accelerated abruptly after World War II. Motor-control neuroscience has a tradition of physicists making major contributions that extends back at least to Borelli (1685). For much of the twentieth century (1909–1964), the influence of Hill (1965) was profound. Theoretical and applied advances, however, in electrical and mechanical engineering had an instant post–World War II impact that was accelerated even further by the dramatic, subsequent developments in computer science. This post-1945 interdisciplinary merger of life and physical science has been a fertile milieu for the ideas of Bernstein.

• The 1967 publication of a group of Bernstein essays augmented Bernstein's international visibility. This publication (Bernstein 1967) consists of papers written by Bernstein between 1934 and 1962. The chapter on the biodynamics of locomotion is prescient of the subsequent 1966–1974 Moscow sequence. Unfortunately, however, an English translation of Bernstein's 1947 monograph (see above) is still not available.

• Lundberg's 1969 treatise gave a particular Western impact to the Moscow group. With its emphasis on cellular (versus systems) neuroscience, and the properties and separate CNS actions of muscle receptors (versus their motor-control function), the post-Sherringtonian training of Western neuroscientists in the late 1940s through late 1960s did not feature the work and ideas of the Bernstein school. Examples of the great success and impact of this training are Denny-Brown, Eccles, Lloyd, and Granit. Rather, this development was left to the subsequent generation of Sherringtonian neuroscientists. Lundberg stands out not only for his instant recognition of the value of the initial controlled locomotion findings but also for several other actions. First, he brought to attention the importance of the far earlier (1911–1922) work of Graham Brown about spinal locomotion (Wetzel and Stuart 1976), removing it from nearly complete obscurity. Second, Lundberg integrated his laboratory's own kinematic and EMG analysis of cat locomotion with findings about the unitary properties of the segmental motor mechanisms (Engberg and Lundberg 1969). Third, he laid the groundwork from the late 1950s to the present for the ever-continuing analysis of interactions at the segmental level between descending command signals, spinal pattern-generating circuits, alternative reflex pathways, and sensory input (Lundberg 1969b, 1981; McCrea 1992; Stuart and McDonagh 1997).

All four of the above features are found in Lundberg's (1969a) widely read monograph *Reflex Control of Stepping*. It made reference to the Shik et al. (1966b) work and provided unpublished additional figures from the report of Severin et al. (1967a) (i.e., Lundberg's figures 4 and 5 are like Severin et al.'s figures 5 and 6).

Following Lundberg's (1969a) example, other Western references to the initial (1966–1967) Moscow reports became progressively more commonplace. In North America, Stuart et al. (1972, their figure 13) summarized the Moscow approach at an international muscle spindle meeting held in 1970. Of much greater moment was reference to the work in Matthews' widely read, 1972 research monograph *Mammalian Muscle Receptors and Their Central Actions* (its pages 568–569). Subsequent impactful citations of the initial 1966–1970 Moscow sequence are too innumerable to document. For example, see the summary bibliography in Desmedt (1973). The group itself began to publish in the West in 1972 beginning with Arshavsky et al. (1972a). This expanded considerably their international readership and the impact of their findings.

• In 1971, Shik's first presentation in the West heightened interest in controlled locomotion. To Western motor-control neuroscientists, it was most heartwarming to hear Shik present his work about controlled locomotion in Munich in 1971. This talk (Shik 1971) was of historic interest not only because it was the first talk in the West by any of the Moscow investigators involved in the 1966–1970 sequence but also because the audience included a young Swedish neuroscientist, Grillner. After completing his PhD under the mentorship of Lundberg (Grillner 1969), Grillner became the first Western scientist to work in Moscow (1971) where he investigated aspects of controlled locomotion (Grillner and Shik 1973). Subsequently, Grillner (1975) achieved a complete incorporation of the 1966–1974 Moscow sequence into Western thought on locomotion (see below). Also in the audience was a young Japanese neuroscientist, Mori. After completing his Ph.D. in Japan in 1964 under Fujimori (Ph.D. dissertation in Japanese, components in Fujimori et al. 1966), Mori had a substantial postdoctoral experience in the U.S.A. with Brookhart (1965–68), where he worked on postural control in conscious dogs (Mori and Brookhart 1968). Subsequently, Mori worked in Moscow in 1973–1974 (Mori et al. 1977) and again in 1983 (Arshavsky et al. 1985), investigating supraspinal aspects of controlled locomotion. His own group continues to make important contributions to this area, particularly concerning the integrated control of posture and movement (Mori 1987, 1995; Mori and Ohta 1986). Finally, a young French neuroscientist, Perret, also heard Shik (1971). Perret completed his Ph.D. under Buser (Perret 1973, 1976) and subsequently had a particularly productive period in Moscow (1975–76) with Orlovsky (Arshavsky et al. 1980, 1988; Deliagina et al. 1981). In retrospect, Grillner, Mori, and Perret were particularly privileged. Between 1970 and 1990, they were the only Western scientists who had the opportunity to work in Moscow with the group that effected the 1966–1974 Moscow sequence. In contrast, over the same time frame, Western laboratories of comparable stature attracted literally hundreds of capable young scientists eager to advance their theoretical and technical skills. Also in contrast is the sorry fact that these gifted Moscow workers had no opportunity to work in Western laboratories during the same time frame. Rather, their contacts with Western workers were largely limited to a Bulgarian symposium series (see below).

• Emergence began in the 1970s of a Bernsteinian emphasis in North American motor-control symposia. For systems motor-control neuroscientists and their trainees, the first North American symposium to include significant reference to the controlled locomotion findings of the Moscow group was a widely attended meeting about *Control of Posture and Locomotion* held in Edmonton, Alberta in the early 1970s (Stein et al. 1973). Two reports were particularly relevant to this chapter's theme. First, after Davis summarized what was then known about the neural control of the lobster swimmeret system (Davis 1973, his figure 8), he compared this control system to that for cat locomotion, with due emphasis on the Moscow sequence. Shortly thereafter, Grillner summarized the then-current understanding of spinal locomotion in the cat. He included a table (Grillner 1973, his table 1). It compared traditional Western electrophysiological approaches to the spinal cord, as usually undertaken in anesthetized preparations, for the controlled locomotion of high-decerebrate cats versus DOPA-driven spinal locomotion. In retrospect, these two reports deserve special historical emphasis in North American motor-control training programs.

The success of the 1973 conference was one of several stimuli for another group to initiate a 10-year symposium cycle about the neural control of locomotion. The first meeting, *Neural Control of Locomotion*, was held at Valley Forge, Pennsylvania in 1975. This meeting's primary theme was the identification of common prin-

ciples of organization in the neural circuits controlling locomotion in both invertebrates and vertebrates (Herman et al. 1976b). A number of the Moscow group were
invited to this meeting but were not granted permission by the U.S.S.R. to attend.
For the North American reaction, see Camhi et al. (1975). A Bernsteinian emphasis
nonetheless prevailed. A series of papers about human solutions to locomotion
(Herman et al. 1976a; Craik et al. 1976; Cook and Cozzens 1976) replaced the hoped
for presentation of Gurfinkel (see below). One report about controlled locomotion
by Grillner (1976) replaced Shik, and another (Grillner and Kashin 1976) about fish
locomotion replaced Kashin. Finally, for a talk about the modulation of central
control mechanisms, a talk by Perret (1976) replaced Orlovsky. Certainly, the Moscow
group was present in Valley Forge in spirit if not in body.

Two subsequent symposia in the Valley Forge tradition have occurred. The 1985
Stockholm version, *Neurobiology of Vertebrate Locomotion*, emphasized common
principles of neuronal organization across all the vertebrates (Grillner et al. 1986).
Again, the Moscow invitees were denied permission to attend, but articles by three
of them appeared in the symposium volume (Arshavsky and Orlovsky 1986;
Berkinblit et al. 1986; Shik 1986). Their full work was thoroughly discussed throughout the meeting and documented in the symposium volume (Grillner et al. 1986).
In 1995, the venue shifted to Tucson, Arizona. The focus of the symposium, *Neurons, Networks, and Motor Behavior,* was on cellular, network, and behavioral levels
of organization in the neural control of locomotion and several other forms of
movement, again in both invertebrates and vertebrates (Stein et al. 1997). Most
heartwarmingly, five of the Moscow group were in attendance at this meeting:
Arshavsky, Deliagina, Feldman, Orlovsky, and Panchin (Stein et al. 1995), as had
been planned by the same organizers both 20 and 10 years earlier.

Also advancing Western appreciation of Eastern bloc motor-control science (and
vice versa) has been a sequence of Bulgarian meetings initiated by the late Alexander
Gydikov in 1969, which will continue for the foreseeable future. For a brief history,
read the preface in Gantchev et al. (1996). An article that included a summary of the
controlled locomotion work of the Moscow group appeared in the volume associated with the second (1972) meeting in the series (Gurfinkel and Shik 1973).

• Several 1970s reviews cemented the impact of the Moscow controlled locomotion studies on Western neuroscience. During the same time frame as the above-
mentioned 1973–1975 symposia, the publication of five 1973–1976 reviews en-
sured the permanence of the 1970–1974 Moscow sequence in the theoretical
armamentarium of Western motor-control neuroscience. In retrospect, it is intriguing that all five reviews have an idiosyncratic quality, thereby illustrating the range
of topics addressed in any review of the neural control of locomotion. They all focus
on the 1966–1974 Moscow sequence and the manifold problems involved in achieving a functional interface between cellular neurophysiology and biomechanics (Hasan
et al. 1985; Hasan and Stuart 1988).

The 1973 review of Gurfinkel and Shik had a retrospective impact because it was
published in a Bulgarian symposium volume (Gydikov et al. 1973) that was not

widely read in the West until the symposium sequence achieved higher visibility more than a decade later (see the prefaces in Gantchev et al. 1987, Stuart 1996). The review gave equal weight to the role of the stretch reflex and vision in the control of posture, and work of the Moscow group about the control of locomotion in unrestrained dogs and high-decerebrate cats. Interestingly, a unifying theme was the idea that posture, like locomotion, was controlled in a Bernsteinian fashion by interactions between descending command signals, a brain stem-spinal cord control program, and sensory feedback. This viewpoint is still conceptually attractive, but it has proven to be more difficult to test experimentally for posture than it has for locomotion.

The 1975 review of Grillner was widely read. It deserves special emphasis for the nearly equal weight given to work undertaken on invertebrates and vertebrates and its strong emphasis on modeling pattern-generating circuitry. These features were also present in the immediately following reviews of Shik and Orlovsky (1976), Orlovsky and Shik (1976), and Wetzel and Stuart (1976) but to differing degrees and intent. The Shik and Orlovsky (1976) synthesis is still of value. It focuses on the challenge of unraveling the details of how reticulospinal neurons interact in their integration of posture and locomotion control and the delivery of power to the musculature for locomotion at variable speed. The Orlovsky and Shik (1976) article is a particular challenge for the late 1990s because it expands on the quote presented above concerning an "absence of a generator of a locomotor rhythm independent of the movements actually performed" (Shik et al. 1966b; Orlovsky and Shik 1976, pages 290–291). Their argument that "It seems possible that afferent signals can trigger (or facilitate) some parts of the central program, and, on the contrary, the central program can strongly influence the afferent signals," is still under intense investigation (Stein et al. 1997). The Wetzel and Stuart (1976) review, too, had a somewhat different emphasis in its focus on the details of the 1966–1974 Moscow sequence. For interdisciplinary motor-control trainees, Wetzel and Stuart (1976) heavily emphasized the history of the study of locomotion.

Western Corecognition of Bernsteinian Thought and the Interface Between Vertebrate and Invertebrate Neuroscience

An intriguing feature of Western motor-control neuroscience is that the same symposia, symposium volumes, and reviews that emphasized the importance of the 1966–1974 Moscow sequence about controlled locomotion also emphasized the significance of selected generalizations about the neural control of movement. These were applicable to invertebrates, nonmammalian vertebrates, and mammalian vertebrates, including humans. For example, among the prominent, widely attended, international symposia discussed above, there was no such interphyletic awareness at the 1965 Stockholm meeting (Granit 1966). In contrast, it was a prominent feature at the 1973 Edmonton meeting (Stein et al. 1973) and the primary focus of the 1975 Valley Forge (Herman et al. 1976b), 1985 Stockholm (Grillner et al. 1986), and 1995 Tucson (Stein et al., 1995, 1997) meetings. It would be presumptuous indeed to claim that this development, at the systems level of analysis, is unique to motor-control science. For example, read about the interphyletic developments concerning olfaction in Hildebrand and Shepherd (1997) and visually guided head-neck movements in Strausfeld (1997). However, the study of locomotion neurobiology, in general, and controlled locomotion, in particular, have been a particularly nurturing area for this development to occur. In a similar vein, as Western appreciation grew for the Bernsteinian approach in the early 1970s, much informal discussion occurred about the interdisciplinary expertise that was so evident in Bernstein (1967), the 1966–1974 Moscow sequence, and, most significantly, Gelfand et al. (1971). From such discussions came the desire to develop interdisciplinary centers in the West where scientists with a variety of expertise in the physical and life (including medical) sciences could collaborate in a similarly optimal fashion. Such a development has occurred in several Western countries, and it should serve motor-control science particularly well for the foreseeable future.

Current Issues in Locomotion Neurobiology

It is fitting to conclude this tribute to Bernstein and subsequent members of the Moscow group with some reflections about current issues in locomotion neurobiology (Prochazka 1996; Stein et al. 1997) and their relation to the ideas resulting from the 1966–1974 Moscow sequence. The following viewpoints are, like the neural control of locomotion itself, organized in tripartite fashion, that is, the current understanding about descending command signals, spinal pattern-generating circuitry, and the role of sensory feedback.

• Three aspects of the descending command are to the experimental forefront. The understanding of goal-directed locomotion (Grillner et al. 1997), the locomotor role of the basal ganglia (Garcia-Rill and Skinner 1986; Reese et al. 1995), and posturo-locomotor integration in the brain stem and the reticulospinal pathway (Mori 1995; Grillner et al. 1997) has advanced considerably since the 1966–1974 Moscow sequence. Nonetheless, much work is still required on state transitions from the locomotor region of the midbrain to the reticulospinal pathway, in particular. Since it is now known that the locomotor role of these structures generalizes across vertebrates (Grillner et al. 1997), the possibility exists to advance understanding of posturo-locomotor integration and state transitions in the reticulospinal pathway at the cellular and network level of analysis. Progress has been made, in particular, in studies on nonmammalian vertebrates. For advances about the lamprey, read Grillner et al. (1997). Knowledge about the precise role played by the cerebellum in enhancing interlimb coordination has not advanced significantly since the 1966–1974 sequence. The same Moscow group has found it more fruitful to undertake analogous cerebellar and cerebellar pathway experiments on animals exhibiting scratching movements (Arshavsky and Orlovsky 1986). Again, resorting to intracellular experimentation on nonmammalian vertebrates should prove fruitful in the immediate future. Finally, a fascinating new aspect of the central command is now on the verge of investigation. It is the interface between mechanisms of supra–brain stem and brain stem contributions to the locomotor command and to the control of multijointed arm movements. A formulation of this issue was recently provided by Grillner et al. (1997). Clearly, the general understanding of voluntary mechanisms in the control of movement will be advanced if this interface can be effected.

• Three aspects of spinal pattern-generating circuitry are to the forefront. The 1995 Tucson symposium (Stein et al. 1995, 1997) brought out the extent to which knowledge of pattern-generating circuitry has advanced since the 1966–1974 Moscow sequence and its review in the mid-1970s (Grillner 1975; Orlovsky and Shik 1976; Shik and Orlovsky 1976; Wetzel and Stuart 1976). Across invertebrate and vertebrate phyla, a library of cellular, synaptic, and network mechanisms has been delineated. Individual phyla apparently exhibit combinations of these building blocks (Getting 1986) that are appropriate for their individual needs and biomechanical constraints. This theme and the task dependency of pattern-generator operation (Stein and Smith 1997) fully deserve their present-day attention. Literally scores of laboratories are advancing understanding of the cellular, network, interactive, and subcomponent aspects of central pattern generators. The full composition of the spinal and ganglionic pattern generators for vertebrate and invertebrate locomotion, respectively, is still a matter of lively debate (Pearson and Ramirez 1997). This debate keeps at the forefront issues about the mechanisms of sensory feedback interaction with central pattern generators that were raised by Shik et al. (1966b) in the report that initiated the 1966–1974 Moscow sequence. A still valuable review of this issue is that of Orlovsky and Shik (1976).

• Modern engineering control theory should guide future research about sensory feedback. This viewpoint is presented here in response to the Shik et al. (1966b) emphasis on an "absence of a generator of locomotor rhythm independent of the movements actually performed" (Orlovsky and Shik 1976; Pearson and Ramirez 1997). As reviewed elsewhere (Stuart and McDonagh 1997), it is largely due to Lundberg, Jankowska, and their trainees that during the last four decades such a wealth of information has become available about the spinal actions of limb afferents (McCrea 1992) and the connectivity patterns of spinal interneurons (Jankowska 1992). These advances have required great technical ingenuity and perseverance, largely using anesthetized cats. Within the last two decades, similarly virtuosic efforts were required to obtain information about the firing patterns of limb afferents and efferents during the natural movements of conscious cats. The time has now come to interface this body of information with current understanding of the role of sensory feedback in the control of locomotion. To this end, Prochazka (1996) has emphasized the potential value of modern engineering control theory in providing a rubric for future research. Such theory includes finite state (conditional), adaptive (self-organizing), and predictive control; neural networks; and fuzzy logic. Prochazka (1996) has emphasized that all six of these approaches are now used in varying combination in prosthetics research, with one such example provided in figure 2.4.

The authors believe that Bernstein would have liked figure 2.4. Furthermore, the subsequent Moscow group should appreciate its interphyletic emphasis. If, indeed, such engineering control theory is subsequently shown to provide the rubric for advancing understanding of the interplay between sensory feedback and spinal (and ganglionic) pattern-generating mechanisms, then the impact of Bernstein and subsequent members of the Moscow group will extend well into the twenty-first century.

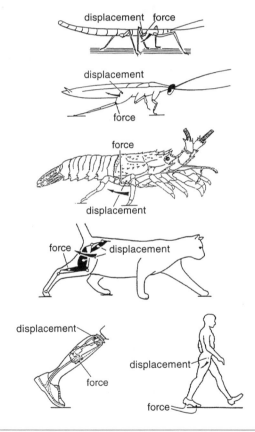

Figure 2.4   An example of finite state control across species in the elaboration of a stepping movement. Shown are schematics of a stick insect, locust, lobster, cat, an active leg prosthesis, and a human with a functional electrical stimulation device. In each case, pairs of sensory variables are indicated that when used in a conditional way, initiate the swing phase of the step. That is, if displacement exceeds threshold and force has declined below threshold, flexion is then initiated. Approximate positions of identified sensors (natural and artificial) are shown.
Reprinted from Prochazka. © 1993 IEEE.

Acknowledgments

The authors would like to thank Dr. Paul S.G. Stein for reviewing a draft of this manuscript and for his stimulating review of some of this material prior to its presentation at the international conference *Bernstein's Traditions in Motor Control*, Pennsylvania State University, University Park, PA, August 22–25, 1996. Preparation of the article was supported, in part, by U.S.P.H.S. grants NS 20577 and NS 07309 and NS 01686. Its contents are solely the responsibility of the authors and do not necessarily represent the views of the awarding agencies.

References

Adrian, E.D., Bronk, D.W. (1928) The discharge of impulses in motor nerve fibers. I. Impulses in single fibres of the phrenic nerve. *Journal of Physiology (London)*, 66: 81–101.

Armstrong, E.M. (1986) The motor cortex and locomotion in the cat. In *Neurobiology of Vertebrate Locomotion*. S. Grillner, P.S.G. Stein, D.G. Stuart, H. Forssberg, R.M. Herman (Eds.), London: MacMillan Press, 121–137.

Arshavsky, Y.I., Berkinblit, M.B., Fukson, O.I., Gelfand, I.M., Orlovsky, G.N. (1972a) Recordings of neurones of the dorsal spinocerebellar tract during evoked locomotion. *Brain Research*, 43: 272–275.

Arshavsky, Y.I., Berkinblit, M.B., Fukson, O.I., Gelfand, I.M., Orlovsky, G.N. (1972b) Origin of modulations in neurones of the ventral spinocerebellar tract during locomotion. *Brain Research*, 43: 276–279.

Arshavsky, Y.I., Berkinblit, M.B., Gelfand, I.M., Orlovsky, G.N., Fukson, O.I. (1972c) Activity of the neurones of the dorsal spinocerebellar tract during locomotion. *Biophysics*, 17: 506–514.

Arshavsky, Y.I., Berkinblit, M.B., Gelfand, I.M., Orlovsky, G.N., Fukson, O.I. (1972d) Activity of the neurones of the ventral spinocerebellar tract during locomotion. *Biophysics*, 17: 926–941.

Arshavsky, Y.I., Berkinblit, M.B., Gelfand, I.M., Orlovsky, G.N., Fukson, O.I. (1972e) Activity of the neurones of the ventral spinocerebellar tract during locomotion of cats with deafferented hind limbs. *Biophysics*, 17: 1169–1176.

Arshavsky, Y.I., Berkinblit, M.B., Gelfand, I.M., Orlovsky, G.N., Fukson, O.I. (1973) Activity of the neurones of the cuneocerebellar tract during locomotion. *Biophysics*, 18: 132–138.

Arshavsky, Y.I., Berkinblit, M.B., Gelfand, I.M., Orlovsky, G.N., Fukson, O.I. (1974) Differences in activity of spinocerebellar tracts during artificial stimulation and locomotion. *Proceedings of the Fifth Symposium on Problems in General Physiology*. Nauka, Leningrad, 99–105.

Arshavsky, Y.I., Gelfand, I.M., Orlovsky, G.N. (1986) *Cerebellum and Rhythmical Movements*. Berlin: Springer-Verlag. (Original Russian version in 1984).

Arshavsky, Y.I., Kots, Y.M., Orlovsky, G.N., Rodionov, I.M., Shik, M.L. (1965) Investigation of the biomechanics of running by the dog. *Biophysics*, 10: 737–746.

Arshavsky, Y.I., Mori, S., Orlovsky, G.N., Pavlova, G.A., Popova, L.B. (1985) Effects of stimulation of midline inhibitory area within the pontine tegmentum on scratch reflex. *Brain Research*, 343: 360–362.

Arshavsky, Y.I., Orlovsky, G.N. (1986) Role of the cerebellum in the control of rhythmic movements. In *Neurobiology of Vertebrate Locomotion*. S. Grillner, P.S.G. Stein, D.G. Stuart, H. Forssberg, R.M. Herman (Eds.), London: MacMillan Press, 677–689.

Arshavsky, Y.I., Orlovsky, G.N., Pavolva, G.A., Perret, C. (1980) Activity of neurons of cerebellar nuclei during fictitious scratch reflex in the cat. 2. The interpositus and lateral nuclei. *Brain Research*, 200: 249–258.

Arshavsky, Y.I., Orlovsky, G.N., Perret, C. (1988) Activity of rubrospinal neurons during locomotion and scratching in the cat. *Behavioural Brain Research*, 28: 193–199.

Arvanitaki, A. (1936). Variations lentes de potentiel associées au fonctionnement rythmique des nerfs non myélinisés isolés. *Journal de Physiologie et de Pathologie Génerale*, 34: 1182–1197.

Barker, D. (Ed.). (1962) *Symposium on Muscle Receptors*. Hong Kong: Hong Kong University Press.

Berkinblit, M.B., Gelfand, I.M., Feldman, A.G. (1986) A model for the aiming phase of the wiping reflex. In *Neurobiology of Vertebrate Locomotion*. S. Grillner, P.S.G. Stein, D.G. Stuart, H. Forssberg, R.M. Herman (Eds.), London: MacMillan Press, 217–227.

Bernstein, N. (1967) *The Co-ordination and Regulation of Movements*. New York: Pergamon.

Bernstein, N.A. (1947) *On the Construction of Movements*. Moscow: Medgiz (In Russian).

Binder, M.D., Heckman C.J., Powers R.K. (1996) The physiological control of motoneuron activity. In *Handbook of Physiology*. Sec. 12. *Exercise: Regulation and Integration of Multiple Systems*. L.B. Rowell, J.T. Shepherd (Eds.), New York: Oxford University Press, 3–53.

Björk, A., Kugelberg, E. (1953) Motor unit activity in the human extraocular muscles. *Electroencephalography and Clinical Neurophysiology*, 5: 271–278.

Borelli, G.A. (1685) *De Motu Animalium,* 2d ed. Leyden: Gaesbeeck. (A small section has been translated into English. Foster, M. (1901) *Lectures on the History of Physiology*. Cambridge: Cambridge).

Brazier, M.A.B. (1959) The historical development of neurophysiology. In *Handbook of Physiology, Neurophysiology*. Vol. I. J. Field, H.W. Magoun, V.E. Hall (Eds.), Washington, DC: American Physiological Society, 1–58.

Brownstone, R.M., Jordan, L.M., Kriellaars, D.J., Noga, B.R., Shefchyk, S.J. (1992) On the regulation of repetitive firing in lumbar motoneurones during fictive locomotion in the cat. *Experimental Brain Research*, 90: 441–455.

Budakova, N.N., Shik, M.L. (1970) Effect of brain-stem stimulation evoking locomotion on ascending reflexes in the mesencephalic cat. *Bulletin of Experimental Biology and Medicine*, 69: 5–8. (English translation from the Russian journal of the same title).

Camhi, J.M., Hoy, R., Ritzman, R., Paton, J. (1975) Lost opportunity. *Science*, 190: 422.

Cook, T., Cozzens, B. (1976) Human solutions for locomotion: The initiation of gait. In *Neural Control of Locomotion*. R.M. Herman, S. Grillner, P.S.G. Stein, D.G. Stuart (Eds.), New York: Plenum Press, 65–76.

Craik, R., Herman, R., Finley, F.R. (1976) Human solutions for locomotion: Interlimb coordination. In *Neural Control of Locomotion*. R.M. Herman, S. Grillner, P.S.G. Stein, D.G. Stuart (Eds.), New York: Plenum Press, 51–64.

Creed, R.S., Denny-Brown, D., Eccles, J.C., Liddell, E.G.T., Sherrington, C.S. (1932) *Reflex Activity of the Spinal Cord*. Oxford: Clarendon Press. (Reprinted in 1972).

Critchlow, V., von Euler, C. (1963) Intercostal muscle spindle activity and its motor control. *Journal of Physiology (London)*, 168: 820–847.

Davey, M.R., Taylor, A. (1966) Activity of cat jaw muscle stretch receptors recorded from their cell bodies in the mid-brain during spontaneous jaw movements. *Journal of Physiology (London)*, 185: 62P.

Davis, W.J. (1973) Neuronal organization and ontogeny in the lobster swimmeret system. In *Control of Posture and Locomotion*. R.B. Stein, K.G. Pearson, R.S. Smith, J.B. Redford (Eds.), New York: Plenum Press, 437–455.

Deliagina, T.G., Orlovsky, G.N, Perret, C. (1981) Efferent activity during fictitious scratch reflex in the cat. *Journal of Neurophysiology*, 45: 596–604.

Desmedt, J. (Ed.). (1973) *New Developments in Electromyography and Clinical Neurophysiology*. Vol. 3. Basel, Switzerland: S. Karger.

Duensing, F., Schaefer, K.-P. (1957) Die neuronenaktivität in der formatio reticularis des rhombencephalons beim vestibulären nystagmus. *Archiv für Psychiatrie und Zeitschrift für das gesamten Neurologie*, 196: 265–290.

Eccles, J.C., Schadé, J.P. (1964) Physiology of spinal neurons. *Progress in Brain Research,* Vol. 12. Amsterdam: Elsevier Publishing Company.

Eccles, R.M., Sears, T.A., Shealy, C.N. (1962) Intra-cellular recording from respiratory motoneurones of the thoracic spinal cord. *Nature*, 193: 844–846.

Eklund, G., von Euler, C., Rutkowski, S. (1963) Intercostal γ-motor activity. *Acta Physiologica Scandinavica*, 57: 481–482.

Eklund, G., von Euler, C., Rutkowski, S. (1964) Spontaneous and reflex activity of intercostal gamma motoneurones. *Journal of Physiology (London)*, 171: 139–163.

Engberg, I., Lundberg, A. (1969) An electromyographic analysis of muscular activity in the hindlimb of the cat during unrestrained locomotion. *Acta Physiologica Scandinavica*, 75: 614–630.

Feigenberg, I.M., Latash, L.P. (1996). N.A. Bernstein: The reformer of neuroscience. In *Dexterity and its Development*. M.L. Latash, M.T. Turvey (Eds.), Mahwah, NJ: L. Erlbaum and Assoc., 247–275.

Feldman, A.G., Orlovsky, G.N. (1975) Activity of interneurons mediating reciprocal Ia inhibition during locomotion. *Brain Research*, 84: 181–194.

Field, J., Magoun, H.W., Hall, V.E. (Eds.). (1959) *Handbook of Physiology, Neurophysiology*. Vol. I. Washington, DC: American Physiological Society.

Field, J., Magoun, H.W., Hall, V.E. (Eds.). (1960a) *Handbook of Physiology, Neurophysiology*. Vol. II. Washington, DC: American Physiological Society.

Field, J., Magoun, H.W., Hall, V.E. (Eds.). (1960b) *Handbook of Physiology, Neurophysiology*. Vol. III. Washington, DC: American Physiological Society.

Fujimori, B., Kato, M., Matsushita, S., Mori, S., Shimamura, M. (1966) Studies on the mechanisms of spasticity following spinal hemisection in the cat. In *Muscular Afferents and Motor Control*. R. Granit (Ed.), New York: Wiley, 397–413.

Gantchev, G.N., Gatev, P., Tankov, N., Draganova, N., Dunev, S., Popivanov, D. (1987) Role of the visual feedback for stabilization of vertical human posture during induced body oscillations. In *Motor Control*. G.N. Gantchev, B. Dimitrov, P. Gatev (Eds.), New York: Plenum Press, 129–134.

Gantchev, G.N., Gurfinkel, V.S., Stuart, D.G., Wiesendanger, M. (Eds.). (1996) *Motor Control VIII*. Sophia, Bulgaria: Academic Publishing House, Bulgarian Academy of Sciences.

Garcia-Rill, E., Skinner, R.D. (1986) The basal ganglia and the mesencephalic locomotor region. In *Neurobiology of Vertebrate Locomotion*. S. Grillner, P.S.G. Stein, D.G. Stuart, H. Forssberg, R.M. Herman (Eds.), London: MacMillan Press, 77–103.

Gelfand, I.M., Gurfinkel, V.S., Tsetlin, M.L., Shik, M.L. (1971) Some problems in the analysis of movements. In *Models of the Structural-Functional Organization of Certain Biological Systems*. I.M. Gelfand, V.S. Gurfinkel, S.V. Fomin, M.L. Tsetlin (Eds.), Cambridge, MA: MIT Press, 329–345.

Gelfand, I.M., Tsetlin, M.L. (1971) Mathematical modeling of mechanisms of the central nervous system. In *Models of the Structural-Functional Organization of Certain Biological Systems*. I.M. Gelfand, V.S. Gurfinkel, S.V. Fomin, M.L. Tsetlin (Eds.), Cambridge, MA: MIT Press, 1–22.

Gesell, R., Magee, C.S., Bricker, J.W. (1940) Activity patterns of the respiratory neurons and muscles. *American Journal of Physiology*, 128: 615–628.

Getting, P.A. (1986) Understanding central pattern generators: Insights gained from the study of invertebrate systems. In *Neurobiology of Vertebrate Locomotion*. Wenner-Gren International Symposium Series. Vol. 45. S. Grillner, P.S.G. Stein, D.G. Stuart, H. Forssberg, R.M. Herman (Eds.), London: MacMillan Press, 231–244.

Gill, K.P., Kuno, M. (1963) Excitatory and inhibitory actions on phrenic motoneurones. *Journal of Physiology (London)*, 168: 274–289.

Goslow, G.E., Reinking, R.M., Stuart, D.G. (1973) The cat step cycle: Hind limb joint angles and muscle lengths during unrestrained locomotion. *Journal of Morphology*, 141: 1–41.

Graham Brown, T. (1914) On the nature of the fundamental activity of the nervous centres; together with an analysis of the conditioning of rhythmic activity in progression, and a theory of the evolution of function in the nervous system. *Journal of Physiology (London)*, 48: 18–46.

Granit, R. (1966) *Muscular Afferents and Motor Control*. New York: Wiley.

Gray, J. (1950) The role of peripheral sense organs during locomotion in vertebrates. *Symposium for the Society for Experimental Biology*, 4: 112–126.

Grillner, S. (1969) Supraspinal and segmental control of static and dynamic gamma-motoneurons in the cat. *Acta Physiologica Scandinavica*, 327: 1–34.

Grillner, S. (1973) Locomotion in the spinal cat. In *Control of Posture and Locomotion*. R.B. Stein, K.G. Pearson, R.S. Smith, J.B. Redford (Eds.), New York: Plenum Press, 515–535.

Grillner, S. (1975) Locomotion in vertebrates: Central mechanisms and reflex interaction. *Physiological Reviews*, 55: 247–304.

Grillner, S. (1976) Some aspects on the descending control of the spinal circuits generating locomotor movements. In *Neural Control of Locomotion*. R.M. Herman, S. Grillner, P.S.G. Stein, D.G. Stuart (Eds.), New York: Plenum Press, 351–375.

Grillner, S., Georgopoulos, A.P., Jordan, L.M. (1997) Selection and initiation of motor behavior. In *Neurons, Networks, and Motor Behavior.* P.S.G. Stein, S. Grillner, A. Selverston, D.G. Stuart (Eds.), Boston: MIT Press. 3-19.

Grillner, S., Kashin, S. (1976) On the generation and performance of swimming in fish. In *Neural Control of Locomotion*. R.H. Herman, S. Grillner, P.S.G. Stein, D.G. Stuart (Eds.), New York: Plenum Press, 181–201.

Grillner, S., Shik, M.L. (1973) On the descending control of the lumbosacral spinal cord from the mesencephalic locomotor region. *Acta Physiologica Scandinavica*, 87: 320–333.

Grillner, S., Stein, P.S.G., Stuart, D.G., Forssberg, H., Herman, R.M. (Eds.). (1986) *Neurobiology of Vertebrate Locomotion*. Wenner-Gren International Symposium Series. Vol. 45. London: MacMillan Press.

Grossman, R.G. (1958) Effects of stimulation of non-specific thalamic system on locomotor movements in cat. *Journal of Neurophysiology*, 21: 85–93.

Gurfinkel, V.S., Shik, M. L. (1973) The control of posture and locomotion. In *Motor Control.* A.A. Gydikov, N.T. Tankov, D.S. Kosarov (Eds.), New York: Plenum Press, 217–234.

Gydikov, A.A., Tankov, N.T., Kosarov, D.S. (Eds.). (1973) *Motor Control*. New York: Plenum Press.

Haertig, E.W., Masserman, J.H. (1940) Hypothalamic lesions and pneumonia in cats. With notes on behavior changes. *Journal of Neurophysiology*, 3: 293–299.

Hasan, Z., Enoka R.M., Stuart D.G. (1985) The interface between biomechanics and neurophysiology in the study of movement: Some recent approaches. In *Exercise and Sport Sciences Reviews*. Vol. 13. R.L. Terjung (Ed.), New York: MacMillan Press, 169–234.

Hasan, Z., Stuart, D.G. (1988) Animal solutions to problems of movement control: The role of proprioceptors. *Annual Review of Neuroscience*, 11: 199–223.

Herman, R., Wirta, R., Bampton, S., Finley, F.R. (1976b) Human solutions for locomotion: Single limb analysis. In *Neural Control of Locomotion*. R.M. Herman, S. Grillner, P.S.G. Stein, D.G. Stuart (Eds.), New York: Plenum Press, 13–49.

Herman, R.H., Grillner, S., Stein, P.S.G., Stuart, D.G. (1976a) *Neural Control of Locomotion*. New York: Plenum Press.

Hess, W.R. (1957) *The Functional Organization of the Diencephalon*. New York: Grune & Stratton.

Hildebrand, J.G., Shepherd, G.M. (1997) Mechanisms of olfactory discrimination: Common principles across phyla. *Annual Review of Neuroscience,* 20: 595-631.

Hill, A.V. (1965) *Trails and Trials in Physiology*. Baltimore: Williams & Wilkins.

Hinsey, J.C., Ranson, S.W., McNattin, R.F. (1930) The role of the hypothalamus and mesencephalon in locomotion. *Archives of Neurology and Psychiatry*, 23: 1–43.

Hoffer, J.A., Sugano, N., Loeb, G.E., Marks, W.B., O'Donovan, M.J., Pratt, C.A. (1987) Cat hindlimb motoneurons during locomotion. II. Normal activity patterns. *Journal of Neurophysiology*, 57: 530–553.

Hughes, G.M., Wiersma, C.A.G. (1960) The co-ordination of swimmeret movements in the crayfish, *Procambarus clarkii* (Girard). *Journal of Experimental Biology*, 37: 657–670.

Jankowska, E. (1992) Interneuronal relay in spinal pathways from proprioceptors. *Progress in Neurobiology*, 38: 335–378.

Jankowska, E., Jukes, M.G.M., Lund, S., Lundberg, S. (1967) The effect of DOPA on the spinal cord 5. Reciprocal organization of pathways transmitting excitatory action to alpha motoneurones of flexors and extensors. *Acta Physiologica Scandinavica*, 70: 369–388.

Jordan, L.M. (1983) Factors determining motoneuron rhythmicity during fictive locomotion. *Society for Experimental Biology Symposium*, 37: 423–444.

Jordan, L.M. (1991) Brainstem and spinal cord mechanisms for the initiation of locomotion. In *Neurobiological Basis of Human Locomotion*. M. Shimamura, S. Grillner, V.R. Edgerton (Eds.), Tokyo: Japan Scientific Societies Press, 3–20.

Jordan, L.M., McCrea, D.A. (1997) Location, organization and properties of spinal and brainstem neurons producing fictive locomotion. *Progress in Neurobiology*. (In press).

Kulagin, A.S., Shik, M.L. (1970) Interaction of symmetrical limbs during controlled locomotion. *Biophysics*, 15: 171–178.

Loeb, G.E. (1984) The control and responses of mammalian muscle spindles during normally executed motor tasks. In *Exercise and Sports Sciences Reviews*. Vol. 12. R.L. Terjung (Ed.), Lexington, MA: D.C. Heath and Company, 157–204.

Loeb, G.E., Bak, M.J., Duysens, J. (1977) Long-term unit recording from somatosensory neurons in the spinal ganglia of the freely walking cat. *Science*, 197: 1192–1194.

Lundberg, A. (1969a) Reflex control of stepping. *Nansen Memorial Lecture to Norwegian Academy of Sciences and Letters*. Oslo: Universitetsforlaget.

Lundberg, A. (1969b) Convergence of excitatory and inhibitory action on interneurones in the spinal cord. In *The Interneurone*. M.A.B. Brazier (Ed.), Los Angeles: University of California Press, 231–256.

Lundberg, A. (1981) Half-centres revisited. In *Regulatory Functions of the CNS: Motion and Organization Principles*. Vol. 1. J. Szentagothai, M. Palkovits, J. Hamori (Eds.), New York: Pergamon Press, 155–167.

MacDonald, J.S., Reid, E.W. (1898) Electromotive changes in the phrenic nerve. A method of investigating the action of the respiratory center. *Journal of Physiology (London)*, 23: 100–111.

Matthews, P.B.C. (1972) *Mammalian Muscle Receptors and Their Central Actions*. London: Arnold.

Maynard, D.M. (1965) Integration in crustacean ganglia. In *Nervous and Hormonal Mechanisms of Integration*. Symposia of the Society for Experimental Biology. No. XX. G.M. Hughes (Ed.), Cambridge: Cambridge University Press, 111–149.

McCrea, D.A. (1992) Can sense be made of spinal interneuron circuits? *Behavioral and Brain Science*, 15: 633–643.

McCrea, D.A., Pratt, C.A., Jordan, L.M. (1980) Renshaw cell activity and recurrent effects on motoneurons during fictive locomotion. *Journal of Neurophysiology*, 44: 475–488.

Mori, S. (1987) Integration of posture and locomotion in acute decerebrate cats and in awake, freely moving cats. *Progress in Neurobiology*, 28: 161–195.

Mori, S. (1995) Neuroanatomical and neurophysiological bases of postural control. *Advances in Neurology*, 67: 289–303.

Mori, S., Brookhart, J.M. (1968) Characteristics of the postural reactions of the dog to a controlled disturbance. *American Journal of Physiology*, 215: 339–348.

Mori, S., Ohta, Y. (1986) Interaction of posture and locomotion and initiation of locomotion in decerebrate cats and freely moving intact cats. In *Neurobiology of Vertebrate Locomotion*. S. Grillner, P.S.G. Stein, D.G. Stuart, H. Forssberg, R.M. Herman (Eds.), London: MacMillan Press, 55–71.

Mori, S., Shik, M.L., Yagodnitsyn, A.S. (1977) Role of pontine tegmentum for locomotion control in mesencephalic cat. *Journal of Neurophysiology*, 40: 284–295.

Muybridge, E. (1957) *Animals in Motion*. New York: Dover.

Orlovsky, G.N. (1969a) Electrical activity in brainstem and descending path in guided locomotion. *Sechenov Physiological Journal of the USSR*, 55: 437–444. (In Russian).

Orlovsky, G.N. (1969b) Spontaneous and induced locomotion of the thalamic cat. *Biophysics*, 14: 1154–1162.

Orlovsky, G.N. (1970a) Connexions of the reticulo-spinal neurons with the locomotor sections of the brain stem. *Biophysics*, 15: 178–186.

Orlovsky, G.N. (1970b) Work of the reticulo-spinal neurones during locomotion. *Biophysics*, 15: 761–771.

Orlovsky, G.N. (1970c) Influence of the cerebellum on the reticulospinal neurones during locomotion. *Biophysics*, 15: 928–936.

Orlovsky, G.N. (1972a) Activity of vestibulospinal neurons during locomotion. *Brain Research*, 46: 85–98.

Orlovsky, G.N. (1972b) Activity of rubrospinal neurons during locomotion. *Brain Research*, 46: 99–112.

Orlovsky, G.N. (1972c) The effect of different descending systems on flexor and extensor activity during locomotion. *Brain Research*, 40: 359–372.

Orlovsky, G.N. (1972d) Work of the Purkinje cells during locomotion. *Biophysics*, 17: 935–941.

Orlovsky, G.N. (1972e) Work of the neurones of the cerebellar nuclei during locomotion. *Biophysics*, 17: 1177–1185.

Orlovsky, G.N., Feldman, A.G. (1972a) Role of afferent activity in the generation of stepping movements. *Neurophysiology*, 4: 304–310.

Orlovsky, G.N., Feldman, A.G. (1972b) Classification of lumbosacral neurons by their discharge pattern during evoked locomotion. *Neurophysiology*, 4: 311–317.

Orlovsky, G.N., Pavlova, G.A. (1972a) Effect of removal of the cerebellum on vestibular responses of neurons in various descending tracts in cats. *Neurophysiology*, 4: 235–240.

Orlovsky, G.N., Pavlova, G.A. (1972b) Vestibular responses of neurons of the descending tracts during locomotion. *Neurophysiology*, 4: 241–245.

Orlovsky, G.N., Pavlova, G.A. (1972c) Response of Deiter's neurons to tilt during locomotion. *Brain Research*, 42: 212–214.

Orlovsky, G.N., Severin, F.V., Shik, M.L. (1966a) Effect of speed and load on coordination of movements during running of the dog. *Biophysics*, 11:414–417.

Orlovsky, G.N., Severin, F.V., Shik, M.L. (1966b) Effect of damage to the cerebellum on the coordination of movement in the dog on running. *Biophysics*, 11:578–588.

Orlovsky, G.N., Shik, M.L. (1965) Standard elements of cyclic movement. *Biophysics*, 10: 935–944.

Orlovsky, G.N., Shik, M.L. (1976) Control of locomotion: A neurophysiological analysis of the cat locomotor system. In *Neurophysiology II*. Vol. 10. International Review of Physiology. R. Porter (Ed.), Baltimore: University Park Press, 291–317.

Pearson, K.G. (1993) Common principes of motor control in vertebrates and invertebrates. *Annual Review of Neuroscience*, 16: 265–297.

Pearson, K.G., Ramirez, J.-M. (1997) Sensory modulation of pattern generating circuits. In *Neurons, Networks, and Motor Behavior*. P.S.G. Stein, S. Grillner, A. Selverston, D.G. Stuart (Eds.), Boston: MIT Press, 225-235.

Perret, C. (1968) Relations entre activités efférentes spontanées de nerfs moteurs de la patte postérieure et activités de neurones du tronc cérébral chez le chat décortiqué. *Journal of Physiology (Paris)*, 60 (Suppl. 2): 511–512.

Perret, C. (1973) Analyse des mécanismes d'une activité de type locomoteur chez le chat. *Unpublished doctoral dissertation*. Paris: University of Paris 6.

Perret, C. (1976) Neural control of locomotion in the decorticate cat. In *Neural Control of Locomotion*. R.M. Herman, S. Grillner, P.S.G. Stein, D.G. Stuart (Eds.), New York: Plenum Press, 587–615.

Perret, C., Cabelguen, J.M. (1980) Main characteristics of the hindlimb locomotor cycle in the decorticate cat with special reference to bifunctional muscles. *Brain Research*, 187: 333–352.

Pitts, R.F. (1943) The basis for repetitive activity in phrenic motoneurons. *Journal of Neurophysiology*, 6: 439–454.

Prochazka, A. (1993) Comparison of natural and artificial control of movement. *IEEE Transactions on Rehabilitation Engineering*, 1: 7–17.

Prochazka, A. (1996) Proprioceptive feedback and movement regulation. In *Handbook of Physiology*. Sec. 12. *Exercise: Regulation and Integration of Multiple Systems*, L.B. Rowell, J.T. Shepherd (Eds.), New York: Oxford University Press, 89–127.

Prochazka, A., Westerman, R.A., Ziccone, S.T. (1976) Discharges of single hind limb afferents in the freely moving cat. *Journal of Neurophysiology*, 39: 1090–1104.

Prosser, C.L. (1934a) Action potentials in the nervous system of the crayfish. I. Spontaneous impulses. *Journal of Cellular and Comparative Physiology*, 4: 185–209.

Prosser, C.L. (1934b) Action potentials in the nervous system of the crayfish. II. Responses to illumination of the eye and caudal ganglion. *Journal of Cellular and Comparative Physiology*, 4: 363–377.

Reese, N.B., Garcia-Rill, E., Skinner, R.D. (1995) The pedunculopontine nucleus—auditory input, arousal and pathophysiology. *Progress in Neurobiology*, 47: 105–133.

Salmoiraghi, G.C., Burns, B.D. (1960) Notes on mechanism of rhythmic respiration. *Journal of Neurophysiology*, 23: 14–26.

Sears, T.A. (1962) The activity of the small motor fibre system innervating respiratory muscles of the cat. *Australian Journal of Science*, 25: 102.

Sears, T.A. (1963) Activity of fusimotor fibres innervating muscle spindles in the intercostal muscles of the cat. *Nature*, 197: 1013–1014.

Sears, T.A. (1964a) Efferent discharges in alpha and fusimotor fibres of the intercostal nerves of the cat. *Journal of Physiology (London)*, 174: 295–315.

Sears, T.A. (1964b) The slow potentials of thoracic respiratory motoneurones and their relation to breathing. *Journal of Physiology (London)*, 175: 404–424.

Selionov, V.A. Shik, M.L. (1992) Responses of medullary and spinal neurons to simultaneous stimula-
    tion of two locomotor points. *Neurophysiology*, 24: 471–481.
Selionov, V.A., Shik, M.L. (1990) Two types of responses of medullary neurons to microstimulation of
    locomotor and inhibitory points of the brain stem. *Neurophysiology*, 22: 257–266.
Selionov, V.A., Shik, M.L. (1991) Convergence of influences from locomotor points of the midbrain and
    medulla and from an inhibitory pontine point onto medullary neurons. *Neurophysiology*, 23:
    297–306.
Severin, F.V. (1970) The role of the γ-motor system in the activation of the extensor α-motor neurones
    during controlled locomotion. *Biophysics*, 15: 1138–1145.
Severin, F.V., Orlovsky, G.N., Shik, M.L. (1967a) Work of the muscle receptors during controlled
    locomotion. *Biophysics*, 12: 575–586.
Severin, F.V., Shik, M.L., Orlovsky, G.N. (1967b) Work of the muscles and single motor neurones during
    controlled locomotion. *Biophysics*, 12: 762–772.
Shik, M.L. (1971). The controlled locomotion of the mesencephalic cat. *Abstracts, XXV International
    Congress of Physiological Science, Munich*, 8: 104–105.
Shik, M.L. (1986). An hypothesis on the bulbospinal locomotor column. In *Neurobiology of Vertebrate
    Locomotion*. S. Grillner, P.S.G. Stein, D.G. Stuart, H. Forssberg, R.M. Herman (Eds.), London:
    MacMillan Press, 39–49.
Shik, M.L., Orlovsky, G.N. (1965). Co-ordination of the limbs during running of the dog. *Biophysics*,
    10: 1148–1159.
Shik, M.L., Orlovsky, G.N. (1976). Neurophysiology of locomotor automatism. *Physiological Reviews*,
    56: 465–501.
Shik, M.L., Orlovsky, G.N., Severin, F.V. (1966a). Organization of locomotor synergism. *Biophysics*, 11:
    1011–1019.
Shik, M.L., Orlovsky, G.N., Severin, F.V. (1968) Locomotion of the mesencephalic cat elicited by
    stimulation of the pyramids. *Biophysics*, 13: 143–152.
Shik, M.L., Severin, F.V., Orlovsky, G.N. (1966b). Control of walking and running by means of electrical
    stimulation of the mid-brain. *Biophysics*, 11: 756–765.
Shik, M.L., Severin, F.V., Orlovsky, G.N. (1967). Structures of the brainstem responsible for evoked
    locomotion. *Sechenov Physiological Journal of the USSR*, 53: 1125–1132 (In Russian).
Stein, P.S.G., Smith, J.L. (1997) Neural and biomechanical control strategies for different forms of
    vertebrate hindlimb motor tasks. In *Neurons, Networks, and Motor Behavior*. P.S.G. Stein, S.
    Grillner, A. Selverston, D.G. Stuart (Eds.), Boston: MIT Press, 61-73.
Stein, P.S.G., Grillner, S., Selverston, A., Stuart, D.G. (Eds.). (1995) *Neurons, Networks, and Motor
    Behavior*. Proceedings of an International Symposium. Tucson, AZ, 8–11 November. Tucson:
    The University of Arizona. (See also URL<:http://www.physiol.arizona.edu/CELL/Department/
    Conferences.html>).
Stein, P.S.G., Grillner, S., Selverston, A., Stuart, D.G. (Eds.). (1997) *Neurons, Networks, and Motor
    Behavior*. Boston: MIT Press.
Stein, R.B., Pearson, K.G., Smith, R.S., Redford, J.B. (Eds.). (1973) *Control of Posture and Locomotion*.
    New York: Plenum Press.
Strausfeld, N.J. (1997) Oculomotor control in insects: From muscles to elementary motion detectors. In
    Neurons, Networks, and Motor Behavior. P.S.G. Stein, S. Grillner, A. Selverston, D.G. Stuart
    (Eds.), Boston: MIT Press, 277-284.
Stuart, D.G. (1985) Summary and challenges for future work. In *Motor Control: From Movement
    Trajectories to Neural Mechanisms. Short Course Syllabus*. P.S.G. Stein (Org.), Bethesda, MD:
    Society for Neuroscience, 95–105.
Stuart, D.G. (1996) *Motor Control VII*. Tucson, AZ: Motor Control Press.
Stuart, D.G., Binder M.D., Botterman B.R. (1979) Features of segmental motor control revealed in
    single–unit recordings during natural movements. In *Posture and Movement*. R.E. Talbott, D.R.
    Humphrey (Eds.), New York: Raven Press, 281–294.
Stuart, D.G., Callister, R.J. (1993) Afferent and spinal reflex aspects of muscle fatigue: issues and
    speculations. In *Neuromuscular Fatigue*. A.J. Sargeant, D. Kernell (Eds.), Amsterdam: Royal
    Netherlands Academy of Arts and Sciences, 169–180.

Stuart, D.G., McDonagh, JC. (1997) Muscle receptors, mammalian, spinal actions. In *Encyclopedia of Neuroscience* [CD-ROM; book form in 1998]. G. Adelman, B. Smith (Eds.), Amsterdam: Elsevier Science B.V. (In press).

Stuart, D.G., Mosher C.G., Gerlach R.L. (1972) Properties and central connections of Golgi tendon organs with special reference to locomotion. In *Research in Muscle Development and the Muscle Spindle.* B.Q. Banker, R.J. Przybylsky, M. Victor, J.M. Meulen (Eds.), New York: Exerpta Medica, 437–462.

Székely, G. (1965) Logical network for controlling limb movements in urodela. *Acta Physiologica Hungarica*, 27: 285–289.

Taylor, A., Davey, M.R. (1968) Behavior of the jaw muscle stretch receptors during active and passive movements in cat. *Nature (London)*, 220: 301–302.

Viala, G., Buser, P. (1965) Décharges efférentes rythmiques dans les pattes postérieures chez le Lapin et leur mécanisme. *Journal of Physiology (Paris)*, 57: 287–288.

von Holst, E. (1932) Untersuchungen über die funktionen des Zentralnervensystems beim regenwurm. *Zoologische Jahrbuecher. Abteilung fuer Allgemeine Zoologie und Physiologie der Tiere*, 54: 157–179.

von Holst, E. (1933) Weitere versuche zum nervösen mechanismus der bewegung beim regenwurm. *Zoologische Jahrbuecher. Abteilung fuer Allgemeine Zoologie und Physiologie der Tiere*, 53: 67–100.

von Holst, E. (1938) Über relative koordination bei säugern und beim menschen. *Pflügers Archiv*, 240: 44–59.

Waller, W.H. (1940) Progression movements elicited by subthalamic stimulation. *Journal of Neurophysiology*, 3: 300–307.

Weiss, P. (1941) Self-differentiation of the basic pattern of coordination. *Comparative Psychological Monographs*, 17: 1–96.

Wetzel, M.C., Stuart, D.G. (1976) Ensemble characteristics of cat locomotion and its neural control. *Progress in Neurobiology*, 7: 1–98.

Wiersma, C.A.G. (1947) Giant nerve fiber system of the crayfish. A contribution to comparative physiology of synapse. *Journal of Neurophysiology*, 10: 23–38.

Wilson, D.M. (1961) The central nervous control of flight in a locust. *Journal of Experimental Biology*, 38: 471–490.

Wilson, D.M. (1965) Central nervous mechanisms for the generation of rhythmic behavior in arthropods. In *Nervous and Hormonal Mechanisms of Integration. Symposia of the Society for Experimental Biology. No. XX.* G.M. Hughes (Ed.), Cambridge: Cambridge University Press, 199–228.

Zajac, F.E., Young, J.L. (1980) Discharge properties of hindlimb motoneurons in decerebrate cats during locomotion induced by mesencephalic stimulation. *Journal of Neurophysiology*, 43: 1221–1235.

3

CHAPTER

Automation of Movements: Challenges to the Notions of the Orienting Reaction and Memory

Lev P. Latash
Chicago, IL, U.S.A.

Introduction: Bernstein's Principle of Activity in Brain Processes

N.A. Bernstein is known mostly for his studies of biomechanics and motor control that formed the theoretical and experimental foundations for these areas of neuroscience. However, his ideas spread far beyond movement science. Studies by Bernstein led to the formulation of the basic problems of and defined the development routes for contemporary integrative neuroscience as a whole. His studies signified the liberation of neuroscience from the reflex tradition. Its reformation led to a new paradigm based on notions from the control theory and certain areas of mathematics. At present, theoretical neuroscience may be considered an independent branch of brain science. The importance of the studies by N.A. Bernstein in this respect may be compared to the importance of Maimonides' reform of Judaism, Luther's reform of Christianity, or Maxwell's revolution in physics.

Bernstein's deep analysis of the regularities and mechanisms of controlling purposeful movements led to the formulation of the principle of activity in brain processes. The key notion of an active (rather than reactive) interaction of an organism with the environment penetrates all the categories of brain processes. It includes the organization of movement, perception, and mentality.

An organism's active interaction with the environment includes overcoming the environmental resistance to realize different goals, programs, and attitudes. The complexity of movement construction (control), as demonstrated by Bernstein (1927, 1935, 1947, 1967, 1990), is apparently related to the fact that the system that controls voluntary movements provides the main means of realizing the organism's most vital needs associated with its physical interaction with the environment. The activity of this system induces changes in the organism-environment interactions and requires high accuracy and reliability. Such actions by a healthy person always imply obtaining a significant result.

A number of steps, conscious and unconscious, can be identified within movement construction. These include, in particular, the formation of a "motor task" (its neural equivalent); the elimination of the redundant degrees of freedom (DOFs) at the level of both brain substrate and peripheral kinematics (the Bernstein problem) (Latash 1993; Turvey 1990); mastering or overcoming unexpectedly changing external and reactive forces; organization of a multilevel, cyclic interaction of different structures of the control hierarchy using continuously functioning feedforward and feedback loops; and assessment of the efficacy (e.g., accuracy, quickness, expediency, and reliability) of the control strategies and tactics, which is likely to be particularly important for motor skill acquisition. This array of complex problems makes quantitative characterization of the neural substrate of motor-control processes nontrivial. The complexity and importance of these processes require the participation of a large number of neurons organized into numerous chains and nets with a degree of redundancy. This assures reliable movement realization even in cases of a lesion or functional unavailability. On the other hand, the complexity of the motor-control system in conditions of numerous simultaneous and successive motor acts requires control frugality with the maximal possible decrease in redundancy.

The requirement of eliminating redundancy primarily refers to brain processes and mechanisms at the higher level of the control hierarchy. Unique opportunities are presented by control of repetitive movements or repetitive fragments involved in different movements. One may assume that the control system solves the problem of redundancy at different hierarchical levels using a complex of interrelated methods. These methods may be based on the internal design of the system for movement production, which allows a substantial decrease in the number of independently controlled variables (i.e., variables supplied by hierarchically higher control levels). For example, the actively discussed λ-model (Berkinblit et al. 1986; Feldman 1979; Latash 1993) has received both theoretical and experimental support in human studies, at least with respect to single joint movements. According to the λ-model, central commands during voluntary movements are not formulated in terms of activation levels of α-motoneurons innervating participating muscles. Instead, a hypothetical equilibrium point is defined through the tonic stretch reflex mechanism including γ-motoneurons and interneurons. Another example is the special organization of a lower control level that allows a high degree of autonomy so that descending signals act as a trigger whose quantitative level can lead to different locomotor patterns such as walking, running, trotting, and galloping (Sherrington 1910; Shik and Orlovsky 1976). Similar findings were described for certain insects whose very high frequency wing beats do not allow a descending (from the head ganglion) control of each beat because of the time deficit (Bassin et al. 1966; Bernstein 1947). Very high possibilities of autonomous control of movements by lower centers were described in the spinal frog. It wipes off a nociceptive skin stimulus in different conditions using a wide range of hind limb movements, including those that had never been used in its individual activity prior to the experiment. In these experiments, the transfer to new control tactics and new, effective motor coordination occurred virtually instantaneously (Berkinblit et al. 1986). These findings show that the organization of a lower level of movement control may be viewed as a set of rules and conditions of coordination switching, a kind of "control matrix" (Bassin et al. 1966; Gelfand and Tsetlin 1962; Gelfand et al. 1971). This coordination switching requires an action by a higher level only when peripheral conditions exceed the abilities of a current control matrix and a change in coordination is required, including the control matrix. These examples illustrate a decrease in the system's redundancy induced by a decrease in the complexity of descending commands and a decrease in their number.

All the mentioned principles of motor control (and maybe some others), which reduce redundancy, belong to the internal resources of the motor system. They apparently characterize mostly inborn methods of control or those developed during early postnatal ontogenesis (e.g., vertical posture and equilibrium control). For movements constructed in later ontogenesis, particularly those developed as a result of numerous repetitions, an important way of harnessing redundancy is movement automation. A crucial role in movement automation is played by "whole brain mechanisms." This term is less compromised than "nonspecific brain systems." It refers to a complex of structures with central roles played by the reticular formation of the brain stem and diencephalon, thalamic nuclei, and limbic system. Characteristic features of the whole brain mechanisms (WBM) include their participation in the generation and regulation of brain electrical rhythms, sleep and wakefulness, emotional and motivational processes, and memory. Obviously, the influence of conscious awareness and will occurs through these mechanisms. There are reasons to believe that during movement automation, these mechanisms use internal resources of the motor system as well.

Movement Automation and the Orienting Reaction

The Major Features of Voluntary Movement Automation

Voluntary movement automation is traditionally viewed as a fixation of the results of a learning process, which takes place in the motor system during repetition of a motor act. Such a learning process leads to the formation of motor skill whose automation produces a machine-like stereotype of the movement. Resulting identical trajectories of the movement assure standard quickness and accuracy in each repetition without conscious concentration of attention every time. For a long time, conditioning (associative learning) was considered the major form of everyday motor learning. Hence, the hypothetical neural mechanisms of movement automation were considered analogous to the neural mechanisms of conditioned reflexes. These involved slow formation ("beating a trail") and memorizing neuronal connections between appropriate neocortical sensory and motor centers. The positive role of numerous repetitions was related to the weakness of a trace effect from a single movement performance and to an amplification or summation of such effects on neural conductivity, until the summed effect reached a threshold required for the formation and fixation of a "beaten trail." Then, the movement became automated.

N.A. Bernstein studied athletic and labor movements whose automation looked like an important method to increase efficacy and reduce redundancy. The results of his studies required complete reconsideration of the described views. Bernstein found out that in motor skill acquisition, the variability of movement trajectories and other characteristics was not eliminated. Movements do not become identical or machine-like, although the ultimate motor outcome becomes highly reproducible. For example, with movements like shaving with a sharp razor blade or shooting at a target, success depends on fractions of a millimeter or of an angular minute. Only high variability in automated movements allows the reaching of such a high accuracy when unexpected forces intrude. Sensory corrections and their interaction with different levels within the control hierarchy appeared central to movement automation. These observations led Bernstein to a very concise and accurate definition of movement automation:

"Automation of a motor act is . . . switching over a number of components of a movement to lower levels of control, i.e. switching over a number of coordinative corrections of the motor act to afferentation pertaining to the lower levels which are most adequate for these particular types of corrections" (Bernstein 1947, page 183).

The importance of movement repetitions for automation has nothing to do with "beating trails" or summation of memory traces. The importance of sensory repetitions is to assess the significance of changes in the environment for a planned motor act and lawful relations among them that can be established after only one to two trials. Motor repetitions include a multistep search of the most effective ways of control, including the creations of the mentioned control matrices, together with their sensory corrections at lower levels of the motor-control hierarchy.

Bernstein (1947) meticulously analyzed the problem of movement automation, primarily the role of internal resources of central control of the motor system. Bernstein made the following three important points. First, the pronounced variability of even highly automated movements reflects the search and usage of the most effective ways of motor control. It leads to outcome stability in the unpredictable environment. Second, the specificity of movement control leads to a reorganization of the whole movement even in response to a very small change in conditions. (Movements never react to a change in a detail of the conditions by a change in a detail of the control.) Third, some skills emerge by a jump (e.g., swimming and riding a bicycle) and are retained for the whole life. Automation can be very quickly restored after a prolonged period of disuse that may lead to some deautomation. These three points suggest an important role of external resources in movement automation, particularly those related to the processes and mechanisms of the orienting reaction (OR) and memory. It is possible to assess some aspects of WBM involvement in movement. On the other hand, the role of OR and memory, particularly of the long-term memory (LTM), in movement automation presents a serious challenge to many key concepts in current theories of the mechanisms and functioning of these fundamental, integrative brain phenomena.

Studies by Bernstein and most of his followers created a tradition of isolated investigations of internal resources of the motor control system. On the one hand, a particular style of detailed motor control research with an extensive usage of mathematical modeling methods has emerged. On the other hand, a deficit has occurred in the knowledge of motor control system interaction with other brain systems during movement construction and automation. The discovery of the functions of the reticular and limbic formations signified the beginning of the studies about the role of WBM in motor control. The influence of cybernetics and, later, of cognitive psychology has resulted in the emergence of concepts that allow expansion of the definition of movement automation to an action in general. A special role was attributed to the concept of decision making. It implied special processes of active choice, mostly unconscious, for triggering of certain types of movements and movement control tactics when alternatives exist. Psychophysiological analysis of decision making is important for understanding brain mechanisms underlying the active nature of brain processes.

The notion of decision making can also be applied to the process of transferring control to "lower" levels during early phases of movement automation. Further development of automation can lead to a decrease in the role of decision making and, eventually, to its disappearance. These considerations have influenced studies of OR whose main results will be briefly reviewed later. They have also forced reconsideration of the nature and role of OR and of its habituation in the integrative activity of the brain. Special attention will be paid to the development of ties between emotional mechanisms of motivation as manifested in the electrodermal component of OR and to decision-making processes related to motor control and, particularly, to movement automation.

Evolution of the Notion of Orienting Reaction

Studies of OR originate from the description by I.P. Pavlov of a complex of changes in the organism which manifest mainly as changes in behavior, including interruption of the current activity and an adjustment of sensory organs to a new or unexpected signal. This complex was called the novelty reflex or the What Is This? reflex, later renamed the orienting reflex. These studies gave rise to a whole series of investigations that considered this "reflex" as related exclusively to sensory processes. Its functional importance was viewed as the mobilization of attention and sharpening of perception of the sensory systems to unexpected (new) events in the environment in order to identify or explore them, primarily as signals of potential danger. It was believed to facilitate the formation of conditioned reflexes. This direction of studies dominated until the 1960s and is still very influential despite serious contradictions established during the last 35 years. Studies of OR benefited significantly from the introduction of polygraphic methods of recording, in particular electroencephalography (EEG). As a result, the complex nature of the "reflex" and its components (manifestations) in different body systems were discovered. Those include signs of activating different somatic (including the brain) and autonomous systems united by their characteristic dynamics. The dynamics supposedly include emergence of the "reflex" in response to an unexpected stimulus, its attenuation and disappearance during stimulus repetition (habituation), and dishabituation when the stimulus unexpectedly changes. At the same time, the theoretical basis of the mechanisms and functional role of the reaction was developed. (The original term, "reflex", was gradually replaced by "reaction" and "response.") The theoretical basis included the well-known concept of a neuronal (neural) stimulus model (Sokolov 1964). It was largely based on the statistical information theory, and OR was considered a response to the incoming information. Based on this information, the brain develops a neuronal model of the stimulus (or of an event). When the stimulus stops to deliver information (because of its repetition), habituation takes place. When a stimulus enters the brain, a standard procedure is initiated by matching the stimulus with an existing neuronal model. If a mismatch between the two occurs, an OR emerges. Despite all the complexity, this concept preserved the traditional, exclusive attitude to sensory information; the passive, "reflective" nature of OR; and even the term "reflex." Action on a person's mentality was considered only with respect to facilitating completion of a conditioned link.

Significance—A Leading Factor in Orienting Reaction

The study of OR in healthy persons and in neurological patients was performed in the 1960s and allowed the authors to suggest an alternative approach to and a basically different functional interpretation for OR and its components. A detailed description of these studies was published in both Russian (Graschenkov and Latash 1963, 1965; Latash 1968; Latash 1993) and English (Latash et al. 1974; Latash 1990). Only a brief account on the basic concepts and the main results will be presented concerning the relation among some components of OR and motor-control processes, especially as related to automation. A few new results will also be discussed.

This line of research has four important, distinctive features. First, OR emerges only in response to significant changes in a situation, which may refer to changes in the environment or in the subject (his or her attitude). This implies that the emergence of an OR is preceded by brain processes, typically unconscious, dealing with an assessment of the significance (even hypothetical) of current changes with respect to a current hierarchy of needs, motivations, and goals. Therefore, the nature of OR is not reactive but active (in Bernstein's sense), i.e., defined ultimately by internal factors of brain activity. Events that have no representation in the person's experience, i.e., absolutely new events, are not significant and do not elicit OR. Second, OR is not a unitary reaction of a general arousal type, as was traditionally assumed. Instead, it is a complex, polyfunctional activity whose different aspects are reflected in different OR components that are actual *components* rather than simply manifestations of one and the same event in different body systems. These components may change in different directions based on their different functional importance and different underlying brain mechanisms. Third, OR represents the processes of the early stage of organizing a new (nonstandard) action—motor, sensory, or mental. Each time, the action is triggered actively, so that a decision about its initiation has to be taken. Habituation of an OR manifests an attenuation of the active control of an action due to its automation. Thus, the traditional interpretation of OR as a passive, sensory-cognitive phenomenon (a reflex "What Is It?") was substituted with its interpretation as *activity forming an action* (reaction "What and How to Do?"), i.e., as an active phenomenon. Fourth, based primarily on the differences in the conditions of elicitation and the rate of habituation and dishabituation, different functional roles were assigned to processes related to such OR autonomous components as phasic skin galvanic responses (SGR), changes in heart rate, and also to changes in EEG components (alpha-desynchronization and vertex potential). SGR is either an increase in the skin conductivity (SCR) or a shift of the skin potential toward larger negative values (SPR). Both types of SGR represent manifestations of activating the emotional-motivational mechanisms during decision making, typically, without a reflection in consciousness. EEG arousal is connected to mobilization of sensory processes drawing attention to a stimulus (correlated with heart rate deceleration) and a general increase in alertness (correlated with accelerated heart rate).

Decision-Making, SGR, and Movement Automation

The described notion has been supported with respect to the role of the significance of a new stimulus in OR elicitation (Bernstein 1969; Bernstein and Taylor 1979; Grastyan et al. 1965; Kochubey 1979; Maltzman 1990; Maltzman and Raskin 1965) and the polyfunctional nature of OR (Barry 1982; Berlyne and McDonnell 1965; Wedenyapin and Rotenberg 1984). However, the notion is still an object of active discussion. Adepts of the traditional point of view focus their objections to the role of emotional mechanisms in OR (see the comment by Sokolov in Latash 1990). Recently, new data have demonstrated that visual and auditory stimuli reaching the conscious mind cannot be considered as impartial percepts that are later evaluated. They arrive already with an emotional "mark" of the positive or negative sign (like or dislike) that is established by the brain processing mechanisms at the instant of perception without participation of consciousness. This initial assessment may be either preserved or overridden during further processing (Bargh 1995). Although the role of unconscious attitudes based on the subject's experience is obvious, in many cases these attitudes reflect hypothetical extrapolations of the experience that may happen to be wrong. In any case, these data strongly support the active notion of OR.

The presented concept has been based on the results of the following studies. Figure 3.1 shows averaged data about the rate and sequence of habituation of SGR and EEG components of OR for three conditions. A shows the response to an "indifferent" stimulus (a tone). B shows the response to the same stimulus transformed into the signal for a motor action leading to an increase in stimulus significance. C shows a situation when stimulus novelty (unexpectedness) and significance were separated and forced to compete with each other. The subject was instructed to act (press a button) in response to the second signal of a pair while the first signal provided a warning. This made the second signal less unexpected but more significant while making the first more unexpected but less significant. The results unambiguously demonstrated a sharp slowing of the SGR habituation with the characteristic inversion of the order of habituation of OR components when the "indifferent" stimulus' significance was increased by turning it into a signal to act. Replacing the motor action with a mental one (counting the second stimuli in mind, a mental representation of the emotiogenic situation; Simonov et al. 1964) did not change the results. The effect was likely related to the increase in the significance of the signal associated with triggering an action.

Data also showed a relation between SGR and movement itself. For example, a positive correlation occurred between the latencies of the SGR over 2.5 s and of the movement (figure 3.2). If the subjects were instructed to perform the movement not immediately after an imperative signal but after an arbitrary delay, OR emerged in response to both stimuli and to the movement initiation. OR in response to the movement initiation demonstrated the biggest delay in SGR habituation. This OR did not result from a change in the routine proprioception and an ensuing mismatch with a "neural model of the stimulus." It is well known that even the most automatic movements do not repeat each other kinematically and, therefore, produce different, unexpected proprioceptive afferentation each time. If OR were induced by changes in these afferent signals, habituation would not have occurred at all. However, it did occur, albeit slower. In motor tasks associated with a particular significance of proprioceptive information (e.g., walking on a rope), changes in afferent signals would conceivably be able to produce OR. Thus, OR can be induced by a signal about significant changes in the situation and by a movement. However, proprioceptive afferent signals typically do not play a role in the latter effect. Note the prevailing relation of movement to the SGR component of OR. The EEG component showed, in the same experiment, an opposite dynamic characterized by an increase of the habituation. Apparently, alpha-desynchronization, as a component of OR, reflects different, sensory-cognitive aspects of OR since an increase (slowing down of habituation) in alpha-desynchronization was seen in cases of increased complexity of the sensory characterization of the stimulus (Berlyne and McDonnell 1965; Bernstein and Taylor 1979). Apparently, movement automation occurs slower and preserves the feature of active action triggering (see later) as compared with sensory actions, particularly during perception of very simple stimuli.

The cause of the SGR during voluntary movement was unveiled with the help of a rather stable time window within which this OR component was observed in response to an indifferent stimulus. The value is based on over 1500 observations. The magnitude of the latency was virtually always within the range from 1.4 s to 2.5 s after the stimulus.

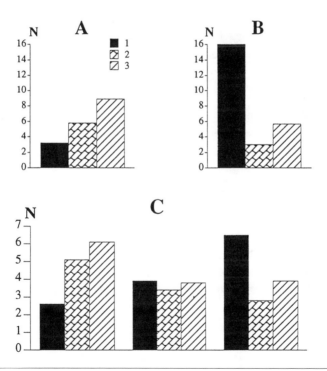

Figure 3.1 Habituation rate of OR components (1) SGR, (2) alpha desynchronization, and (3) vertex-potential in normal subjects. Ordinate is the number of stimulus presentations. *(A)* presentations of "indifferent" tones (instruction to sit relaxed), *(B)* presentations of signals to action (instruction to press a button in response to the same tones), and *(C)* presentations of "indifferent" tones, warning tones, and imperative tones. Mean values for 41 subjects are shown.

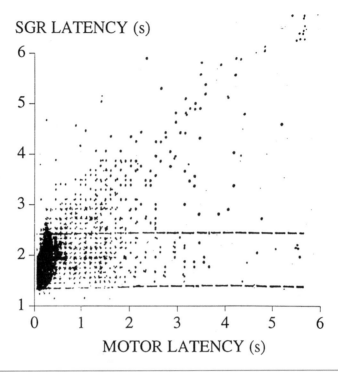

Figure 3.2　The relationships between the latencies of the SGR (ordinate) and motor response to the imperative signal (abscissa). Time in seconds. Note the relationship with SGR latencies over 2.5 s.

Table 3.1 shows that in about 10% of observations, SGR emerged too late after the stimulus to be considered a direct response to the signal and too early to be related to the initiation of the motor act. In most cases (over 80%), SGR could emerge, in different trials, with respect to several (two to four) time moments in the interval from the imperative signal to the movement. This variability of the SGR timing suggests its relation to a process that is tightly coupled neither with perception of the stimulus nor with initiation of the movement. It may therefore be associated with the process of decision making that may reflect the instability of additional information processing and be the only one able to demonstrate timing fluctuations depending on the significance of the action, its complexity, fatigue, and so forth.

The first person to describe SGR was Tarchanoff (1889), who established a relation between SGR in humans and a psychological phenomenon, an emotion. He called the reaction the "psycho-galvanic reflex." This view has been confirmed many times about conscious emotional feelings. During OR, the emergence of an SGR is typically not associated with a conscious emotion. There are two possible explanations. First, a microactivation of emotional processes that is not consciously

Table 3.1   SGRs in Response to the Imperative Signal and to the Movement

Related to . . .				
Total # of SGRs	Signal	"Signal + Movement" complex	Movement initiation	A moment between signal & movement
1864	638	659	398	174

Related to . . .				
# of experiments	4 time moments	3 time moments	2 time moments	Single time moment
131	21	47	40	23

	# of subjects			
62	17	29	12	4

perceived occurs in the brain. Second, during OR, SGR is unrelated to emotional brain mechanisms (Sokolov's view).

Role of Unconscious Microemotion in the Orienting Reaction: Mental Disorders, Self-Stimulation, and Pretuning

The results of the author's studies corroborate the former view, implying the universality of the functional interpretation of SGR suggested by Tarchanoff. Figure 3.3 illustrates the results of a similar study of OR in neurological patients with local lesions of the hippocampus and other limbic structures, frequently associated with Korsakoff's amnesia (Latash 1968; Dallakyan et al. 1970). In cases of emotional inhibition such as apathy or lack of spontaneity (the upper part of figure 3.3), SGR, as an OR component, was either absent or habituated very quickly. It was not amplified when the stimulus turned into a signal to act. In patients who survived the acute phase (typically, in those with nontumorous lesions), emotional disinhibition such as euphoria and confabulations was frequently observed while the memory deficits persisted. In these cases (the lower part of figure 3.3), SGR was restored and amplified in response to a signal to action. Its habituation was markedly slowed. The characteristic reversal of the order of habituation of OR components was also observed.

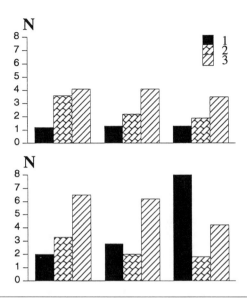

Figure 3.3   Habituation rate of OR components in patients with local lesions of the hippocampus and other limbic formations (usually with Korsakoff's amnesia). Upper part—patients with emotional inhibition; lower part—patients with emotional disinhibition.

Psychopathology is rich with examples of relations between SGR and the emotional state. For example, sharp emotional activation with the development of strong anxiety in acute alcohol withdrawal syndrome is accompanied by an SGR storm (Danilin et al. 1984). Both phenomena were quickly suppressed by benzodiazepines (of bromazepam type). SGR nonresponding was observed in some forms of depression and other psychiatric states. Interestingly, SGR changes are typically not perceived subjectively, i.e., they exist beyond the realm of consciousness. Damasio (1994) also noted this as a kind of alienation of a consciously perceived emotional influence from an absent emotional feeling. Now, the relationship between SGR and emotional activation is acknowledged more and more. SGR is even considered a specific somatic marker of emotions (Damasio et al. 1991). Thus, Sokolov's position (Sokolov, 1990) rejecting this interpretation of SGR as an OR emotional component looks unfounded.

What is the function of the short-lasting, unconscious activation of emotions in the decision making processes? One may assume that, during such processes, a state of microanxiety (microstress) emerges whose magnitude depends on the significance of the movement being organized and the complexity of choice from available alternatives (amplifying function of emotions). Emotional reinforcement can help to make the choice. On the other hand, taking into account the assessment function of emotions related to the availability of their sign, emotional microactivation can reinforce positive choices (rewarding) and block negative choices (punishment). This possibility corresponds well with the views of Routtenberg (1968) about two types of ascending activation, reticular and limbic. This possibility was demonstrated in experiments with brain self-stimulation (SS) of the medial forebrain bundle in the lateral hypothalamus during simultaneous recording of EEG and brain temperature in the rat (Latash and Kovalson 1973).

The upper part of figure 3.4 shows that, when freely moving rats stimulated the positive rewarding zone with moderate current strength (the duration of the stimulation was defined by pressing the lever by the animal itself), during relatively long periods of stimulation, there was a typical rhythmic, synchronized activity in the hippocampus (RSA, hippocampal theta rhythm, since theta frequencies dominated as a component of OR and during REM sleep, Green and Arduini 1954; Kovalson 1971; Vanderwolf 1971; Gottesman 1992) with somewhat increased frequency (8 to 9 Hz), formally related to the alpha-band by spectral analysis. During intervals between successive pressings, one could see either desynchronization or theta-rhythm of a lower frequency. An increase in the strength of the stimulating current by about 50% led to a marked change in the picture (the lower part of figure 3.4).

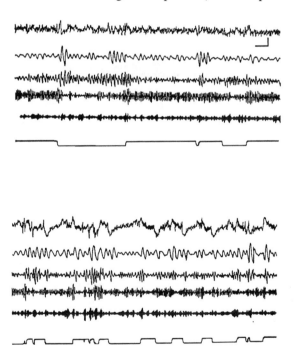

Figure 3.4 EEG in free-moving rats during SS of MFB in the lateral hypothalamus with stimulus train length determined by the animal. In both fragments from top to bottom: dorsal hippocampal EEG; frequency analysis in delta, theta, alpha, and low beta bands; SS marks (solid line shifted upward). Calibration: EEG—50 μV, time—1 s. Upper fragment—SS with moderate current intensity (6 V, 0.1 ms, 100 Hz), RSA during SS; lower fragment—SS with higher current intensity (10 V, 0.1 ms, 100 Hz), RSA in intervals between SS.

Pressings became brief but frequent, RSA increased its frequency to 10 to 11 Hz and moved into intervals between successive pressings, while during pressings, EEG desynchronization was seen. Since elements of the punishing system are scattered within the brain stem and diencephalon and have a higher threshold, moderate stimulation induces largely a rewarding effect. During stronger and iterative stimulation, an increased rewarding effect is accompanied by a steeply rising punishing effect which forces the animal to stop the stimulation. Pressing the lever in the absence of stimulation did not induce either similar hippocampal EEG dynamics or typical sharp increase in the brain temperature, that would be similar to the one seen during REM sleep which is also characterized by dominating RSA, although in the absence of voluntary motor activity and EMG.

These data should not be interpreted as a direct expression of positive and negative emotions in EEG. Rather, it may be concluded that EEG differently reflects incentives for a voluntary movement associated with anticipation of a desired outcome or those associated with anticipation of an undesired outcome requiring an evasive movement. Apparently, unconscious emotional sanctioning of a certain choice of action is a part of the structure of the cognitive unconscious as described by Kihlstrom (1987).

Relations between particular aspects of voluntary movement control and the activity of emotional mechanisms have not been studied systematically. The data in this area are fragmented.

To analyze these relations, the phenomenon of motor "pretuning" in humans was used in accordance with Gurfinkel et al. (1965). This phenomenon represents an increase in the excitability of spinal alpha-motoneurons as reflected by the amplitude increase in the H-reflex of corresponding muscles, which can be seen for a few hundred milliseconds with an additional sharp elevation 60–80 ms prior to movement initiation, i.e., in the latent period after a signal sounds. Changes in the phenomenon during numerous movement repetitions were studied in neurological patients with local lesions of limbic formations and no SGR, and in healthy subjects and patients with a pronounced SGR. No patient displayed any clinically diagnosed motor abnormalities (Latash et al. 1974). The subjects performed repetitive ankle plantar flexions in response to signals, i.e., in conditions implying developing movement automation. H-reflexes were induced in m. gastrocnemius. An "indifferent" stimulus, i.e., one that did not require the movement, induced a moderate increase in the H-reflex amplitude, which could be seen for about 500 ms. Repetitions of the "indifferent" stimulus led to a relatively quick habituation of this effect, although not as quickly as the habituation of the SGR. Thus, the H-reflex elevation in this case behaved like a component of OR. Figure 3.5 shows the dynamics of the phenomenon prior to the voluntary movement in the two subject groups. Each subject performed about 85 to 90 movements at time intervals ranging from 10 s to 15 s. Comparisons were made between the average indices for the H-reflexes observed during the first 40 to 45 movements (the first half of the experiment) and during the rest of the movements (the second half of the experiment). Subjects with SGR showed, after the signal, a 15% to 20% increase (compared with the control

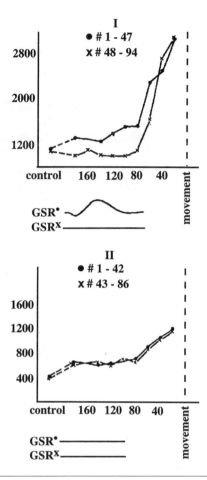

Figure 3.5 The phenomenon of "motor pretuning" in normal subjects with distinct SGR (above) and in patients with local limbic lesions and no SGR (below). Ordinate—H-reflex amplitude in μV; abscissa—time before movement.

value) of the H-reflex amplitude during the first half of the experiment (the "pretuning fringe"). This increase emerged after the first signal; it was relatively stable and showed an additional increase in the H-reflex amplitude of about 150% that began 70–80 ms prior to each movement. The second half of the experiment was characterized by a small but significant decrease in the H-reflex amplitude during the "pretuning fringe" phase, while the additional increase persisted, was more sharp, and was more tightly tied to the movement initiation.

Patients with limbic brain lesions and no SGR demonstrated different dynamics. Baseline H-reflex amplitude was about 30% of the values for the first group. The "pretuning fringe" increase in the reflex amplitude during the first half of the experiment was larger, and the same absolute values corresponded to about 50% increase.

The additional increase was smaller, about 100%. No changes occurred in the dynamics of these indices in the second half of the experiment. In particular, the H-reflex amplitude did not decrease during the "pretuning fringe."

The findings in the first group of subjects can be interpreted as correlates of movement automation based on an economical strategy characterized by a more precise timing of the additional increase in the motoneuronal excitability. The patients with limbic lesions were unable to generate SGR because of the lesion in the brain substrate of emotions. They demonstrated a different movement automation strategy. The long-lasting increase in the motoneuronal excitability persisted during the "pretuning fringe" in the absence of its attenuation and with a smaller additional increase. In this case, decision making was apparently limited to choosing between relatively close alternatives. Gelfand and Tsetlin (1962; see also Gelfand et al. 1971) suggested that this type of motor task solution could be done with only a "local search" without an extrapolation "step along the ravine." As a result of the automation strategies used, the excitability range prior to a movement gets more narrow while the variability of the movement latencies increases. An effective participation of emotional processes in decision making allows the use of a narrow range of latencies with a short period of higher excitability and more economic control strategies without involving the higher levels of the control hierarchy. Individual movements become more standardized. This is a rather speculative conclusion that seems plausible enough to be used as a working hypothesis for future studies.

This is how the author sees the role of one of the WBM, the mechanism of emotional activation in movement automation. At present, it is hard to speculate on the degree of specificity of this role partly because of the proximity or even overlap of the brain substrate for movement control and for OR (subcortical nuclei and dopaminergic mediation with a particular role of D4 receptors). On the other hand, similar relations exist between OR mechanisms and, particularly, SGR and the organization of some nonmotor actions. Studies of mnemic actions in patients with Korsakoff's amnesia demonstrated the following phenomena. In these patients, the duration of short-term memory (STM) traces was assumed to be much shorter, leading to an impairment in the transfer of the traces into LTM. A test suggested by Konorski (1961) was used: The subjects were instructed to press a button in response to the second stimulus of a pair but only when both signals were identical. Time intervals between the stimuli could vary. The maximal interval at which the test yielded results implying remembrance of the first stimulus was viewed as a measure of the duration of storage in STM. Subjects without memory disorders and with SGR in response to a signal to action could perform the test with any intervals between the stimuli, up to tens of minutes. In patients with amnesia and no SGR, the test could not be performed properly with interstimuli intervals more than 8–10 s. This was shown in experiments with a quasi-random presentation of pairs of identical and different stimuli at different time intervals. If the interval increased slowly, at small increments of 1–2 s (figure 3.6, upper part), the maximal intervals at which the test was performed correctly increased nearly twofold, sometimes exceeding 20 s. In other words, the results of the test depended not only on the absolute interval duration but also on the type of its time change. If the interval increased at steps of no more than 4 s (figure 3.6, lower part), the results improved significantly, i.e., the intervals at which the test was correctly performed increased. These data do not question the crude memory disorder following a bilateral lesion of the limbic system but suggest its complex nature. This phenomenon seems closely related to the "local search" suggested for the behavior of the "pretuning" in the same categories of patients. It suggests intricate roles for the WBM in the functioning of special brain subsystems, and provides an appropriate link to the next section.

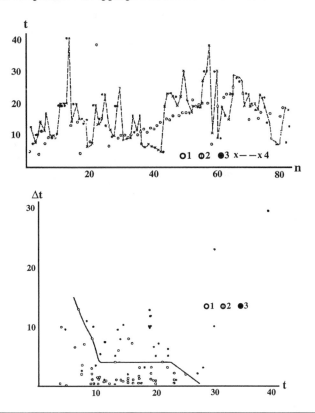

Figure 3.6 Influence of the time interval increase within a pair of stimuli upon the Konorski test performance by patients with Korsakoff's amnesia due to limbic lesions (no SGR). Instruction—to press a button to the second signal only if the first and second signals are the same. (1) empty circles, correct performance of positive signal combination (the same); (2) empty circles with vertical line, correct performance of negative signal combination (different); (3) filled circles, error; (4) dotted line, time intervals between pairs. Upper part—successive test presentations. Ordinate—time in seconds; abscissa—ordinal number of signal pairs. Lower part—relations between magnitude of step size of time interval increase within each pair and maximal interval length with correct performance of the test. Ordinate—delta t (step size); abscissa—time in seconds. The solid line shows the zone with a prevalence of correct performances (below the line).

Movement Automation and Memory

The important role of neural memory in movement automation is defined by two factors. First, according to Bernstein, the main functional purpose of automated movement repetition is the creation and testing of optimal control tactics. The well-known Latin proverb, "Repetition is the mother of learning," should apparently be rephrased as, "Repetition of solving is the mother of creativity." A process that, in the future, would allow the movement to be performed without control from the "higher" levels of the control hierarchy implies fixation of intermediate results of skill acquisition in memory. This is particularly obvious when skill acquisition requires prolonged periods of time, e.g., automation of erect walking, piano playing, car driving, and so forth. Second, fixation of the result of automation, frequently for the whole duration of an individual's life, requires an ultimate fixation in LTM. This is true both for automation of motor coordination occurring without participation of consciousness (the so-called memory of implicit learning) and for the results of consciously acquired skills (explicit memory).

Presently, it is very hard to suggest a satisfactory explanation of the mechanisms and processes within the CNS that define the interaction between skill acquisition and memory. In particular, it is not clear which of the intermediate products and in which form get stored in LTM, what stimulates continuation of the search for better solutions, or what leads to its termination. The main difficulties, however, are due to the state of knowledge about the mechanisms of memory in CNS. Considerable progress in this area has led, in particular, to establishing the special role of the hippocampal formation in the organization of mnemic activity and to the identification of major types of memory: STM, LTM, working or operative memory, and explicit and implicit memories. However, a number of fundamentally important problems are still without well-based answers. Among them are the following:

- Which of the external (with respect to the CNS) events, in which form, and in which composition are being remembered?

- Which of the brain processes occurring during perception of external events, including results of consciously perceived or not perceived creative processes, are being fixed in memory?

- Are all or only some of neural elements (neurons, their fragments, and constellations, i.e., nets) able to fix memory traces?

- What are the neural substrates of STM and LTM, and what are the mechanisms of information exchange between them?

- What is being affected during brain injuries leading to memory disorders? Is it particular mechanisms of memory fixation, storage, and retrieval, or whole brain mechanisms (WBM) that participate in the organization of mnemic activity, such as general arousal, consciousness, attention, motivation, incentives, classification, and encoding?

These questions do not even have firmly established negative answers. Existing experimental data and related concepts are, as a rule, very much open for criticism and cannot form the foundation for a consensus. However, the following LTM characteristics are relatively well established and commonly accepted. First, storage continues throughout the whole life of a healthy person, while forgetting is a result of problems with retrieval. Second, storage is associated with structural changes in a neural substrate without participation of neurodynamics, i.e., neuronal firing or other processes that require additional energy expenditure, though these processes may be vital for fixation, retrieval, and modification of LTM. Third, storage is not affected despite the continuing renewal of the substance forming the structure ("the skeleton") of neurons. There are serious objective problems with separating processes directly related to memory, because any mnemic activity includes processes related to WBM. On the other hand, virtually any integrative brain activity proceeds in an interaction with memory or is based on memory.

Studying the processes of movement automation may be interesting with respect to particular problems these processes pose to neural memory in general and to some of its aspects in particular. Such studies may potentially generate answers to some of the mentioned questions. One of the most important questions for this analysis is the question of LTM substrate in CNS.

Hypotheses of the Synaptic Nature of Memory Traces and the Neuron-Net Hypothesis

Studies of movement automation have shown that the widespread views about the role of selective fixation of facilitated synaptic connections between interacting neurons (the "beating the trail" concept) in recording and storing LTM traces meet insurmountable difficulties. As Bernstein (1967) showed, unexpected changes in the conditions of movement execution, particularly due to changes in passive and reactive forces, led to irreproducible kinematic and neural patterns during repetitions of an automated movement. This means that, each time, the movement control reorganizes, even if differences in kinematics are small. This fact renders fixation of an automated movement as rigidly fixed connections of neural elements senseless. Taking into account the pronounced, unexpected variability of movements, zillions of fixed facilitated connections would be needed, probably with numerous parallel loops for reliability. In this case, the problem of redundancy would have turned into an insurmountable obstacle. Besides that, particular automated movement realizations frequently have no analogs in past experience and therefore cannot be based on connections fixed earlier in the memory. Movement automation would turn from a means of solving the problem of redundancy into a factor aggravating the problem. The only acceptable alternative is remembering principles, sets of rules, or "control matrices" realized by neural nets, i.e., everything that displays processes of brain activity self-organization (Freeman 1990). No rigidly fixed connections can exist among neural elements. The hypothetical net must be able to reorganize not only by including new elements and connections but largely by changing relations among already involved elements, modifying the spatial and temporal configuration of neural nets. Lashley (1950) came to these conclusions based on his famous studies. He noted that each memorizing neuron can potentially take part in many memories (in all the memories in the limit case), while every memory is represented everywhere over the brain hemispheres. These ideas formed the foundation of the quasi-holographic principle of the memory organization in the brain (Osovets et al. 1983; Pribram 1971), of the role of nonlinear dynamic properties of the brain electrical rhythms in the LTM mechanisms (Osovets et al. 1983), and of the role of statistical, spatial-temporal configurations of neuronal ensembles (John and Schwartz 1978), as contrasted by the commutative theories of brain memory. In different brain formations, neural nets (ensembles) are assumed to possess certain structural features that impose certain restrictions and play the role of operators (Bassin et al. 1966).

All the mentioned considerations are based on a general idea that input of information into LTM and its storage imply the emergence in the CNS substrate (structures) of something that was not present earlier. Differences in opinions are related to the nature of changes realizing LTM. According to one school of thought, these changes are localized in the apparatus of interneuronal connections, in synapses, mostly implying their presynaptic structures. A memory can be retrieved with the help of signals only from the sources that were used for recording the memory. According to competing views, these changes are stored in intracellular structures, such as the nucleus, protoplasm, endoplasmic reticulum, ribosomes, and in others within the soma and dendrites. Such a storage mechanism allows retrieval under the influence of signals from various sources with a possibility of interaction that can also take place during recording. An important role in LTM recording and retrieval can be played by intracellular computations and the activity of particular neural circuits.

A marked disbalance exists in the number of experimental studies, involved research groups, and publications supporting the mentioned hypotheses (synaptic and neuron-net). The synaptic hypothesis has been much more vigorously explored. Progress in this direction has been well disseminated through scientific publications, so this chapter will mention these data only briefly and present a critical analysis of its major ideas. Studies supporting the neuron-net hypothesis will be described in more detail.

LTP Is not a Storage for STM and LTM

The central position in the experimental support of the synaptic hypothesis is taken by studies about the phenomenon of long-term potentiation (LTP; for reviews, see Baudry and Davis 1994; Kandel and Hawkins 1992; Petri and Mishkin 1994; and Zalutsky and Nicoll 1990). LTP was discovered in neurons of the hippocampus as a relatively long-lasting increase in the synaptic transmission strength after a brief period of a high-frequency stimulation of presynaptic axons (Bliss and Lomo 1973). Such an increase can last seconds, minutes, and hours in a urethane-anesthetized rabbit, days and weeks in alert, free-moving animals.

Soon, the primary role in this aftereffect was attributed to processes taking place in the presynaptic part of the synapse. LTP was considered a mechanism and index of neural memory. During the last 10–15 years, an avalanche-like increase has occurred in the number of studies of LTP at the cellular, subcellular, and molecular levels. These studies have led to the accumulation of an enormous number of facts related to the properties of LTP, the sequence of biochemical processes, the role of primary and secondary transmitters, special enzymes, protein molecules, ions, and ion channels in the genesis of LTP (for details see Kandel and Hawkins 1992; Manabe and Nicoll 1994; Min Zhao et al. 1993; Petit et al. 1994; Petri and Mishkin 1994; Zalutsky and Nicoll, 1990).

Note that all these sophisticated studies of the hippocampus were performed under an implicit consensus that hippocampal LTP was identical to traces of stored memories. This belief is so strong that the underlying assumption of a relation between LTP and memory processes has never been actually discussed, until now. The idea of a relation between the hippocampus and neuronal mechanisms of memory is based on the following four presumptions. First, the phenomenon is observed in the hippocampus, whose role in memory processes has been firmly established by clinical and psychological studies. Second, a long-lasting aftereffect exists, the LTP, and its timing characteristics generally correspond to the time of the existence of memory in the shape of neurodynamical processes, such as STM and the period of trace consolidation. This has been demonstrated in experiments with electroconvulsive shock and the action of protein synthesis inhibitors (Duncan 1949; McGaugh 1966, 1989). Third, neurodynamical processes underlying LTP, which are mostly presynaptic, explain the selective relation between processes of information recording and retrieval. They both use the same axonal terminals and a pre-postsynaptic interaction, postulated by Hebb (1949), as a precondition for fixation of a trace in a synapse. Fourth, a certain similarity exists among subcellular and molecular changes in neurons of Aplysia during learning and changes in hippocampal neurons during LTP.

The presumption of LTP localization in the hippocampus appears to be most important. Presently, studies of amnesia after hippocampal injuries in humans, monkeys, and rodents (Cohen and Eichenbaum 1993; Milner et al. 1968; Mishkin et al. 1984; Petri and Mishkin 1994; Squire 1992; Squire and Zola-Morgan 1991; Warrington and Weiskrantz 1970, 1978) have led to a conclusion that despite the memory deficit related to the inability to assimilate recent memories, no disorders of either LTM or STM storage occur in such cases meaning that these memories are not stored locally in the hippocampus. Thus, prior to linking hippocampal LTP to storage of memory traces, a brain object for the storage must be identified. The function of the hippocampus is commonly assumed to be related to fixation of new traces in LTM, while the LTM is supposedly stored outside the hippocampus, i.e., the function is related to the processes of memory organization and transformations (its consolidation) rather than to its storage. However, increasingly sophisticated studies of LTM in patients with amnesia have demonstrated progressively larger amounts of newly arrived information stored in LTM ("shrinking the deficit"). In particular, it appeared possible to retrieve from LTM in amnesiacs new motor skills (Milner et al. 1968), events with strong emotional contents such as information about an upcoming surgery (Latash 1968), image recognition based on fragments (Warrington and Weiskrantz 1970, 1978), categorization using a template set of objects while remembering a particular object from a set was impossible (Knowlton and Squire 1993), and so forth. These observations have implied a limited but very important and complex, special role of the hippocampus in the transfer of new traces into LTM. They formed the basis for an assumption of a multiple (nonunitary) nature of memories in CNS and of their mechanisms. An essential role was played by the mentioned separation of memories into declarative (explicit, flexible) and nondeclarative (implicit, procedural) (Cohen and Eichenbaum 1993; Mishkin et al. 1984; Schachter et al. 1993; Squire 1992; Squire and Zola-Morgan 1991). The first group of memories involved objects, facts, and events ("knowing that"). It emerges quickly and is characterized by the possibility of consciousness participating in memorization and retrieval (recollection) and by an important role of the hippocampus in LTM fixation. The second group of memories stored acquired motor skills (including products of movement automation, according to the descriptions), habits, and simple conditioning ("knowing how"). It slowly elaborated, requiring many repetitions without participation of consciousness and without involving hippocampal formations into its brain mechanisms. However, classifying memories into explicit and implicit has not always been simple. Such a separation is straightforward when comparing how children acquire the abilities to walk and to produce speech movements during speech development, or how one remembers newly met persons or new surroundings. It is not clear, however, how to classify cases such as learning how to drive a car or play piano. The role of consciousness in skill acquisition here is obvious while the skill is acquired slowly. However, after the skill has been acquired, retrieval of stored coordinations can frequently be done without participation of consciousness. Another example, which is hard to classify, is learning motor coordinations while stepping on a moving escalator. These coordinations are elaborated rather quickly and with participation of consciousness. However, after being acquired they can act against control of consciousness, e.g., motor discoordinations observed when a person steps on a motionless escalator.

Two methods (mechanisms) seem to be used during memory transfer into LTM. Some of the memorized events and relations are being treated separately, while others are being treated in parallel, e.g., memorizing objects as such or by identifying common features that allow categorization (Knowlton and Squire 1993). Still other events and relations may be memorized by alternating recruitment of different mechanisms for the fixation of different components (aspects) of an event at different stages of forming an ultimate trace. One may view the processes of recording and/or retrieving the two types of memories as basically different because of the participation (for explicit memories) or nonparticipation (implicit memories) of consciousness in humans or of its analog in laboratory animals, or of only certain components of consciousness, such as particular emotional processes, arousal, genesis of brain rhythmic activity, and so forth. However, long-term storage of the ultimate LTM traces can be realized by identical or very similar changes in the neuronal substrate.

Assuming a special role for the hippocampus in the trace fixation of explicit LTM makes the problem of LTP's role related to neurophysiological manifestations of the participation of consciousness in this type of memory. The problem is motivated by the absence of sufficient data that would suggest an existence in the hippocampus of storage of any STM and/or LTM traces.

Much attention has been paid during the last 15 years to another possible mechanism to record and retrieve distributed LTM traces in representative systems of neocortex. This mechanism is more closely related to WBM and the functioning of consciousness. It is reflected in the brain EEG rhythms, and their patterns can be interpreted according to regularities of nonlinear dynamics (cf. "deterministic chaos" and "strange attractors" in Babloyantz 1990; Freeman 1990, 1991; Osovets et al. 1983; and Skarda and Freeman1987). Several hypotheses have been offered based on new functional interpretations of EEG rhythms generated by dynamical systems.

One of the first hypotheses was related to the mechanisms of recording and retrieving LTM in the brain, presumably realized with participation of consciousness (Osovets et al. 1983). The hypothesis was based on the assumption that in neocortical substrates, memorization included an interaction of two information flows. The first contains specific impulses, and the second contains rhythmical processes expressed in EEG wave patterns. The hypothesis explains quasi-holographic features of LTM like distributivity and associativity (Lashley 1950; Pribram 1971; Sperry 1947), offers an efficient and frugal mechanism of retrieval from LTM (without any address indexation and comparators for matching with "neural models of stimuli"), and implies close ties between the processes of memory and conscious awareness. These ties are reflected in special EEG wave patterns created by nonlinear interactions of thalamic autogenerators among themselves and with messages from reticular and limbic formations changing under the influence of specific signals. This convergence of two flows is assumed to be an obligatory condition for LTM recording with participation of consciousness. The wave pattern displays the microstate of the whole brain and can alone facilitate readout of the LTM. The phenomenon of state dependence that was so impressively illustrated in Chaplin's film *City Lights* finds its natural explanation in the framework of the hypothesis. For details, see the English translation of the Osovets et al. (1983)paper in Sov. Phys. Usp.

Another hypothesis about the role of chaotic dynamics of EEG rhythms in the mechanisms of consciousness and memory has been offered by Freeman (1990, 1991). It represents development of the notion of cooperative processes in the brain as the apparent basis of higher forms of mental activity. Within this hypothesis, the 40 Hz rhythm dynamics is claimed to be most important while Osovets et al. (1983) delegate the main role to commonly dominating thalamo-cortical rhythms of the alpha-beta band because of their more evident relations to different states of consciousness.

Within the presented hypothesis, the role of the hippocampus may be related to its significant and diverse effects upon the interaction among EEG autogenerators.

This assumption is related to the important role of the hippocampal RSA in the WBM and in the processes of consciousness, particularly with its volitional function. Attempts have been made to link RSA to STM. Within the synaptic hypothesis and ideas about the role of LTP in storage of memory traces, the existence of hippocampal RSA finds no place. However, its relation to emotional processes and incentives toward behavioral activity (particularly toward motor activity) and its role in alert states and in REM sleep make its participation in all neurodynamical stages of explicit memory realization rather plausible, taking into account the particular relation of this kind of memory to consciousness.

Hypotheses about a relation of LTP to certain processes relevant to aspects of consciousness are rather speculative and require experimental confirmation. The same can be said, as has just been illustrated, about the assumed role of the hippocampus in the storage of memory traces. However, attempts at discovering storage of STM and LTM traces in the hippocampus failed. In contrast, participation of the hippocampus and all the limbic system in WBM has been proven experimentally, including such important components of consciousness as arousal and activation of emotional processes.

For quantitative characteristics of LTP, these were obtained during an artificial stimulation of axon tracts terminating on hippocampal neurons, frequently in anesthetized animals, and also in studies of hippocampal slices *in vitro*. They were not obtained in controlled conditions including natural processes of memory. Therefore, quantitative indices, including timing indices, may only be used to confirm theoretical possibilities and not parameters of an actual function displayed in normal conditions. The following questions emerge. Do the frequencies, durations, and intervals of experimental impulse discharges correspond to those in natural conditions? What can be the effects of the removal of the WBM?

Recording and retrieval selectivity also remains a problem within the synaptic hypothesis despite the claim of dominance of presynaptic processes. The primary role of postsynaptic processes and their ability to display temporal and spatial summation are well known. In some cases, shifts in postsynaptic potentials, PSPs, are shown to be necessary and sufficient factors for the generation of LTP (Petit et al. 1994).

The similarities with the neural memory processes in Aplysia are not extended enough. Processes observed in Aplysia during the elaboration of a simple form of classical conditioning correspond to implicit memory. They should not be compared with LTP mechanisms in the hippocampus of higher animals, which characterize explicit memory. All the above show that, presently, no persuasive proofs exist of links between the hippocampal LTP and substrate changes that realize LTM storage.

Neurophysiological Analysis of the Spinal Memory

A direct attempt at testing the "synaptic" hypothesis was undertaken by Eccles (1965), one of its pioneers, who developed the hypothesis as a concept of the role of synaptic "use" and "disuse" in the formation of improved (or deteriorated) synaptic transmission. Experiments on spinal cats, with selective stimulation of muscle afferents, whose synaptic connections with spinal motoneurons had been well documented, using different fixed postures of the limb, led to a negative result. Eccles concluded that the spinal cord was an "inappropriate" object to study the neuronal mechanisms of memory in the CNS.

It so happened that, at the same time, Gerard and colleagues (Chamberlain et al. 1963a, b) in Ann Arbor, MI started studies of the spinal memory using an experimental model suggested in the 1930s by DiGiorgio. Half of the cerebellum was ablated in an animal (DiGiorgio studied dogs, while Gerard and his colleagues studied rats). After the animal had recovered from the anesthesia, a postural asymmetry was observed, mostly seen in the hind limbs. One hind limb was flexed, while the other was extended. Apparently, these effects resulted from the hemicerebellectomy that had created an asymmetry of the descending signals to the spinal cord and, as a result, an asymmetry of neuronal excitability leading to a postural asymmetry. The following complete transection of the spinal cord at a thoracic level (above the site of postural asymmetry) eliminated the asymmetrical descending inflow. It induced different effects upon the postural asymmetry depending on the time period of its existence. If the spinal cord transection was performed after not more than 30 min, the postural asymmetry disappeared. If more than 30–45 min elapsed, the postural asymmetry persisted all the time the animal survived (up to five days). Since protein synthesis inhibitors eliminated the long-lasting asymmetry, it has been concluded that, after 30–45 min, the state of long-lasting asymmetry may be considered as LTM with trace storage in neuronal substrate. The period of postural asymmetry of less than 30–45 min was associated with STM and trace consolidation. Later, these results were reproduced in other laboratories (Patterson et al. 1984; Steinmetz et al. 1982).

In the second half of the 1960s and first half of the 1970s, the author's group attempted an electrophysiological study of the described phenomenon of spinal memory (Latash 1979; Tikhomirova 1969; Tikhomirova and Latash 1976; Tikhomirova et al. 1976). The purpose of the study was to learn the role in LTM of different neural elements within neuronal formations in which the long-term storage of trace changes definitely takes place. The relatively simple neuronal organization of the spinal cord and some of the well-studied functional connections within the segmental reflex apparatus fit well this purpose. If the role of the neuronal formations were discovered, it could be used as a model to study neural memory in higher animals.

Half of the cerebellum was surgically removed in decerebrated, curarized cats. After the surgery, an asymmetry of monosynaptic reflex responses recorded in a cut ventral root (VR MSRR) at the lumbar-sacral level was observed. As seen in figure 3.7, the hemicerebellectomy led to an expressed asymmetry of the amplitudes of VR MSRRs in the low-lumbar and high-sacral segments. The VR MSRRs were induced by the bilateral stimulation of the central end of a cut muscle nerve (usually, n. tibialis). As a rule, MSRRs with a higher amplitude were seen on the ipsilateral to the ablation side. The original amplitude of MSRRs was considered 100%. A background asymmetry of the MSRR amplitudes could be up to 10–15%. Testing started 1 h after anesthesia and was repeated every 2.5 min.

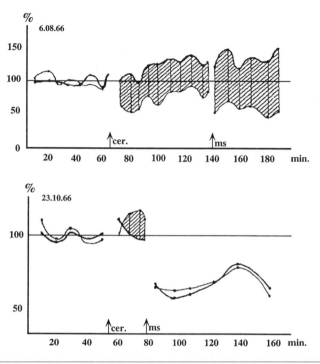

Figure 3.7 "Spinal memory" in decerebrate cats. Amplitude asymmetry of VR MSRR at the S1 segment was caused by hemicerebellectomy. Ordinate—VR MSRR amplitude in percent to a background value (100%); abscissa—time in minutes. Filled circles connected with thick line—VR MSRR on the ipsilateral to hemicerebellectomy side, with thin line—on contralateral side. (Cer) removal of hemicerebellum, (ms) spinal cord transsection at L2 level. Upper part—the fixation time equaled 72 min. VR MSRR asymmetry persisted after ms. Lower part—the fixation time equaled 10 min. VR MSRR disappeared after ms.

Spinal cord transection was later performed at an upper lumbar level. If the VR MSRR asymmetry duration prior to the spinal cord transection was over 30 min (figure 3.7, upper part), the asymmetry persisted and lasted until the end of the acute experiment (5–8 hours). If the asymmetry duration prior to the spinal transection was smaller, the asymmetry disappeared (figure 3.7, lower part). Thus, the phenomenon of spinal memory described by Gerard and others was reproduced using monosynaptic reflex arc activation in conditions when the gamma loop was cut. Therefore, memory trace fixation could happen only in elements of the arc or in interneurons controlling these elements.

The following studies were performed on nondecerebrated rats in order to have direct comparison with the observations by Gerard. Otherwise, the procedure was identical to that described for the cat experiments. The results were basically the same. The hemicerebellectomy induced an amplitude asymmetry of VR MSRRs that, after 50 min, persisted following a transection of the spinal cord at a lower-thoracic–upper-lumbar level (figure 3.8, upper part). After the spinal cord transection, unlike in cats, rats demonstrated a dramatic increase in the amplitude of VR MSRRs at the side of hemicerebellectomy, up to 550%. Since the rats were not decerebrated, these observations suggested a tonic inhibitory influence of the neocortex upon the excitability of spinal neurons that was removed by the spinalization. The much higher increase in the ipsilateral VR MSRR amplitude suggests that the cerebellum exerted an inhibitory influence upon segmental VR MSRRs that could, however, be masked by the effects from the neocortex. The abrupt increase in the amplitude of the ipsilateral responses was temporary and later decreased, on average, to 152% of the background level at the ipsilateral side and to 138% at the contralateral side. If the spinal cord was cut 15 min after the emergence of the VR MSRR asymmetry (figure 3.8, lower part), the asymmetry disappeared and led to a symmetrical increase in the amplitude of the responses.

To analyze the ability of different elements of the monosynaptic arc (the motoneuron and presynaptic terminals of the afferent neuron) to "memorize," a unilateral, electrical, rhythmic stimulation of the central end of the cut muscle nerve was performed in spinal rats. The parameters of the stimulation were chosen to re-

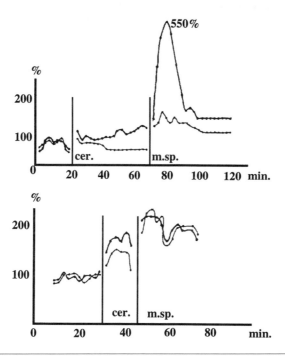

Figure 3.8 "Spinal memory" in nondecerebrate rats. VR MSRRs are at the L6 segment. Symbols and abbreviations are the same as in figure 3.7. Upper part—the fixation time equaled 45 min. VR MSRR persisted after m.sp. Lower part—the fixation time equaled 15 min. VR MSRR disappeared after m.sp.

semble those of the descending inflow. Based on some general considerations, the intensity was close to the threshold or 3–4 times the threshold, the frequency was 10 Hz, 50 Hz, and 100 Hz, and the duration was 40–60 min. None of the stimulation regimes was able to induce an amplitude asymmetry of VR MSRRs. The amplitudes did not differ from those of the background responses (figure 3.9, lower part), which is similar to the findings by Eccles (1965). However, if the same stimulation was used in a decerebrate rat, a clear amplitude asymmetry emerged with the higher amplitude at the side of nerve stimulation (figure 3.9, upper part). The following transection of the spinal cord at an upper level, after an appropriate time delay, did not lead to the elimination of the asymmetry, which is similar to the findings after hemicerebellectomy. Thus, elements of the spinal monosynaptic reflex arc were unable to "memorize" repeated stimuli within the arc but displayed memory when the stimuli were transmitted along spino-bulbar-spinal pathways. Naturally, a hypothesis emerged implying that a spinal interneuron was the site of LTM fixation and storage in the spinal cord. Four series of experiments were performed to test this hypothesis.

Within the first series, the effects of cooling upon the amplitude asymmetry induced by hemicerebellectomy were studied (figure 3.10). Brief periods of cooling (up to 10 min) applied to the dorsal surface of the lumbar-sacral spinal segments led to a deterioration of conduction along sensory pathways. A quick bilateral disappearance of VR MSRRs was followed by a similarly quick restoration of the asymmetrical responses during warming. If the cooling was longer lasting (15–20 min), during the warming up symmetrical VR MSRRs were restored first, followed by a later restoration of their asymmetry.

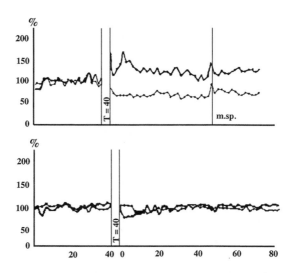

Figure 3.9   Effect of long-lasting (40 min), unilateral stimulation of tibial nerve (f = 100 Hz) on the VR MSRR amplitude. Other symbols and abbreviations are the same as in figure 3.7. Lower part—in the spinal rat; upper part—in the decerebrate rat.

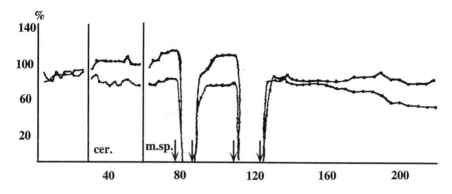

Figure 3.10 The effect of local cooling at L6–S1 segments on "spinal memory" in the rat. Each pair of vertical bars—application and removal of cooling, 10 min and 16–17 min. Other symbols and abbreviations are the same as in figure 3.7.

The second series used pharmacological agents suppressing interneuronal activity in dosages subthreshold for effects on motoneurons. Figure 3.11 (upper part) shows that a single injection of meprobamate (mianezine) in twin solution induced a short-lasting reduction of the VR MSRRs asymmetry due to a reduction in the amplitude of the larger response. Multiple injections led to a reduction followed by a suppression of the asymmetry. If a concentration was reached that was effective for motoneurons as well, VR MSRRs were completely eliminated (figure 3.11, lower part).

Within the third series, the long-lasting stimulation of the flexor reflex afferents (n. suralis) was used to induce activation of interneurons in spinal rats. The procedure and parameters of the stimulation were the same as in the experiments with long-lasting stimulation of muscle afferents. The stimulation of the cutaneous nerve led to an asymmetry of VR MSRRs (figure 3.12) by increasing the amplitude of the ipsilateral responses (to the side of stimulation) and decreasing the amplitude of the contralateral responses.

The fourth series studied the question of what was actually "memorized" by interneurons controlling elements of the monosynaptic reflex arc, activation or inhibition of inhibition. It involved studying the role of monoaminergic (primarily catecholaminergic) descending inhibitory systems of reticulospinal neurons described by Lundberg (1966). Rats received reserpine for 1–2 days (10 mg/kg in a 25% solution), and the experiment with hemicerebellectomy was then performed. In such experiments, the asymmetry either did not occur or was reduced. Intravenous injection of L-dopa (100 mg/kg) led to an emergence of or an increase in the asymmetry. Probably, catecholaminergic processes participate in the realization of the effects of hemicerebellectomy. However, its relation to the spinal storing substrate remains unclear.

Thus, for the first time, a study of the fixation, storage, and retrieval of trace changes in the neural substrate of higher animals was performed. It did not assume

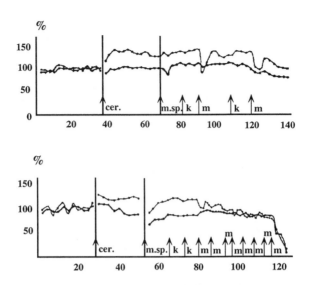

Figure 3.11   The effect on "spinal memory" of intravenous injections of meprobamate (m). (K) control injection of equal amounts of physiological solution with twin. Upper part— two injections, one of 40 mg/kg and the other of 20 mg/kg. Lower part—repetitive injections (40 mg/kg and 15 mg/kg).

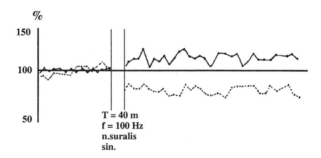

Figure 3.12   A VR MSRR amplitude asymmetry in a spinal rat elicited by stimulation of the left sural nerve (40 min, 100 Hz). Other symbols and abbreviations are the same as in figure 3.7. Dotted line—amplitude of VR MSRR contralateral to stimulated sural nerve.

phenomena of memory based on more or less plausible speculations and indirect, ambiguous evidence but clearly and unambiguously demonstrated their existence. The substrate was rather simply organized including a two-neuron reflex arc and interneurons controlling the arc. The phenomena of memory were demonstrated using indices that did not coincide with an assumed memory mechanism. This contrasted the studies of the hippocampal LTP that played both functions in the analysis, thus making separate analysis of each of the functions very hard if not impossible.

It has also been clearly demonstrated that retrieving LTM can be done using inputs that had not been used during recording. Not all the neural elements were able to "memorize" their interaction. Apparently, input neurons (the first sensory neuron) and output neurons (motoneurons) do not have this ability. This makes sense taking into account their functions: undistorted transmission of various, changing environmental information and the "final common pathway" for the variety of motor commands. One is tempted to suggest the following: "What is present in all the neurons of the CNS cannot by itself be the carrier (storage substrate) of LTM." The interneuronal apparatus plays an important role in LTM storage. This apparatus represents an important tool for the formation of neural nets in the brain. Hypothetically, interneurons mediating inhibitory influences have a particularly important role. Finally, WBM, whose effects are mediated by reticular systems, participate in memory at the level of spinal interneurons.

Spinal memory does not seem to be an experimental artifact. Conceivably, spinal memory is used in the process of fixation of movement automation which is characterized by frequent use of acquired standardized coordinations forming the basis of postural control, locomotion, professional motor skills, and, possibly, changes similar to adaptive plasticity of the spinal stretch reflex described by Wolpaw et al. (1983) in intact monkeys.

By many features, "spinal memory" may be considered an implicit memory. Comparisons of the results of experimental studies of explicit memory (represented by the analysis of the role of the hippocampus) and of implicit memory (represented by the studies of the "spinal memory") suggest that both types of memory can be unified into one common notion LTM based on the "neuron-net" hypothesis rather than on the "synaptic" hypothesis. To paraphrase one of the well-known quotations by Bernstein, the chances of finding phenomena of neural memory in synaptic changes are roughly equal to the chances of stumbling over a meridian while walking in your backyard (Bernstein 1947, p. 13).

Conclusions

Movement automation is considered a modification (a reorganization) of the central motor control with a purpose to eliminate the redundancy of neural processes underlying construction of the movement. Processes related to eliminating the redundancy take place at all levels of the control hierarchy. However, the primary goal of automation is to eliminate redundancy at the "higher" levels. These are levels whose major functions involve planning and initiating new motor (as well as sensory and intellectual) acts related to the most complex problems and incentives. This view preserves the accurate and deep, classical definition of automation given by Bernstein, as transfer of neural control of repeated movements from a "higher" level of control to a "lower" level. However, contemporary practice of studying the mechanisms underlying the movement automation processes is mostly directed at processes within the system of motor control in its narrow sense. Relatively little attention is being paid to "whole brain mechanisms" (WBM), which are responsible for such means of brain self-organization and self-regulation as alertness-arousal, motivation, emotion, attention, sleep-wakefulness, memory, and consciousness. The main purpose of this chapter has been to analyze the relations among these mechanisms and movement automation. On the one hand, such relations allow an insight into certain aspects of the mechanisms and dynamics of automation. On the other hand, certain specific requirements typical for the processes of movement automation force researchers to reconsider widespread theories of the functioning of some of the WBM, particularly of processes and mechanisms defined as the orienting reaction (OR) and of memory.

The role of the WBM in motor control in certain cases is so important and related to the specificities of the motor-control system that these mechanisms need to be included in definitions of movement construction stages. First of all, this is true with respect to mechanisms of decision making, i.e., of an active choice of tactics and realization of a movement from a set of alternatives that typically takes place outside the sphere of consciousness. One can naturally assume that these processes characterize the construction of new movements and their first realizations, when coordinations are formed by a "higher" level of control. In the process of movement automation, decision-making processes are reduced and even disappear. They may reemerge in cases of movement disruption and deautomation. In other words, the level of activity of the decision-making processes is an important index of the degree of movement automation.

A detailed study of OR in healthy persons as well as in neurological and psychiatric patients has demonstrated an important relation of the habituation of the skin galvanic component (SGR) of OR to the significance of changes in the situation that implies the necessity of action organization. A study of the effects of a special motor instruction associated with an increase in the stimulus significance has shown a relation of this component to movement. Other components of OR did not show such relations. These observations lead to a reconsideration of traditional views about the functional role of OR represented in a theory of "the neural model of the stimulus" (Sokolov 1964). A new concept has been suggested based on

1. the active (rather than reactive) nature of OR with a leading role being attributed to significance of changes in the situation;

2. the polyfunctional (rather than unitary) nature of the observed changes in OR components that usually display nonparallel dynamics; and

3. a relation of OR and its habituation to the organization and automation of a new action (motor, sensory, or intellectual).

An analysis of conditions leading to an increase or a decrease in the SGR component of OR (slowing down or acceleration of its habituation) lead to a conclusion that this component is a reflection of emotional processes that are typically not perceived consciously and are related to decision making. Habituation of this component is considered a reduction in decision making and in an accompanying emotional activation when a movement becomes progressively more automated. This activation of emotional processes can be considered a reflection of a microanxiety when one needs to choose between alternatives or as an emotional reinforcement of a choice.

In rats, the dynamics of hippocampal RSA (which are rather close to the dynamics of SGR in humans) are related to the sign of an emotional effect during self-stimulation. This observation suggested reinforcing effects of emotional activation. This mechanism was assumed to be defective in patients with local damages within the limbic system and suppressed SGR. These patients did not demonstrate disorders of certain motor acts. However, movement automation in these patients was based on different, less effective tactics as compared with healthy subjects with unchanged SGR. In the former group (patients), a so-called "local search" was observed. It was associated with an increased readiness to move as reflected by an increased excitability of a corresponding group of spinal motoneurons seen throughout all the latent period. In the latter group (healthy subjects), a more economical strategy, "ravine search," (Gelfand and Tsetlin 1962) was associated with a decrease in the excitability of motoneurons during virtually the whole latent period and its increase just prior to the movement.

Phenomena of movement automation pose a number of problems for neuronal theories of memory due to two basic factors. First, movement kinematics and, therefore, neural executive processes are continuously variable even in the most standard, sequential, automatic movements. Second, control of such movements is transferred from "higher" levels to "lower" ones. Both factors pose insurmountable problems for views claiming that LTM is stored in synapses of the CNS. These views assume a particular role of presynaptic changes that allow a rigid possibility of LTM retrieval only through inputs that were used during memorizing. The first factor (movement variability) makes such rigid relations negative phenomena requiring a huge redundancy of neuronal connections. The second factor (transfer of control) requires an ability to retrieve memories through inputs that were not used during memorizing. This is because the "lower" control mechanism, such as a control matrix (Bassin et al. 1966; Gelfand et al. 1971), must be formed using both descending and peripheral signals. Retrieval, however, should proceed without descending influences (if there are no disruptions) or with the help of alternating descending and peripheral signals (in cases of deautomation and reautomation). Besides that, peripheral signals should also have an ability to switch over different coordinations within a control matrix depending more on the nature of the incoming information than on the nature of the conducting pathways.

However, researchers currently focus their attention on the role of synaptic changes in the hippocampus manifested by the LTP phenomenon as a mechanism of LTM storage. A review of studies based on this hypothesis suggests that the hypothesis lacks appropriate experimental foundation. The function of the hippocampus is still unknown. One may say, however, that it is not related to storage of either STM or LTM since the storage is not affected by hippocampal disorders. The facts show only an unclear, but important, role of the hippocampus in forming memories that are stored in other brain formations. The author would rather draw attention to the possibility that the hippocampus participates in WBM playing a crucial role in realizing the influence of consciousness, especially its aspects related to motivation and emotional activity, on processes of organizing and modifying new memories. RSA and even LTP may be considered indices of specifics of forming explicit memories that occur with participation of consciousness (possibly also of REM sleep). The approximate timing correspondence of LTP and STM/LTM (the latter based on clinical and psychological studies) can easily get an alternative explanation.

Within the synaptic hypothesis of memory, the problem of selective retrieval from LTM is solved with the help of identical neuronal inputs for recording and retrieval. This explanation becomes less impressive when taking into account the increase in resulting redundancy and also the recently demonstrated role of postsynaptic processes in LTP. Note that these processes are characterized by interaction of different inputs. The problem of selectivity of recording and retrieval of LTM is much better solved by the hypothesis about the role of EEG rhythms in these processes based on the regularities of nonlinear dynamics (deterministic chaos). All these support the neuron-net hypothesis of LTM. It considers a whole neuron (including the intracellular substrate) as a structural unit of memory (see also recent data by Barnea and Nottebohm 1996) and a net as a functional unit. In other words, a neuron "remembers" codes of neural nets that recruit it using special combinations of input and output signals.

Direct experiments searching for memory based on a hypothesis of an amplification of synaptic connections led to negative results, testing the effects of use and disuse of synapses (Eccles 1965). Commonly during experimental studies of memory, no unambiguous proofs show that memory is in a tested brain substrate. In order to avoid this problem, we studied the phenomenon of "spinal memory." Until that time, it had been studied only by its behavioral characteristics such as the emergence of a stable postural limb asymmetry following hemicerebellectomy. The asymmetry could persist for hours and days after a total spinal cord transection at a higher level. Since the phenomenon was reproduced as an amplitude asymmetry of VR MSRRs, the search zone of traces of spinal memory was limited to a two-neuron arc of the phasic myotatic reflex and interneurons controlling this arc. Besides that, processes of memory storage and realization are separable in the substrate (unlike LTP). A number of facts are incompatible with the "synaptic" hypothesis of memory while corresponding well to the "neuron-net" hypothesis. Among these facts are

1. the existence of neurons unable to remember;
2. the possibility of memory retrieval through inputs that were not used in the recording process; and
3. the participation of WBM (reticular descending systems) that were responsible for the LTM storage via an important role played by controlling interneurons.

The described data about "spinal memory" were obtained using the lowest levels of the neuromotor system. Thus, the mechanisms underlying these data are assumed to be close to memory mechanisms participating in movement automation. It may be argued that the authors analyzed processes of implicit memory that form the basis of motor skills, proceed without participation of consciousness, and are not mediated by hippocampal structures. They are therefore basically different from the mechanisms of explicit memory. This argument, however, cannot be used to support the role of hippocampal synapses in storing explicit LTM. Moreover, there are reasons to believe that explicit and implicit memories can exist independently but can function in parallel or as a sequence of phases. This happens, for example, while automating professional and some everyday movements. Such an interaction favors a logical assumption that, if different mechanisms of recording and retrieval are used on the level of object's images or descriptions, a single form of LTM storage exists related to a more abstract object representation. Transitions between the memory forms are largely controlled by WBM. Automation of movements of an adult person may be more commonly based on such LTM transitions than may seem at a first glance.

References

Babloyantz, A. (1990) Chaotic dynamics in brain activity. In *Chaos in Brain Function*. E. Basar (Ed.), Berlin: Springer, 42–49.

Bargh, J. (1995) Presentation to annual meeting of American Psychol. Association. *New York Times,* August 8: B5 and B9.

Barnea, A., Nottebohm, F. (1996) Recruitment and replacement of hippocampal neurons in young and adult chickadees: An addition to theory of hippocampal learning. *Proceedings of the National Academy of Sciences,* 93: 714–718.

Barry, R.J. (1982) Novelty and significance. Effects of fractionation of phasic OR measures. *Psychophysiology,* 19: 28–35.

Bassin, P.V., Bernstein, N.A., Latash, L.P. (1966) On the problem of relations between brain structure and function in its contemporary understanding. In *Physiology in Clinical Practice*. N.I. Graschenkov (Ed.), Moscow: Nauka, 38–71 (In Russian).

Baudry, M., Davis, J.L. (Eds.). (1994) *Long-Term Potentiation*. Vol. 2. Cambridge: MIT Press.

Berkinblit, M.B., Feldman, A.G., Fukson, O.I. (1986) Adaptability of innate motor patterns and motor control mechanisms. *Behav. Brain Sci.,* 9: 585–638.

Berlyne, D.E., McDonnell, P. (1965) Effect of stimulus complexity and incongruity on duration of EEG desynchronization. *Electroencephalog. Clin Neurophysiol.,* 18:156–161.

Bernstein, A.S. (1969) To what does the orienting response respond? *Psychophysiology,* 6: 338–350.

Bernstein, A.S., Taylor, K.W. (1979) The interaction of stimulus information with stimulus significance in eliciting the skin conductance orienting response. In *The Orienting Reflex in Humans*. H.D. Kimmel, E.H. van Olst, F. Orlebeke (Eds.), Hillsdale, N.J.: Erlbaum, 499–519.

Bernstein, N.A. (1927) New in methods for studying labor movements. In *Psychophysiology of the Labour.* Vol. 2. K.C. Kekcheev, V.T. Rabinovich (Eds.), Moscow-Leningrad, 70–81 (In Russian).

Bernstein, N.A. (1935) The problem of interrelation between coordination and localization. *Arch. Biol. Sci.,* 38: 1–34 (In Russian).

Bernstein, N.A. (1947) *On the Construction of Movements*. Moscow: Medgiz (In Russian).

Bernstein, N.A. (1967) *The Coordination and Regulation of Movements*. Oxford: Pergamon Press.

Bernstein, N.A. (1990) *Movement Physiology and Activity*. In the series *Classics of Science*. Moscow: Nauka (In Russian).

Bliss, T.V.P., Lomo, T. (1973) Long-lasting potentiation of synaptic transmission in the dentate area of the anaesthetized rabbit following stimulation of the perforant path. *Journal of Physiology,* 232: 331–356.

Chamberlain, T.J., Halick, P., Gerard, R.W. (1963a) Fixation of experience in the rat spinal cord. *Journal of Neurophysiology,* 26: 662–673.

Chamberlain, T.J., Rothschield, G.H., Gerard, R.W. (1963b) Drugs affecting RNA (ribonucleic acid) and learning. *Proceedings of the National Academy of Sciences,* 49: 918–924.

Cohen, N.J., Eichenbaum, H. (1993) *Memory, Amnesia, and the Hippocampal System.* Cambridge: MIT Press.

Dallakyan, I.G., Latash, L.P., Popova, L.T. (1970) Some regular relationships between the galvanic skin response expression and EEG changes in local lesions of the limbic (rhinencephalic) formations of the human brain. *Proc. USSR Acad. Sci.,* 190: 991–994 (In Russian).

Damasio, A.R. (1994) *Descartes's Error. Emotion, Reason, and the Human Brain.* N.Y.: Avon Books.

Damasio, A.R., Tranel, D., Damasio, H. (1991) Somatic markers and the guidance of behavior: Theory and preliminary testing. In *Frontal Lobe Function and Dysfunction.* H.S. Levin, H.M. Eisenberg, A.L. Benton (Eds.), New York: Oxford University Press, 217–229.

Danilin, V.P., Krylov, E.N., Latash, L.P., Magalif, A.Yu., Rait, M.L. (1984) Some aspects of the psychophysiology of alcohol withdrawal syndrome. In *Somatic and Neurologic Aspects of Alcoholism and Alcohol Psychoses.* V.V. Kovalev (Ed.), Moscow: Moscow Psychiatric Research Institute, 46-55.

Duncan, C.P. (1949) The retroactive effect of electroshock on learning. *J. Comp. Physiol. Psychol.,* 42: 32–34.

Eccles, J.C. (1965) Possible way in which synaptic mechanisms participate in learning, remembering and forgetting. In *The Anatomy of Memory.* Kimble (Ed.), Palo Alto, CA: Science and Behavior Books, 12–87.

Feldman, A.G. (1979) *Central and Reflex Mechanisms of Motor Control.* Moscow: Nauka (In Russian).

Freeman, W.J. (1990) On the fallacy of assigning an origin to consciousness. In *Machinery of the Mind..* R.E. John (Ed.), Boston: Birkhauser, 14–26.

Freeman, W.J. (1991) *Development of new science of brain dynamics with guidance from the theory of nonlinear dynamics and chaos.* A paper presented at the 8th International Conference on Biomagnetism. Munster, Germany, 18–24 August.

Gelfand, I.M., Gurfinkel, V.S., Fomin, S.V., Tsetlin, M.L. (Eds.). (1971) *Models of Structural-Functional Organization of Certain Biological Systems.* Cambridge: MIT Press.

Gelfand, I.M., Tsetlin, M.L. (1962) Some methods of controlling complex systems. *Adv. Math. Sci.,* 17: 3–25 (In Russian).

Gottesman, C. (1992) Theta rhythm: The brain stem involvement. *Neurosci. & Biobehav. Rev.,* 16: 25–30

Graschenkov, N.I., Latash, L.P. (1963) On the active character of the orienting reaction. In *Brain Reflexes. Abstracts of International Symposium.* Moscow: USSR Academy of Sciences, 18-19.

Graschenkov, N.I., Latash, L.P. (1965) The role of orienting reaction in organizing for action. *Probl. Psychol.,* 1: 21–41 (In Russian).

Grastyan, E., Czopf, I., Anagyan, L., Szabo, J. (1965) The significance of subcortical motivational mechanisms in the organization of conditioned connections. An attempt of the physiological interpretation of the basic mechanisms of motivation. *Acta Physiol. Acad. Sci. Hung.,* 26: 9–46.

Green, J.D., Arduini, A.A. (1954) Hippocampal electrical activity in arousal. *Journal of Neurophysiology,* 17: 532-577.

Gurfinkel, V.S., Kots, Y.M., Shik, M.L. (1965) *The Regulation of the Posture in Humans.* Moscow: Nauka (In Russian).

Hebb, D.O. (1949) *The Organization of Behavior: A Neuropsychological Theory.* New York: Wiley.

John, R.E., Schwartz, E. (1978) The neurophysiology of information processing. *Annual Review of Psychology,* 29: 1–29.

Kandel, E.R., Hawkins, R.D. (1992). The biological basis of learning. *Scientific American,* 267: 78–87.

Kihlstrom, J.F. (1987) The cognitive unconscious. *Science,* 237: 1445–1452.

Knowlton, B.J., Squire, L.R. (1993) The learning of categories: Parallel brain systems for items memory and category knowledge. *Science,* 262: 1747–1749.

Kochubey, B.I. (1979) On definition of the concept of orienting reaction in humans. *Probl. Psychol.,* 3: 35–46 (In Russian).

Konorski, J. (1961) Physiological approach to the problem of recent memory. In *Brain Mechanisms and Learning.* Oxford: Blackwell Scientific Publications, 115–130.

Kovalson, V.M. (1971) Some characteristics of electrical activity of the brain in free-moving rats. *Bull. Exp. Biol. Med.,* 71: 13–18.

Lashley, K.S. (1950) In search of the engram. In *Physiological Mechanisms in Animal Behavior. Symposium of the Society for Experimental Biology.* Cambridge: Cambridge University Press, 454–482.

Latash, L.P. (1968) *Hypothalamus, Adaptive Activity, & Electroencephalogram.* Moscow: Nauka (In Russian).

Latash, L.P. (1979) Trace changes in the spinal cord and some general problems of neurophysiology of memory. In *VII Gagra Talks. Neurophysiological Bases of Memory.* T.N. Oniani (Ed.), Tbilisi, Georgia: Metsniereba, 118–130 (In Russian).

Latash, L.P. (1990) Orienting reaction, organizing for action, and emotional processes. *Pavlovian Journ. Biol. Sci.,* 25: 123–131.

Latash, L.P., Fishman, M.N., Dallakyan, I.G. (1974) Some signs of the activity of emotional-motivation mechanisms and their correlation with the processes of the organization of action. In *The First International Congress on the Higher Nervous Activity.* K. Goldwurm (Ed.), Milan, 18 to 20 October, Pisa: Pacini, 799–814

Latash, L.P., Kovalson, V.M. (1973) LHA self-stimulation effects on EEG and brain temperature in white rats. *Physiol. Behav.,* 10: 651–655.

Latash, M.L. (1993) *Control of Human Movement.* Champaign, IL: Human Kinetics.

Lundberg, A. (1966) Integration in the reflex pathways. In *Nobel Symposium. I. Muscle Afferents and Motor Control.* R. Granit (Ed.), Stockholm: Almquista Wiksell, 275–305.

Maltzman, I. (1990) The OR and significance. *Pavlovian J. Biol. Sci.,* 25: 111–122.

Maltzman, I., Raskin, D.C. (1965) Effects of individual differences in the orienting reflex on conditioning and complex processes. *J. Exp. Res. Personality,* 1: 1–16.

Manabe, T., Nicoll, R.A. (1994) Long-term potentiation: Evidence against an increase in transmitter release probability in the CA1 region of the hippocampus. *Science,* 265: 1888–1892.

McGaugh, J.L. (1966) Time-dependent processes in memory storage. *Science,* 153: 1351–1356.

McGaugh, J.L. (1989) Dissociating learning and performance: Drug and hormone enhancement of memory storage. *Brain Res. Bull.,* 23: 339–345.

Milner, B., Corkin, S., Teuber, H.-L. (1968) Further analysis of the hippocampal amnesic syndrome: A 14-year follow-up study of H. M. *Neuropsychologia,* 6: 215–234.

Min Zhuo, Small, S.A., Kandel, E.R., Hawkins, R.D. (1993) Nitric oxide and carbon monoxide produce activity-dependent long-term synaptic enhancement in hippocampus. *Science,* 260: 1946-1950.

Mishkin, M., Malamut, B., Bachevalier, J. (1984) Memories and habits—Two neural systems. In *The Neurophysiology of Learning and Memory.* J.L. McGaugh, N.M. Weinberger (Eds.), New York: Guilford, 68–88.

Osovets, S.M., Ginsburg, D.A., Gurfinkel, V.S., Zenkov, L.R., Latash, L.P., Malkin, V.B., Melnichuk, P.V., Pasternak, E.B. (1983) Electrical activity of the brain: Mechanisms and interpretations. *Adv. Phys. Sci.,* 141: 103–150 (An English translation is in *Sov.Phys.Usp.,* 26, no. 9 (September 1983), American Institute of Physics, 1984, 801–828).

Patterson, M.M., Steinmetz, J.E., Romano, A.G. (1984) Reflex alterations in the spinal system: Central and peripherally induced spinal fixation in rats. In *Primary Neural Substrates of Learning and Behavioral Change.* D.L. Allison, J. Farley (Eds.), Cambridge: Camridge University Press, 155–167.

Petit, D.L., Perlman, S., Malinow, R. (1994) Potential transmission and prevention of further LTP by increased CaMKII activity in postsynaptic hippocampal slice neurons. *Science,* 266: 1881–1886.

Petri, H.I., Mishkin, M. (1994) Behaviorism, cognition, and neuropsychology of memory. *Amer. Sci.,* 82: 30–37.

Pribram, K.H. (1971) *Languages of the Brain.* Englewood Cliffs, NJ: Prentice Hall.

Routtenberg, A. (1968) The two-arousal hypothesis: Reticular formation and limbic system. *Psychol. Rev.,* 75: 51–80.

Schacter, D.L., Chiu, C.J., Ochsner, K.N. (1993) Implicit memory: A selected review. *Ann. Rev. Neurosci.,* 16: 159–182.

Sherrington, C.S. (1910) Flexion reflex of the limb, crossed extension reflex, and reflex stepping and standing. *Journal of Physiology,* 40: 28-121.

Shik, M.L., Orlovsky, G.N. (1976) Neurophysiology of locomotion automatism. *Physiol. Rev.,* 56: 465–501.

Simonov, P.V., Valuyeva, M.N., Ershov, P.M. (1964) Voluntary regulation of skin galvanic reflex. *Probl. Psychol.,* 6: 45–50 (In Russian).

Skarda, C.A., Freeman, W.J. (1987) How brain makes chaos in order to make sense of the world. *Behav. Brain Sci.,* 10: 161–195.

Sokolov, E.N. (1964) Orienting reflex as information regulator. In *Orienting Reflex and Problems of Reception in Normal and Pathological Conditions.* E.N. Sokolov (Ed.), Moscow: Acad. Pedagogical Sci., 3–20.

Sokolov, E.N. (1990) Comment on Latash's paper. *Pavlov. Journ. Biol. Sci.,* 25:129–130.

Sperry, R.W. (1947) Cerebral regulation of motor coordination in monkeys following multiple transections on sensorimotor cortex. *Journal of Neurophysiology,* 10: 275–294.

Squire, L.R. (1992) Memory and the hippocampus: A synthesis from findings with rats, monkeys, and humans. *Psychol. Rev.,* 99: 195–231.

Squire, L.R., Zola-Morgan, S. (1991) The medial temporal lobe memory system. *Science,* 255: 1380–1386.

Steinmetz, J.E., Cervenka, J., Dobson, J., Romano, A.G., Patterson, M.M. (1982) Central and peripheral influences on retention of postural asymmetry in rats. *J. Comp. Physiol. Psychol.,* 96: 4–11.

Tarchanoff, I. (1889) On galvanic effects in the human skin during stimulation of sensory organs and different forms of mental activity. *Herald Clin. ForensicPsychiat. Neurol.,* 1: 73–81.

Tikhomirova, L.I. (1969) Fixation in the spinal cord of monosynaptic reflex asymmetry caused by extirpation of a half of the cerebellum. *Proc. Acad. Sci. USSR,* 185: 229–233.

Tikhomirova, L.I., Latash, L.P. (1976) On the role of interneurons in storing trace changes in rat's spinal cord. *Proc. Acad. Sci. USSR,* 228: 995–998.

Tikhomirova, L.I., Latash, L.P., Kuman, I.G. (1976) Electrophysiological study of the trace changes in the excitability in the rat spinal cord. *Sechenov Physiol. J. USSR,* 62: 1110–1117.

Turvey, M.T. (1990) Coordination. *Amer. Psychol.,* 45: 938–953.

Vanderwolf, C.H. (1971) Limbic-diencephalic mechanisms of voluntary movement. *Psychol. Rev.,* 78: 83–113.

Warrington, E.K., Weiskrantz, L. (1970) Amnesic syndrome: Consolidation or retrieval? *Nature,* 228: 628–630.

Warrington, E.K., Weiskrantz, L. (1978) Further analysis of the prior learning effect in amnesic patients. *Neuropsychologia,* 16: 169–177.

Wedenyapin, A.B., Rotenberg, V.S. (1984) Skin galvanic reaction as a manifestation of decision-making process in conditions of time deficit. *Pavlov. J. Higher. Nerv. Activity,* 34: 207–211.

Wolpaw, J.R., Seagal, R.F., O'Keefe, J.A. (1983) Adaptive plasticity in primate spinal stretch reflex: Behavior of synergist and antagonist muscles. *Journal of Neurophysiology,* 50: 1312–1319.

Zalutsky, R.A., Nicoll, R.A. (1990) Comparison of two forms of long-term potentiation in single hippocampal neurons. *Science,* 248: 1619–1624.

4

CHAPTER

The Model of the Future in Motor Control[1]

Josef M. Feigenberg

Gilo, Jerusalem, Israel

Nikolai Aleksandrowitsch Bernstein wrote the paper *From the Reflex to the Model of the Future* when he was terminally ill. The paper was published in a popular magazine, *Nedelya (The Week),* in 1966, after Bernstein's death (Bernstein 1966b). Being a physician, Bernstein knew better than anybody else that he had only a few months of life left. He tried during those scarce months to summarize his scientific accomplishments. He was primarily preoccupied with preparing a manuscript, *Essays on the Physiology of Movement and Physiology of Activity,* for publication. This book was published in 1966, after Bernstein had died (Bernstein 1966a). Bernstein did not have time for smaller and less significant projects. However, he wrote this popular article for the general audience. This action suggests that Bernstein considered this article a summary of one of the main achievements of his scientific career. Its title is symbolic and reflects a major idea of the physiology, biology, and psychology of activity.

[1]Translated from Russian by Mark L. Latash.

The Evolution of Bernstein's Idea
of the Model of the Future

The theory of reflexes had developed during a long period of time, from René Descartes to I.P. Pavlov. While developing, this theory embraced and explained more and more functions of the central nervous system. It dominated neurophysiology. Researchers clearly had a tendency to use the theory of reflexes as the basis for some of the directions of psychology research.

The theory of reflexes is based on causality and deals with the past and the present. Everything that happens now has a cause from the past. Physiology and some other biological sciences showed a tendency to follow the routes that had previously led to success in the sciences about inanimate nature. A well-developed science is able to address two basic questions about the objects it studies. First, *how* does a phenomenon occur? Second, *why* does it occur? Formulating the second question already places the phenomenon into the rigid frames of causality. The necessity and sufficiency of answering these questions was obvious for sciences whose theories had reached maturity, particularly physics.

In physiology as well as in other sciences about animate nature, answering these questions was insufficient. Bernstein was the first to emphasize this point. One of the salient features of the phenomena pertaining to the animate nature is their purposefulness, i.e., relation to a certain goal; this feature is inapplicable to phenomena of the inanimate nature. Thus, phenomena of the animate nature raise a third question: *What is the purpose* of a certain phenomenon? or, Which observable problem does it try to solve?

This formulation met strong resistance from some physiologists. A cause seemingly occurs not prior to a consequence but after it. This is certainly not true. Causes still take place prior to consequences. The solution to this problem is provided by a model of the future, which is encoded within the central nervous system of a living organism. Bernstein wrote, "Goal, understood as an encoded in the brain model of desired future, defines processes that should be considered as goal-oriented. . . . The whole dynamics of the purposeful struggle with the help of appropriate mechanisms form a complex which should be united under the term *activity*" (Bernstein 1990, pages 454–455).

The notion of the *model of desired future* was suggested by Bernstein. Actually, an animal or a human cannot formulate a motor task or a motor program without a model of what should result from the planned action. This notion does not violate the principle of causality. Consequences of an action do not define the earlier action as was claimed by Bernstein's opponents who did not understand the idea. Rather, *an earlier model of desired future defines and leads a following action.* In his seminal paper, published in 1935, Bernstein wrote about, "The presence in the central nervous system of a 'motor plan,' its overall formula" (cited in Bernstein 1990, page 282).

In the 1961 paper *Routes and Problems of the Physiology of Activity,* Bernstein analyzed in detail the notion of the model of the future, that is, the model of what should be. He wrote, "Similarly to how the brain forms a *reflection* of the real *external world*, of an actual *present* situation, and of earlier, memorized situations that happened in the *past*, it must also have an ability to 'reflect' (actually, to construct) a situation pertaining to a near *future* which has not yet become reality and which the brain tries to bring to existence led by its biological needs. Only such a clear image of desired future can be used to formulate a problem and to create a program for its solution" (Bernstein 1990, page 416).

Within the same article, Bernstein emphasized an essential difference between the model of the past/present and the model of the future: "The former model is *unambiguous* and *categorical* while the latter one can be based only on extrapolation within a certain *probability*" (Bernstein 1990, page 416). The same problem is revisited in a paper published in 1962, "It is possible to program an action with respect to a certain goal only based on an image or a model of a situation to which this action must lead and with respect to which the action is undertaken. However, since future events can be assessed or predicted only using probabilistic prognosis (a neat term by J.M. Feigenberg), it is clear that analysis of underlying physiological processes must be based on the theory of probabilities including its most recent developments" (Bernstein 1962, page 15). Bernstein notes that *probabilistic modeling of the future* forms the basis of activity of all organisms, starting from the lowest ones.

Further development of this idea can be found in a paper by Bernstein published in 1963. Bernstein emphasized that a beneficial or vitally important action cannot be programmed or realized if the brain has not created a driving, directing force in the form of a model of desired future: "We are confronted by two interrelated processes. One of them is *probabilistic prognosis* based on a perceived current situation, kind of an *extrapolation* for a period of time. . . . In parallel with this probabilistic extrapolation (based on an assumption of "non-intervention"), there takes place a process of *programming* an action whose goal is to bring about realization of the desired future" (Bernstein 1990, page 438).

Probabilistic prognosis is an important component of behavior organization of humans and animals (Feigenberg 1980; Feigenberg and Ivannikov 1971). In some pathological cases, the characteristic disorders of probabilistic prognosis can be observed in humans (Feigenberg 1969, 1971).

Probabilistic prognosis occurs based on memory of past events and information of current events delivered by sensory organs. The possibility of probabilistic prognosis is based on the corresponding memory organization (Feigenberg 1994).

Probabilistic prognosis results from life evolution in the probabilistically organized world. This prognosis represents a subjective model of the environment in an animal-centered reference frame. Thus, a formula of behavior can be expressed as: *to see—to foresee—to act*, i.e., to see an actual situation, to foresee its further development, and to act to achieve a certain goal (Feigenberg 1986).

Movements are programmed and performed in the dynamic, changing environment. Therefore, one has to transform into a desired state (model of desired future) a state that would occur with the highest probability after a certain time period, i.e., when a planned movement becomes reality rather than a state that exists at the time of planning. This is partly due to the fact that outcome of a movement depends on both motor commands and the perpetually changing environment, in particular on changes in forces and motions of bodies in the environment. We mean here changes in forces and motions of bodies that are not under control of a human or an animal. In order to achieve a desired motor goal accurately and within a reasonably short time, one must get ready to counteract (or to use) such external force changes and object motions as well as reactive forces, i.e., forces emerging due to joint coupling.

If one lived in a static environment, the following components would be needed to achieve a certain goal:

- A model of what should be, i.e., a model of the desired future *(Sollwert[2])*
- A model of what is *(Istwert[2])*
- A model of actions whose purpose is to transform *Istwert* into *Sollwert*

In a dynamic world which may change independently of actions by the subject, a model of desired future should be based on a hypothetical situation that would be realized when the planned actions are performed rather than on a situation that exists when motor planning takes place. For example, when someone shoots at a moving target, the person must aim not at the target but at the place the target most probably will be when the bullet arrives.

Therefore, in order to plan actions in a dynamic environment, one needs to be able to predict changes that would most likely take place in the environment at the time of action realization. This level of probabilistic prognosis may be termed *probabilistic prognosis by an external observer*: a subject observes events that are beyond his or her control and whose progress is independent of his or her actions. A typical example is trying to predict the weather.

A higher level of probabilistic prognosis is *predicting the results of actions that actively involve both the subject and the environment*, i.e., when changes in the environment depend on actions by the subject. In this case, prognosis should include the subject's own actions. In other words, the subject should consider what would happen in the environment if he or she does this or that and how these actions would move prospective situations closer to the model of desired future.

Both levels of probabilistic prognosis refer to actions in a passive environment, i.e., in an environment that has no goal of its own that may differ from the goals of the subject. Such an environment is assumed not to try to predict actions by or ruin the predictions of the subject. Theory of games address such situations as "games with the Nature."

One more, even higher level of probabilistic prognosis is when the environment includes another active subject whose goals may differ from the goals of the original

---

[2]N.A. Bernstein used the German terms *"Sollwert"* and *"Istwert"* even in his Russian papers. *Istwert* denotes a magnitude or an image existing at a current instant of time. *Sollwert* means a magnitude or an image of something that should occur in the future as a result of active behavior.

subject. Both subjects try to predict the future actions of each other. They make their prognoses taking into account these future actions and develop programs of actions leading to their own, subjective goals. This is a "game with a Partner" rather than a "game with the Nature." Such situations typically occur in sportive games, military actions, and a whole spectrum of situations that involve at least two human subjects whose goals are different (and frequently opposing). In such a situation, one subject deals with *probabilistic prognosis while "playing with an active Partner"*. One subject needs to predict with the highest probability the actions of the other partner and to prevent the partner from guessing the first subject's own actions (a tricky move in sports).

This is the hierarchy of probabilistic prognoses. Since playing with an active partner is most complex and potentially dangerous, humans tend to consider an unknown situation as one involving an active partner. This might be the origin of anthropomorphic or animal-like gods, reflecting forces of nature in the consciousness of ancient peoples. Such gods may be absolutely unrealistic, but they are always active, and their activity is somehow reflected in the peculiar combinations of human and animal features. Such gods can typically be fooled or bribed.

Plans and prognoses in humans are dramatically different from those in animals. In animals, psychological memory stores only information relevant to the personal experience of the given animal, while plans and prognoses never go beyond its life span. In contrast, human memory goes far back into the remote past, beyond the limits of an individual's life. It includes the experience of previous generations. Correspondingly, human plans and prognoses go beyond one's own life. This may be one of the most essential differences between humans and animals.

Experimental Studies of Reaction Time

Timing and accuracy of motor reactions are major factors in their appropriateness. The following discusses results of some studies investigating how probabilistic prognosis affects reaction time in humans.

What Is Important: The Frequency of a Signal or the Probability of Its Prognosis?

In our experiments, subjects sat in front of a row of electric bulbs with buttons. Each bulb had a corresponding button. Reaction time was measured, that is the time required for a subject to press the button corresponding to a bulb that turned on. If the bulbs were turned on in a random sequence and with an equal probability, reaction time decreased when fewer bulbs were present in the setup. If the probability of certain bulbs being turned on was different, reaction time was shorter for more probable (more frequent) stimuli. Why does a more probable signal induce shorter reaction times? Some researchers thought that different stimuli excited different neural structures. Did the more frequently excited neural structures change their state and become "quicker"? This question was analyzed in a series of studies performed by my colleague M. Tsiskaridze.

A subject saw only two bulbs, which could turn on in a random order, and was instructed to press the corresponding buttons as quickly as possible. The time intervals between the stimulus (a lighted bulb) and the reaction were measured. Two signals, A and B, appeared randomly. However signal A appeared twice as frequently as signal B. Here is an example of such a sequence.

BAABAAAABABBAAAAABABBAAA

In such an experiment, the subject reacts to signal A quicker than to signal B.

After reaction time reached a plateau, the experimenter changed the strategy of signal presentation without telling the subject. A short interval of a random sequence was repeated again and again. This interval is underlined in the previous example. The interval was chosen so that the number of A stimuli was still twice as high as the number of B stimuli. Thus, starting from a certain moment, the sequence of stimuli presentation looked like the following.

AAABABABAAABABAAABABAAABABAAABAB

Reaction times to both stimuli, A and B, started to decrease and approached simple reaction time, i.e., reaction time to a single stimulus, when there is no choice involved.

This result suggests that reaction time depends not on the frequency of signal presentation within a sequence but on the probability with which the subject expects this particular stimulus, i.e., on the probabilistic prognosis. In fact although the frequency of stimulus A in the second sequence was still twice as high as the frequency of stimulus B, just as at the beginning of the experiment, the subject could with certainty (i.e., with the probability of 1) predict the next stimulus. The probability of guessing the next stimulus correctly in this experiment was as high as in experiments with single stimulus presentation. Consequently, reaction time is similar in this experiment to that during simple reaction time tests. These results lead to the assumption that reaction time depends on the probabilistic prognosis of a particular stimulus.

To test the last conclusion, the stimulus frequency and subject's prognosis were separated. An experiment was designed where one of the signals became more frequent, while the subject could predict the other stimulus with an increasing probability. At first, a subject was presented with a sequence where stimulus A appeared more frequently than stimulus B. The probability of stimulus A was 0.9, while the probability of stimulus B was 0.1 $(P_A = 0.9; P_B = 0.1)$. After some time, reaction times reached a plateau with shorter reaction times for A and longer reaction times for B. At that moment, (T_1 in figure 4.1) the experimenter started to present only stimulus B without telling the subject. Thus, signal B became the only stimulus.

However, reaction time did not decrease but even started to increase. If the sequence of B was interrupted by a single A, reaction time to this stimulus was shorter. Such single measures were used to interpolate the dashed curve in figure 4.1.

This result corresponds well with the assumption that reaction time depends on probabilistic prognosis with respect to a stimulus. While successive B stimuli followed one another, the subject predicted stimulus A with an increasing probability. Correspondingly, the reaction time to stimulus B increased, while the reaction time to stimulus A decreased. However, when the number of stimuli B presented in a row continued to increase, the relation between reaction times to stimuli B and to stimuli A changed again. Reaction time to stimuli B decreased, while reaction time to stimuli A increased. This means that the subject stopped waiting for signal A and was predicting stimulus B with a higher and higher probability. The subject had changed his or her hypothesis with respect to the probabilistic structure of the sequence of stimuli. Different persons demonstrated different time courses of changes in hypothesis about the probabilistic structure of the environment at different times.

At a later time (T_2 in figure 4.1), the experimenter returned to the original sequence of signals ($P_A = 0.9$; $P_B = 0.1$). Reaction times to both A and B returned to their values recorded in the first part of the experiment.

This experiment was still suboptimal because, in the course of the experiment, the experimenter switched from a random sequence of stimulus presentation to a fixed sequence. To substantiate the findings, it was necessary to maintain random presentation of stimuli and, simultaneously, to separate stimulus presentation frequency and probabilistic prognosis of their emergence.

Reaction Time (ms)

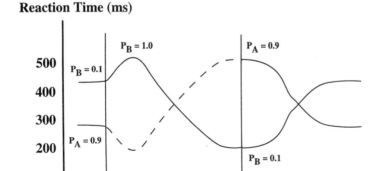

The Order of Signal Presentation

Figure 4.1   A scheme illustrating the dynamics of reaction time change to two signals, A and B, following changes in the probability of their appearance. The total number of signals in the experiment was 240.

Experiments With Sets of Four Stimuli

The following experiment was designed and performed. A subject was confronted with four stimuli *(A, B, C,* and *D)*. Each stimulus had a corresponding button to be pressed as quickly as possible when the stimulus emerged. An essential feature of the experiment was the probability with which different stimuli were presented. The first stimulus was chosen out of four, with a probability of 0.25; the next stimulus was chosen out of the remaining three, with a probability of 0.33; the next stimulus was chosen out of the remaining two, with a probability of 0.5; and then, the remaining, fourth stimulus was presented (with the probability of 1). Thus, in each sequence of four, the probability of each consecutive stimulus increased. The next sequence of four was chosen according to the same principle, and so on. Thus, a long sequence of stimuli (120 in each series) had equal proportions of each of the four stimuli—within each sequence of four, the order of stimuli was random. The sequence of stimuli in this experiment represented a particular case of the Markov chain, i.e., it was a sequence within which the probability of a stimulus depended on the stimuli that preceded it (in contrasts to the Bernoulli sequence within which the probability of each stimulus is independent of other stimuli).

As a result, the average reaction times to each of the stimuli *(A, B, C,* and *D)* were not statistically different, on the order of 360-365 ms ($T_A = T_B = T_C = T_D$). However, the average reaction times to stimuli at different places within the sequences of four was different. The reaction time to the first stimulus was the largest (on average, 460 ms). The reaction time to the second stimulus was somewhat smaller (on average, 424 ms). The reaction time to the third stimulus was even smaller (on average, 351 ms). The reaction time to the fourth stimulus was the smallest (on average, 207 ms). Note that the reaction time to the first stimulus was equal to the reaction time in an experiment with four stimuli in a Bernoulli sequence. The reaction time to the third stimulus was equal to the reaction time measured in an experiment with two stimuli in a Bernoulli sequence presented with equal probability. The reaction time to the last, fourth stimulus was equal to the simple reaction time, i.e., the reaction time to a single stimulus. We did not perform the experiments with three stimuli presented as a Bernoulli sequence with equal probability. Thus, reaction time depended on the probability with which the subject could predict a stimulus but not on which of the stimuli actually was presented.

The mentioned rule about reaction time *($T_1 > T_2 > T_3 > T_4$)* was seen in healthy subjects whether or not they realized the structure of the presented sequence of stimuli. It seems that even when a person does not comprehend the rule underlying a sequence of stimuli, he or she is able to grasp the probabilistic structure of the sequence subconsciously and use it adequately for probabilistic prognosis, preparation, and performance of a motor reaction.

Reaction time therefore depends on probabilistic prognosis, i.e., on a measure of signal expectancy. Stimuli frequency affects reaction time only as long as it affects probabilistic prognosis.

Probabilistic Prognosis and Motor Anticipation

Which psychophysiological events in the organism are induced by the prognosis? To answer this question, another experimental series was performed that used two light signals *(A* and *B)*. The subject was instructed to react to one of the signals by moving the right hand and to react to the other signal by moving the left hand. A few seconds prior to a signal presentation, a tone warned the subject that one of the signals would be presented. Figure 4.2 shows that, if the signals were presented with equal probabilities *(P$_A$ = P$_B$ = 0.5)*, the background electrical activity of forearm muscles increased in both arms prior to the light signal presentation. The bilateral anticipatory changes corresponded to a prognosis of equal probabilities of movements by the right or left hand.

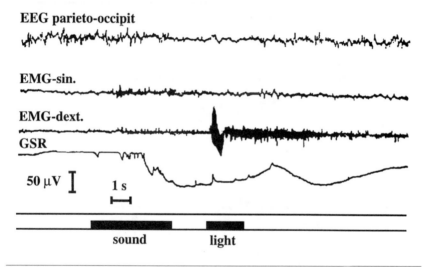

Figure 4.2   In conditions with equal prognoses of signal probability for movements of the right and left arms, the reaction of the right arm (EMG-dext.) is preceded by bilateral anticipatory changes in the EMGs.

Figure 4.3 illustrates a fragment of another experiment. The light signal for right-hand movement was presented more frequently than the other signal. In this case, the probabilistic prognosis led to a unilateral anticipatory change in the muscle activity seen only in the right arm.

In still another experiment, a light signal was presented prior to a "go"-signal (a number, 2 or 3). A red signal *(R)* was used prior to a "go"-signal 2, which corresponded to right-hand movement. A green signal *(G)* was used prior to a "go"-signal 3, which corresponded to a left-hand movement. Figure 4.4 shows that the red signal led to anticipatory changes in the muscle activity in the right arm, while the green signal led to anticipatory changes in the muscle activity in the left arm. In some cases, the experimenter presented a "go"-signal for the right hand after a warning

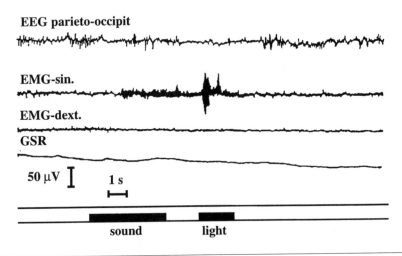

Figure 4.3   If the probabilistic prognosis predicts a signal to move the right arm, the reaction of the right arm (EMG-dext.) is preceded by anticipatory changes in the right arm EMG.

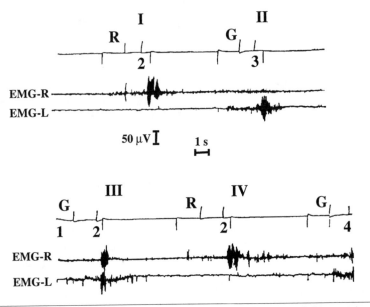

Figure 4.4   (I) and (IV) A reaction by the right hand (EMG-R) preceded by anticipatory changes in the right-hand EMG when a triggering signal for a right-hand movement was expected. (II) A reaction by the left hand (EMG-L) preceded by anticipatory changes in the left-hand EMG when a triggering signal for a right-hand movement was expected. (III) A reaction by the right hand (EMG-R) preceded by anticipatory changes in the left-hand EMG when a triggering signal for a left-hand movement was expected.

green light. In this case, no anticipatory changes occurred in the activity of right arm muscles, and reaction time increased.

Thus, probabilistic prognosis of a signal induces anticipatory preparation of an action (i.e., a preparation prior to a "go"-signal) only in organs that are expected to react to the following "go"-signal. This anticipation decreases the reaction time.

Probabilistic Prognosis of the Signal
or Probabilistic Prognosis of the Response?

Thus, one may hypothesize that prognosis of an appropriate response reaction is essential. Until now, we have discussed the probabilistic prognosis about the emergence of one of two signals where each signal corresponds to a definite reaction by one of the hands. The probability of signal emergence and the corresponding reaction were the same. Additional experiments were required to study the dependence of the reaction time on the probability of an expected reaction. In particular, it was necessary to measure reaction time to signals of equal probability in conditions when probabilistic prognosis for movements by the right hand differed from prognosis for movements by the left hand. The following experiment was designed to answer this question.

The subject faced a display that could present one out of eight signals. The signals are A_1, A_2, A_3, A_4, B_1, B_2, B_3, and B_4. They were presented in a random order with probabilities p_1, p_2, p_3, p_4, q_1, q_2, q_3, and q_4, correspondingly. If a signal from the A group appeared, the subject was supposed to press a lever with the left hand. If a signal from the B group appeared, the subject was instructed to press another lever with the right hand. The subject could predict the probability of pressing the lever with the left hand (P) as the sum of the probabilities of all the signals from the A group $(P = p_1 + p_2 + p_3 + p_4)$. The probability of pressing the lever with the right hand (Q) could be predicted as the sum of the probabilities of the signals from the B group $(Q = q_1 + q_2 + q_3 + q_4)$.

In experiments when the total frequency of the signals from the A group was higher than the total frequency of the signals from the B group, i.e., $P > Q$, an asymmetry of the electromyographic signals and reaction time occurred. The amplitude of electromyographic signals was higher in the left arm, while the reaction time was shorter in the left hand. In this experiment, however, the probability of emergence of one of the signals from the A group (A_1) was equal to the probability of a signal from the B group (B_1), i.e., $p_1 = q_1$. Despite the lack of difference in the probabilities for these stimuli, the reaction time to A_1 in the left hand was shorter than the reaction time to B_1 in the right hand. This means that different reaction times to stimuli of the same probability occur if the probabilities of reactions differ. The probabilistic prognoses of a signal and of a reaction must be distinguished.

Does the last observation mean that reaction time depends *exclusively* on the probability of the response while probabilistic prognosis of signals is important only as a predictor of the probability of reactions? Note that in the earlier experimental series with two signals and two responses, probabilistic prognoses for the signals and for the reactions coincided.

This question required modification of the experiment with eight signals and two reactions. If reaction time depends only on the probability of a corresponding reaction, reaction times to signals A_1 and A_2 should be the same even if the probabilities of these signals are different ($p_1 \neq p_2$) since both signals require the same reaction with the probability P. The experiments showed, however, that reaction time depended on the probability of a signal: the reaction time was shorter when the signal with a higher probability was presented. The hypothesis that reaction time depends only on motor preparation was therefore not confirmed. This suggests that preparation depending on probabilistic prognosis occurs in both motor and sensory domains of the organism.

Probability and Significance of a Stimulus

The described experiments studied the dependence of the reaction time on the probability of a signal or, more correctly, on probabilistic prognosis of its emergence. Signal probability was defined within the experimental procedure, and the reaction time was measured. The significance of all the signals was the same. In real life, however, the significance of a signal also plays an important role. A rare signal informing of a serious danger may induce a very fast reaction.

In experiments when signals A and B had the same probability but quick reactions to signal A were more highly rewarded, the reaction time to signal A was shorter. In other experiments when the probability of signal B was higher while quick reactions to signal A were more highly rewarded, it was possible not only to reach the same reaction times to both signals but also to have a faster reaction to the rare signal.

Reaction time may depend on *mathematical expectation* of a reward (or mathematical expectation of the lack of punishment), that is on the average value of the expected reward.

Therefore, it seems necessary to define quantitatively, within an experiment, the significance of signals and to measure reaction times. However, how can significance be quantitatively defined? Significance is a subjective notion. It varies among individuals and may differ for one and the same person in different conditions. If the significance of a signal is measured by a monetary reward for quick, correct reactions, one may say that a signal rewarded with one dollar is half as important as a signal rewarded with two dollars. However, signals rewarded by one-hundredth of a cent and by two-hundredths of a cent would be equally important because the significance of either of them can be defined as "very small". This is similar to how a person who does not know astronomy reacts to distances of one billion and ten billion light-years. Both distances are assessed as "very large".

Thus, an experimenter cannot define quantitative significance of a signal within an experiment. The experimenter can only define *reward*. No direct proportionality exists between reward and significance. Such a dependence exists only in a rather narrow range of "actual rewards"; this range varies for different persons. It becomes impossible to compare two nonmonetary rewards because their relative signifi-cance is very subjective. To one person, a sandwich is more important than a ticket to a museum; for another, the relation may be the opposite. An experimenter can define "reward" quantitatively and *make a conclusion about the significance only based on the results of the experiment.*

How can an experiment quantitatively compare the significance of two signals for a given person in a particular state? An index of the signal significance may be a change in the dependence of a measurable parameter (for example, reaction time) upon the probability of signal emergence. To compare the significance of two sig-nals, one may start with the following three assumptions:

1. If signals A and B emerge randomly with equal probabilities $(P_A = P_B)$ and the average reaction times to these stimuli are equal $(T_A = T_B)$, one may assume that the significance of signal A equals the significance of signal B $(V_A = V_B)$.

2. If under the same conditions $(P_A = P_B)$, the reaction time to signal A is shorter than the reaction time to signal B $(T_A < T_B)$, one may conclude that the significance of signal A is higher than the significance of signal B $(V_A > V_B)$.

3. Reaction times can be made equal by decreasing the frequency of presenta-tion of the more important signal. If reaction times to both signals are the same $(T_A = T_B)$ while the probability of signal B is N times higher than the probability of signal A $(NP_A = P_B)$, the significance of signal A is N times the significance of signal B $(V_A = NV_B)$.

So, it looks as if the central nervous system has developed a common scale that allows a person to compare the significance of phenomena of very different nature. Without this, expressing a preference or making a choice would be impossible. This subjective scale is very similar to monetary scales that allow comparison of the relative values of different phenomena, such as a leg of lamb, the quality of a house, and the attractiveness of a performance. In this field, psychologists may have ben-efited from recollecting the notion of "ultimate usefulness" (Grenznutzentheorie) of 19th century economists.

Summarizing the Chain of the Experiments

We started with a well-known fact: If two signals that require different motor re-sponses are presented randomly, reaction time is shorter for the signal presented more frequently. However, is the signal *frequency* itself important?

Experiments with repetitions of a short sequence within a long series of signals have demonstrated that frequency itself is not important. Although the unequal ratio of the signal frequencies was preserved, actual reaction times to both stimuli became equal. This result suggests that reaction time depends on the probability the person expects a signal, i.e., on probabilistic prognosis.

To test this hypothesis, we needed to run an experiment within which changes in the frequency of signal presentation and changes in the probabilistic prognosis would have different signs. After a series of experiments with two randomly presented signals of different probabilities, we performed a study with just one signal which showed that reaction time depended on the probabilistic prognosis and not on the signal frequency.

However, in all these studies, we moved during the experiment from a random sequence of signal presentations to a fixed sequence. So it was important to define the dependence of reaction time on probabilistic prognosis in conditions of a random signal sequence within which the probabilistic prognosis of a signal would not depend on the frequency of the signal among other signals. In experiments with combinations of four different signals in conditions of equal frequencies for each of the signals, it was shown that reaction time depended on the probabilistic prognosis with respect to the emergence of a signal at a certain place within a series of four different signals.

Which psychophysiological changes within the body are brought about by probabilistic prognosis? To answer this question the electromyogram (EMG) of muscles expected to be involved in the required motor reaction was recorded. The amplitude of the EMG was higher in muscles that were expected to react to a more probable signal. These muscles showed anticipatory changes in the background activity that reflect preparing to move. Stronger anticipatory changes were associated with quicker reactions. Reaction time may therefore depend on anticipation of a certain motor response.

A series of experiments with eight signals and two possible reactions demonstrated that the reaction times to two signals of equal probabilities that required different reactions could be different. Reaction time was shorter in response to a signal that appeared more frequently, thus having a higher probabilistic prognosis.

So maybe reaction time depends *exclusively* on probabilistic prognosis of a required motor reaction and on motor preparation? This hypothesis was not confirmed. A signal whose emergence was expected with a higher probability than that of the other signal within a pair showed a faster reaction time if both signals required the same response (for example, both signals demanded moving the right arm). This means that both probabilistic prognosis of a signal and probabilistic prognosis of a motor reaction are essential. In other words, *anticipatory changes take place in both motor and sensory domains.*

Reaction time, however, depends not only on probabilistic prognosis but also on the significance of a signal. Significance cannot be defined quantitatively *a priori* the way the probability of a signal can be defined. Significance can be measured only in an experiment. An experiment to measure significance, which, unlike reward, is subjective, was described.

Conclusion

Within the described series of experiments, we tried to get answers to questions on the relations between reaction time and probabilistic prognosis. We have also been interested in the processes of motor preparation (anticipation) and their relation to probabilistic prognosis. However, the problems of sensory preparation (anticipation) have not been touched. We have only come to a conclusion that sensory anticipation is necessary because motor anticipation, by itself, is unable to account for the experimental observations. To study sensory preparatory processes, a different series of experiments should be designed and performed.

References

Bernstein, N.A. (1935) The problem of interrelation between coordination and localization. *Archives of Biological Sciences,* 38: 1–34 (In Russian).

Bernstein, N.A. (1961) Routes and problems of the physiology of activity. *Problems of Philosophy,* 6: 77–92 (In Russian).

Bernstein, N.A. (1962) New directions in the development of contemporary physiology. In *Proceedings of the Conference on the Methods of Physiological Human Studies.* A.A. Letavet, V.S. Farfel (Eds.), Moscow: USSR Academy of Sciences, 15–21 (In Russian).

Bernstein, N.A. (1963) New directions of development of physiology and their relation to cybernetics. In *Philosophical Problems of Physiology of the Higher Nervous Activity and Psychology.* Moscow: USSR Academy of Sciences, 299–322 (In Russian).

Bernstein, N.A. (1966a) *Essays on the Physiology of Movements and Physiology of Activity.* Moscow: Meditsina (In Russian).

Bernstein, N.A. (1966b) From the reflex to the model of the future. *Nedelya (The Week),* #20: 4.

Bernstein, N.A. (1990) *Physiology of Movements and of Activity.* In the series *Classics of Science.* O.G. Gazenko, I.M. Feigenberg (Eds.), Moscow: Nauka (In Russian).

Feigenberg, I.M. (1969) Probabilistic prognosis and its significance in normal and pathological subjects. In *A Handbook of Contemporary Soviet Psychology.* M. Cole, I. Maltzman (Eds.), New York and London: Basic Books, 355–369.

Feigenberg, I.M. (1971) Probability prognosis and schizophrenia. *Soviet Science Review,* March 1971: 119–123.

Feigenberg, I.M. (1980) The ability to look into the future (probabilistic prognosis). *Pavlovian Journal of Biological Sciences,* July/Sept: 135–138.

Feigenberg, I.M. (1986) *To See—to Foresee—to Act.* Moscow: Nauka (In Russian).

Feigenberg, I.M. (1994) Ein Gedächtnismodell auf der Basis der Wahrscheilichkeitsprognostizierung. *Behindertenpädagogik,* 33: 374–389.

Feigenberg, I.M., Ivannikov, V.A. (1971) Relations between conditions and orienting reactions. *Soviet Psychology,* 9: 271–285.

5

CHAPTER

Bernstein's Principle of Equal Simplicity and Related Concepts

Mario Wiesendanger

Laboratory of Motor Systems, Department of Neurology,
University of Berne, Berne, Switzerland

The equal simplicity principle is a cornerstone in the scientific work of Bernstein. It has been discussed in depth in Bernstein's important paper entitled *The Problem of the Interrelation of Co-ordination and Localization*. It was first published in the *Archivy Biologicheskih Nauk* in 1935 (figure 5.1), and later translated into English as chapter 2 in Bernstein (1967 and 1984) and into German in Bernstein (1988). Although the principle also pertains to perception, Bernstein discussed it mainly with regard to performance of intentional, goal-oriented movements. In essence, the principle says that a given objective, for example making circular movements with an outstretched arm, can be performed with equal ease (or simplicity) in front of or sideward of the body (figure 5.2). This is not a trivial observation considering that execution of circular movements in different work spaces requires different sets of muscles. In other words, the ease of execution is effector independent. Similarly, a person can write large characters on the blackboard equally well as small characters on a sheet of paper. In both cases, the individual's writing characteristics are preserved. Bernstein further reasoned that a stored movement engram consciously used to launch the motor act does not include the metric definition of the muscle actions required for goal achievement. According to Bernstein,

Figure 5.1 Cover page of Bernstein's original paper *The Problem of the Interrelation of Co-ordination and Localization,* published in 1935. The reprint is signed by Bernstein and dedicated to Marc Shik who generously gave it to this chapter's author in 1991. The portrait of Bernstein is taken from the German publication of Bernstein 1988, translated by L. Pickenhain, which contains an introduction by V. Gurfinkel (with permission from publisher).

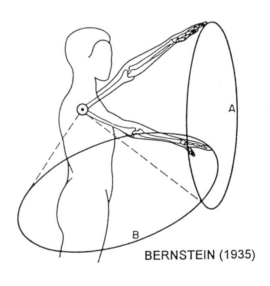

Figure 5.2 Example of circular movements performed in different workspaces to illustrate the principle of equal simplicity (with permission from publisher).

motor control is thus organized in at least two different hierarchical levels: upper, goal-related level(s) and lower level(s) responsible for metric execution.

This chapter's aim is twofold. First, it will show that the ground for many of Bernstein's ideas was prepared during the early part of the 20th century in the United States and particularly in Germany. Second, it will discuss the author's studies in monkeys performing a bimanual task that corroborate the notion of multiple control levels. These studies demonstrate that metric changes in the execution of the individual limbs can be induced by changing constraints or by subjecting the monkey to a brain lesion. In both situations, quantification of bimanual coordination of the two arms for goal achievement was, however, not compromised. The chapter will argue that this demonstrates the equal simplicity principle of Bernstein (1935) and also the motor equivalence principle of Lashley (1930, 1933), which have similar connotations.

Early Ideas Related to the Principle of Equal Simplicity

Terminology

To Bernstein goes the merit of having exposed the principle of equal simplicity in most detail and in conjunction with related concepts. He did this with physiological examples as well as with nonphysiological instruments. He formally showed that the degree of simplicity depends on the construction of the instrument (template, compass, ellipsograph), and the construction of the instrument determines how many variables it can handle. For example, a circle can be drawn with the radius given by the template, with a compass, or with an ellipsograph. The template has no degree of freedom since one can draw a circle only with the fixed template radius and with no eccentricity. Both the compass and the ellipsograph have one degree of freedom. The former with respect to the variable radius, the latter with the variable eccentricity. With a compass, drawing circles with different radii is performed with equal simplicity. With an ellipsograph, ellipses with different eccentricities are also made with equal simplicity. These transitions from one radius to another or from one eccentricity to another occur on lines of equal simplicity, which are on two different planes, as illustrated in figure 5.3.

The heuristic value of the principle is that by investigating lines of equal simplicity for a particular instrument or biological system, one may gain insight into the construction scheme of that system. The term simplicity is used deliberately by Bernstein to stress the general nature for any criteria of performance (e.g., velocity, variability, and precision). In a biological context, the term simplicity could be interchanged with adaptability for a system with changing constraints. The example of figure 5.2 is instructive. It illustrates that the circular trajectories, performed either in front of or laterally from the body, are executed with about equal ease and also precision. From this, Bernstein concludes that the construction scheme for this

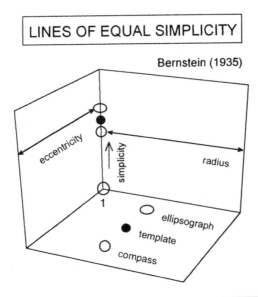

Figure 5.3 Bernstein formally described the principle of simplicity and the lines of equal simplicity by means of three instruments: a circular template, a compass, and an ellipsograph. With all instruments, a circle can be drawn with about equal ease (or simplicity). Whereas the template has no degrees of freedom (only one fixed radius, no eccentricity), both the compass and the ellipsograph have one degree of freedom. The line of equal simplicity for the compass is along one surface and concerns the variable radius, the line of equal simplicity for the ellipsograph along the other surface concerns eccentricity. The dependent variable measures simplicity on the z-axis. Transitions from one possible task to another that would deviate from the lines of equal simplicity are not possible without changes in the operational construction of the instrument. Bernstein suggested that investigations of lines of simplicity indirectly provide hints about the principle of operation of a given system. (Compare Bernstein 1984, page 111 and the present text).

movement task concerns the shape of the trajectories rather than the muscle actions (the muscular scheme).

Topology[1] is a related expression used by Bernstein that he opposes in the expression of metrics. Again, he explains topology in nonphysiological terms. Linear objects have either no configuration of any meaning or they do have a meaning that, however, cannot be expressed in the language of metrics. The latter is illustrated with a group of hand drawn stars of various configurations or with a group of the character A, also hand drawn with different styles and sizes (figure 5.4A) What does this have to do with motricity? Bernstein posits that human skilled performances are characterized by topological classes not only in drawing and writing but also for any

---

[1]Bernstein used the term "topology" as in gestalt psychology, not in the neurobiological definition of peripheral relationships with neural structures as established in mapping studies.

A CLASSES OF TOPOLOGY
   (Bernstein, 1935)

B MOTOR EQUIVALENCE IN HANDWRITING
   (K.S. Lashley, 1942)

blindfolded subjects

	subject 1	subject 2
right hand	*motor equivalence*	*motor equivalence*
left hand	*motor equivalence*	*motor equivalence*
right mirror (reversed)	*motor equivalence*	*motor equivalence*
left mirror (reversed)	*motor equivalence*	*motor equivalence*
teeth	*motor equivalence*	

Figure 5.4   *(A)* Examples of two classes of first-order topologies as metaphors for representations in the central nervous system. The common cognitive denominator for one class is a star, for the other the letter *A*. Note that all drawings have different metrics. *(B)* Handwriting is used as an example of a topology by Bernstein and as an example for the principle of motor equivalence by Lashley (1930). The illustration is from Lashley (1942). It shows preservation of the characteristic style of handwriting (chirography) by two subjects (left and right) when subjects write with the preferred hand, nonpreferred hand, from left to right, from right to left, or with the teeth (in one subject). Mirror writing has been reversed for better comparison.

skills and gestures. In other words, each of a person's skills has its peculiarity or gestalt (topologies of higher order). Thus, topologies of higher order are considered as mental images of purposeful movements without attributes of size or workspace. The term clearly has a cognitive meaning. According to Bernstein, the topological criteria are what is represented in the cerebral cortex and not the operational stages of muscle actions. This is also suggested by the fact that an action is planned in terms of its goal, not in terms of muscle contractions. In Bernstein's words, the topology of a skill (e.g., handwriting) is far away from the topology of muscles and joints controlling the metrics. Cortical representations would be leading engrams, i.e., topologies of higher order. Lower-order representations (subordinate engrams) may then serve to select muscle actions. Inherent in the concept of high-order topology is its invariant nature, whereas the resulting metrics of the movement has a high variability.

The Motor Equivalence Principle of Lashley

Although Lashley's approach to the problem was different from that of Bernstein, the principles of motor equivalence and equal simplicity are related as will become apparent. In a recent interesting essay about motor equivalence, Abbs and Cole (1987) remarked, "While the origin is obscure, for more than 50 years observations of variable means to invariant ends have been termed *motor equivalence (Lashley 1933)*.[emphasis added]" Lashley derived this principle from behavioral experiments in rats and monkeys. In the latter, he investigated the role of the precentral motor cortex in the animal's capability to achieve a manipulatory task. As an experimental psychologist, he was not so interested in metric changes of performance following precentral lesions but rather in measures of goal achievement. The task consisted of opening problem boxes (pull boxes, crank boxes, and hasp boxes), which the monkeys had learned to open to retrieve their food reward. Performance was thus assessed by the success score. Postoperatively, the monkeys had, of course, paretic deficits, and this is briefly mentioned in the paper. However, the main conclusion was that the monkeys quickly adapted to the new situation, sometimes using tricks to open the problem boxes again. In Lashley's words, "After practically complete destruction of both precentral gyri, this animal gave evidence of perfect retention of visual habits and habits of manipulation. Direct adaptive changes in behavior were made to compensate for weakness of the left arm" (Lashley 1924). This was motor equivalence. Interestingly, both Lashley (1933, 1942) and Bernstein (1935; see figure 5.4B) mention handwriting under different conditions (whole arm versus hand, preferred hand versus nonpreferred hand, and so forth) as examples of motor equivalence and equal simplicity, respectively. Lashley used the presence of motor equivalence as a further argument in his conviction that the theory of localization in terms of centers was wrong. Although Bernstein strongly criticized Lashley's equipotentiality theory of the cerebral cortex, he was in other ways quite near to Lashley's ideas. It is not known whether Bernstein was aware of this. In any case, he mentioned Lashley only in the context of the localization issue.

The motor equivalence principle has been reintroduced by Hebb (1949), and it recently has seen a renaissance (Abbs and Cole 1987; Fentress 1989; Jeannerod and Marteniuk 1992; Paulignan et al. 1990; Pavloski 1989; Wright 1990). Motor equivalence was also clearly present in the present author's behavioral studies in monkeys trained to perform a bimanual task (see the second part of this chapter). Note that motor equivalence is tightly linked with the degree of freedom problem often ascribed to Bernstein. Large variability in the metrics of movements appears to be the prerequisite for the possibility of flexibly organizing a motor task to attain the goal. Motor equivalence has, in fact, been observed under changing conditions of environmental constraints, with alterations of the peripheral motor apparatus, and after brain lesions. The existence of multiple solutions (trajectories) for attaining a given goal, i.e., the redundancy in the degrees of freedom, cannot be ignored in motor control models. In fact, Jordan (1990) and Bullock et al. (1993) have incorporated it into connectionist models of motor learning.

Obviously, a number of terms and concepts, such as adaptation, coordination, localization, hierarchies, flexibility, and plasticity, are linked with or are implicit in the principle of equal simplicity and motor equivalence. These terms had surely been discussed well before Bernstein. Most of this literature, published before World War II, is in German and not easily available. Therefore, it is hardly mentioned in relevant recent discussions.

Plasticity, Flexibility, and Adaptation

The physiologist Bethe (1872–1954, father of Hans Bethe, the theoretical physicist and Nobel prize winner who emigrated in 1933 to the U.S.A.) was trained in medicine and also wrote a Ph.D. thesis in biology. He started his academic career in Strasbourg, working at the Physiology Institute directed by Goltz, who was a prominent opponent of the doctrine of centers in the brain. Bethe was no doubt influenced by his master and, like Lashley, was impressed about the adaptive potency of living organisms. Bethe soon became the leader in what became the *Plastizitätslehre* (Bethe 1931). He based it essentially on behavioral observations. At the same time, not accepting the neuron theory, he emphasized the flexibility in the distribution of neural networks. Interestingly, Cajal (1911 and 1955), who of course did not agree with Bethe's premises of diffuse neuronal networks, nevertheless had similar views about central nervous system plasticity. It is now obvious that the modifiability of synaptic strength and connectivity plays an essential role in neural plasticity. Many of the antilocalizers were wrong in their premise but correct in their observations of dynamic representation. Bethe was convinced that the nervous system continuously adapts itself to cope with the environment and to achieve the goal. For example, he observed that when the third pair of legs used by some insects for swimming was amputated, the insect immediately shifted to the middle pair of legs for swimming. On the basis of many such examples, he developed the *Prinzip der gleitenden Koppelung* (literally, the principle of gliding coupling). This expresses the notion of flexibility in neural interactions (principle of flexible coupling would perhaps better transmit this notion). During World War I, Bethe participated, in cooperation with the famous surgeon Sauerbruch, in the development of hand prostheses based on the principle of flexible coupling as illustrated in figure 5.5.

Transplantation of tendons in adult dogs (interchanging flexors and extensors) (Bethe and Fischer 1931) and of whole limbs in developing amphibians (Weiss 1922, 1951) were characterized by a considerable degree of adaptive power and reorganization with restored coordinated locomotion. This work in the early part of the century together with neurological observations (Goldstein 1931) greatly strengthened the notion of plasticity. That is, the central nervous system, including the human brain, has great adaptive powers.

Coordination

Coordination has a long history in theories of motor control. In 1901, Kohnstamm already defined coordination as the process that constrains individual muscles into a spatiotemporal activation pattern (synergy) to achieve common goals. Coordination is thus the essence of goal-oriented intentional and automatic movements. The habilitation thesis of Foerster (1902) deals with the breakdown of coordination in movement disorders resulting from certain brain lesions. In 1922, Wachholder, a physiologist in Breslau, started an extensive series of experiments in healthy human subjects to study the rules of coordination of simple hand movements. A summary of these results, together with many theoretical considerations, appears in a masterly review (Wachholder 1928). For that time, the technical standard was remarkably high thanks to the Rockefeller Foundation, which helped to acquire the most sensitive string galvanometers then available for electromyography. This allowed Wachholder to record simultaneously for the first time, with high temporal resolution and relatively selective intramuscular electrodes, the action currents from an antagonistic muscle pair as well as the position trace. In the context of Bernstein's work, much of Wachholder's contributions, partly in collaboration with Altenburger, is still of high scientific interest. However, with exception of the three-burst pattern of ballistic movements, it has been largely forgotten.[2]

Localization, Movement Melody, and Reafference Principle

As expressed in the title of Bernstein's paper (1935), coordination is interrelated with localization. By localization, Bernstein understands something different than

---

[2]For a more detailed account of the major concepts of Wachholder, compare Wiesendanger (1997).

A. MECHANICAL MODEL

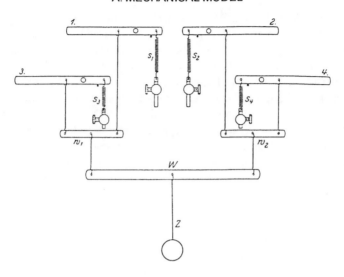

B. HAND PROSTHESES BASED ON PRINCIPLE OF SOFT COUPLING (BELOW)

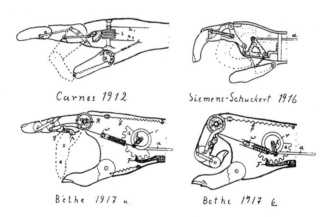

Figure 5.5 *(A)* A mechanical model of Bethe (1931) illustrating the principle of flexible coupling *(Prinzip der gleitenden Koppelung)* in the central nervous system. This allows for constraining muscles into a synergy. Levers 1–4 represent the control structures for the fingers of the hand (except the thumb). Depending on the stiffness of the springs S_1–S_4, a single command Z will produce, via the coupled system of levers, a synergy of the fingers with variable excursions. Bethe postulates that coupling in neural networks is similarly flexible and adaptive, for example to the grasped object. *(B)* Examples of hand prostheses developed during World War I. Note that only in the prosthesis Bethe 1917b, constructed on the principle of flexible coupling, can the index finger be flexed in all three joints by a single pull of an attached functioning tendon.
Reprinted from Bethe 1917.

the keyboard or homunculus type of representation still prevailing in the *Zentrenlehre* of clinical neurology. In fact, Bernstein prefers to use the term topology already referred to in this chapter. A leading topology is envisaged as a dynamic, spatial-temporal mental image (engram) of a behavioral act. This, too, was a commonly used concept in the German literature during the early years of the century. For example, at the 4th Congress of Experimental Psychologists in Innsbruck in 1910, von Monakow (1911) delivered a lecture entitled *Organization and Localization of Movements in Humans.* Mainly based on clinical observations of apractic patients, he conjectured that the cerebral cortex represents purposeful synergies, like reach and grasp synergies. These include the temporal sequencing and patterning of movement elements (what he called the movement melody). He also proposed that the organization of these complex synergies was a matter of widely distributed regions outside the focal motor cortex, particularly with respect to the limb-kinetic succession factors. The choice and definition of a goal is a mental process. As early as 1879, Lincke proposed that voluntary, goal-oriented movements are regulated by neural computation of the discrepancy between goal and actual position of the limb (see also Henn 1971). The mental goal representation is intimately related to the concept of internal body representation or *Körperschema* in the German literature (Schilder 1923), which is needed for this computation. The proposition of Lincke foreshadowed the cybernetic area that became fashionable after World War II when the work of Norbert Wiener became known. The idea was taken and illustrated by von Holst and Mittelstaedt in 1950 in the well-known diagram of the reafference principle with its closed loops (figure 5.6A; note that the author was a pupil of Bethe). Bernstein developed the same idea as von Holst and Mittelstaedt in his 1957 article (figure 5.6B), translated into English as Chapter IV in Bernstein (1967). Interestingly, Bernstein used the terms *Sollwert (*goal) and *Istwert* (actual value), common expressions found in the German literature about control theory.

Another pioneer of the idea of mental movement representations was Wachholder (1928). He insisted on the importance of the *Bewegungsentwurf,* i.e., the movement project, as a leading principle of goal achievement. Finally, a spatiotemporal leading topology also has much in common with the debated concept of the movement program. In his essay *Reflections on Motricity,* Granit (1980) describes a motor program as "an organized anticipation" of the forthcoming movement. Using functional imaging techniques, the metabolic correlate of a mental movement project can now be visualized (e.g., Kawashima et al. 1994). Cortical activation patterns are much in line with those proposed by von Monakow more than 80 years ago (see above).

The Principles of Equal Simplicity and Motor Equivalence Are Revealed in Bimanual Coordination

Aim and Task Description

In the author's laboratory, the mechanisms underlying bimanual coordination are studied. Many human skills require both hands, which cooperate beautifully for a

MOTOR CONTROL BASED ON GOAL DEFINITION AND ERROR CORRECTIONS

A. REAFFERENCE PRINCIPLE

B. SIMPLEST CLOSED LOOP CONTROL OF MOVEMENTS

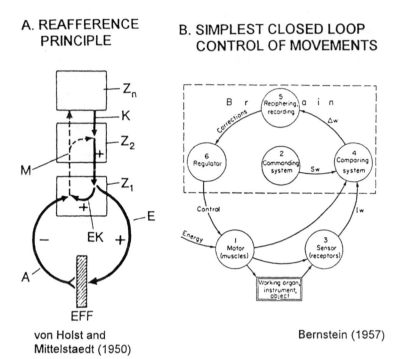

von Holst and Mittelstaedt (1950)

Bernstein (1957)

Figure 5.6 Motor control based on goal definition and error corrections. *(A)* Reafference principle of von Holst and Mittelstaedt (1950). Z1, Z2, and Z3 are hierarchic control levels (centers) of the nervous system. (K) command signal, (E) efference signal, (EFF) effector, (A) reafference signal, i.e., proprioceptive signals produced by self motion (which the authors sharply distinguish from exafference signals produced by external perturbations), (EK) efference copy, (M) difference between EK and A, i.e., the error signal transmitted back to the central control levels. *(B)* Similar scheme for closed-loop error correction by Bernstein (originally published in Russian in 1957, translated into English in chapter 4 of Bernstein, 1967, with permission from Publisher). Note the terminology used in the German literature of control theory: (Sw) *Sollwert* = efference copy of command signal, and (Iw) *Istwert* = reafference signal from moving limb. In this scheme, the brain center is differentiated in a command system, a comparator, a computational correcting system, and the regulator. Bernstein considers this the "simplest possible" control system of movements.

unified goal. Which cortical area is implicated in this higher-order topology of a bimanual action is also studied. The literature hints that the supplementary motor area (SMA) on the mesial surface of the frontal lobe is a bilaterally organized system (Penfield and Jasper 1954; Travis 1955) and thus might be implicated in bimanual coordination (Brinkman 1984; see also Wiesendanger et al. 1994 for review).

The following experiments in monkeys were performed to study bimanual co-ordination in a purposeful, naturalistic task and address two questions. First, can a quantifiable invariance of bimanual coordination be determined? Second, is the SMA an essential component in the control of a learned and well-coordinated bimanual task? To this end, monkeys were trained to perform a bimanual, asym-metric pull and grasp task using a drawer manipulandum. The setup and protocol were designed to allow changes in task constraints or to place brain lesions in order to evaluate possible changes in task performance. Details of setup, protocol, and data analysis, as well as results from unlesioned animals have been published (Kazennikov et al. 1994). In brief, monkeys were sitting behind a screen waiting until two windows opened, providing access for the two arms to reach separately for the drawer manipulandum. The leading left hand grasped the handle of the drawer, the right hand followed to pick up, with the precision grip, a piece of food from the small well of the drawer. During picking, the left hand had to maintain the spring-loaded drawer in the fully opened position (figure 5.7). A number of sensors provided signals for marking events on separate channels concerning the left and right hands. For the left, the sensors recorded movement onset, touching of drawer handle, and drawer fully opened. For the right hand, they recorded movement onset

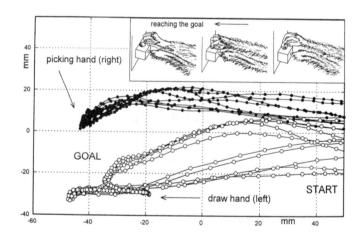

Figure 5.7   Inset—Three frames of the bimanual pull and grasp paradigm (drawn from video frames), as the two hands reach the goal. Goal achievement occurs in the middle frame when the drawer is just completely opened by the pulling hand, and the right index finger reaches into the small food well to pick up the food morsel with the precision grip. Trajectories of five superimposed trials—Note the variability of individual trajectories that converge precisely at the goal. One reflective (passive) marker was stuck on the middle phalanx of the right index finger and on the base phalanx of the left index finger (ELITE-system, 2D-recording, sagittal plane). This performance was equally well performed in complete darkness (see figure 5.8).

and insertion of the index finger into the food well. From these digitized values, intermanual time differences and linear correlation coefficients of movement times were calculated. The well-trained monkeys proficiently performed the task with unfailing motivation for more than 100 trials per daily session. Acquisition of this naturalistic skill by monkeys may be compared to well-practiced manual skills of everyday life in humans. In fact, human performance of the same task using a manipulandum adapted to human geometry was similar to that in monkeys, as reported in Kazennikov et al.

Demonstration of Variable Means to Invariant Ends

A video analysis of the movement sequences already suggested invariant synchronization of the two hands as they reached the goal. This was confirmed objectively by measuring the time structure of the bimanual synergy. Figure 5.8 illustrates the basic results obtained in the control situation (histograms upward, open symbols), as well as in complete darkness (histograms down, filled symbols). In figure 5.8A, the time histograms of left and of right hand events are plotted separately on the left and right, respectively. The histograms of intermanual intervals, as the hands reach the goal (right to left), are plotted on the left of figure 5.8B; the regression plot of left hand and right hand arrival times at the goal is displayed on the right of figure 5.8B. Two important results are readily observed. First, a large trial-by-trial variability occurs in the single-hand components of the synergy (figure 5.8A). Even so, the task had been practiced for a very long time and would have been declared as overtrained by experimental psychologists. Second, this prominent variability of the individual hands contrasts markedly with the intermanual parameters illustrated in figure 5.8B. The right to left intervals for goal-reaching events were near zero with a small variance, i.e., the two hands were well synchronized at the goal and the linear correlation coefficient was high. See the plots for numerical results.

Are There Lines of Simplicity in the Bernsteinian Sense?

The above results obtained under normal light conditions will be compared with those when monkeys performed the task in complete darkness (figure 5.8A; histograms down, filled symbols). Obviously, the individual hand movements were significantly delayed when visual guidance was not possible. This effect in the individual hands contrasts strikingly with the unchanged invariance of intermanual synchronization and the high correlation coefficients shown in figure 5.8B. This means that lack of visual guidance did not perturb goal coordination and that the monkeys easily adapted to the changed situation. The transition from light to dark was thus accomplished with equal simplicity and motor equivalence. Another conclusion was that the goal parameters were excellent descriptors of the quality of bimanual coordination.

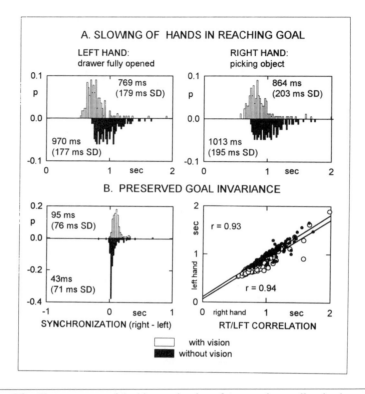

Figure 5.8   Time structure of the bimanual task performance in a well-trained monkey. Data, collected over several weeks, were generated by a number of sensors providing the event markers for each limb. In the four plots, comparison is made between performance under normal light condition with view on the workspace (histograms up, open symbols) and performance in complete darkness (histograms down, filled symbols). Note shift to the right in the time histograms of the left and the right hand, and preserved goal invariance (synchronization, left to right correlation coefficients), demonstrating motor equivalence. Reprinted from M. Wiesendanger 1997.

Where in the Brain Is the Leading Topology of Bimanual Coordination Represented?

The author set out to test the hypothesis that the SMA plays an essential role in constraining the two arms for a unified bimanual synergy. Two well-trained monkeys were subjected to a bilateral and a unilateral lesion of the SMA, respectively[3]. This report is confined to the effects on movement changes in the individual limb components and on bimanual coordination. Surgery was performed under deep Nembutal anesthesia and aseptic conditions. The results of the bilateral lesion are

---

[3]The full account of these experiments will be published elsewhere; preliminary data are reported in Wiesendanger et al. (1994).

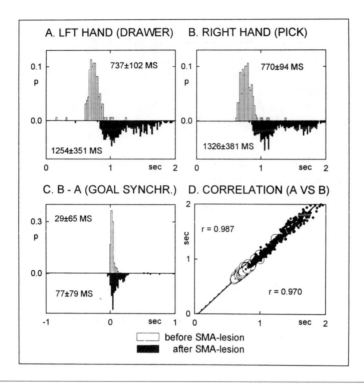

Figure 5.9 Effect of bimanual task performance before and after bilateral ablation of the SMA in a monkey. Same presentation style as in Figure 5.8. Data shown were obtained during the monkey performing in the dark; similar results were also seen when the performance occurred under normal light condition. Note the massive prolongation of the left and right hand in arrival times at the goal. In spite of this, bimanual coordination remains largely intact. The longer goal synchronization values are from the immediate postoperative period; the correlation coefficient remains very high. This is again a demonstration of motor equivalence.

presented in figure 5.9 using the same style as figure 5.8. The preoperative control data are shown with open bars and symbols, the postlesion data with filled bars and symbols. Obviously, the lesion had dramatic effects on both the left and the right limbs, with a shift to longer time values in the histograms, a bimodal distribution, and a large increase in variance (figure 5.9A, B). Surprisingly, however, the goal invariance was preserved (figure 5.9C, D; the slight increase in synchronization variance was from data obtained during the first postoperative sessions). These clear-cut but unexpected results unambiguously demonstrate the presence of two different control levels. Only the control structure for the individual hand actions was disturbed, not (or only minimally and transiently) the control structure for interlimb coordination. In Bernstein's perspective, the SMA is not a high-order topology crucial for bimanual coordination. Formulated in Lashley's terminology, the results show clear motor equivalence. To the author's knowledge, this is the first quantitative demonstration of motor equivalence in a lesion study. Remember that Lashley based the motor equivalence principle on his observations in monkeys with precentral cortex lesions, which is a similar paradigm as the present one. However, he evaluated only the score (success or no success) of opening problem boxes, i.e., goal achievement. He did not evaluate the low-level metrics of limb performances, which were qualitatively described as paretic.

What Is the Heuristic Value of Motor Equivalence and Equal Simplicity Principles?

The limitations of the principles will be considered first. Using the concrete example just described, one can only conclude that the SMA is not crucial for bimanual coordination since the SMA possibly participates in the control of bimanual coordination in cooperation with other cortical or subcortical structures. Following Bernstein's argumentation, the present results positively indicate that the SMA is more closely related to the control scheme of the individual arms and perhaps even of the arm muscles (the muscle formula). In accord with this are the now well-established direct SMA connections with the spinal cord (Macpherson et al. 1982a, b). Most recent tracing studies even indicate direct corticomotoneuronal connections (Rouiller et al. 1996). Generally speaking, differentiating hierarchical control levels in lesion studies (Bernstein's graded topologies) may help to understand better the "Interrelation between coordination and localization" (Bernstein 1935).

Besides theoretical considerations, the notion of multiple control levels also has possible practical implications. The first concerns the amazing confusions existing in the literature about lesion effects in a given brain structure. Even for lesions of massive structures of the human brain, such as the cerebellum or the pyramidal tract, the reported effects vary from no disturbances to massive ones. See for example the article by Bucy (1957) with the provocative title *Is There a Pyramidal Tract?* This unfortunate situation may have many reasons, but one is almost certainly linked with the issue of control levels. To give an example, neglecting to differentiate the two control levels in the above bimanual task experiment may have led to the false statement of, "No effects produced by an SMA lesion." Discussing the issue of stratified control levels in biological systems, Weiss (1951) appropriately commented, "Since coordination exists at all levels, disturbances of coordination at any one level need not imply disturbance or disruption of coordination at others. Disregard of this fact accounts for much of the confusion in past discussions of functional regulation, re-integration, re-education, compensation, etc."

The second issue is the possible practical consequence of correctly identifying control levels of dysfunction in neurological patients as a guide for rehabilitation. Simplifying, rehabilitation strategies will have to stress cognitive aspects of goal-oriented tasks rather than deal with muscle tone, force, and reflexes if the pathology of the motor disorder concerns high-order topologies.

Conclusion

Without doubt, Bernstein has greatly influenced research in motricity during the last 20 years. Phillips (1986) draws attention to the important message of Bernstein who was able "To set limits to the ways in which the cerebral cortex could plausibly operate in producing skilled movement, and to criticize theories based on results of electrical stimulation of the motor areas in monkeys and men." As also rightly remarked by Phillips, Bernstein was a pioneer in measuring and analyzing trajectories (cyclograms) of rhythmic movements of everyday life, such as locomotion and diverse manipulations. The many original and thought-provoking scientific achievements of Bernstein will not be further discussed here. Professor Gurfinkel, who, together with the Bernsteinian *Moscow school,* has brilliantly continued the legacy of Bernstein, is in a much better position than the author to do that.

This chapter attempted to show that from the beginning of the 20th century, ideas emerged, particularly in Germany, that ran contrary to established views. First, recall some of those: the keyboard notion and, emerging from that, the homunculus idea, which was an attractive but too simple concept of cortical function; and functional localization constrained too exclusively as a hierarchical system of brain centers *(Zentrenlehre).* Von Monakow, Bethe, Goldstein, and Lashley were all critical with regard to the doctrine of brain centers. Their criticism made these authors vulnerable, especially Lashley who went too far with his equipotentiality idea. Their skepticism, however, was well founded on results from animal experiments and clinical experience that revealed brain plasticity. The quarrel about localization issues was more about its organization and whether it is distributed in interacting networks or in more isolated and hierarchically organized centers. The former assumption seems to prevail today. The adaptive capacity of the brain was one of the new leading ideas that, for example, led Lashley directly to his principle of motor equivalence. This chapter argued that Bernstein's principle of equal simplicity has much in common with the latter. Both deal with the capacity of an organism to adapt to variable environmental or internal conditions (such as brain lesions). In order to assess this adaptation, objective measurements were necessary. Early pioneers who measured and studied coordination (such as Foerster, Kohnstamm, Bethe, and Wachholder) contributed significantly to the new deal, i.e., to knowledge about the organization of natural movements.

Additionally, motor-control research was caught in a more and more detailed investigation of spinal reflexes, mostly studied in the tradition of Sherrington, i.e., in reduced preparations. This was useful and necessary work. However, it diverted from the fact that motricity of intact animals and human beings is not exclusively stimulus bound but primarily goal oriented with adjustments that anticipate the movement's result. All these more cognitive aspects of motor control, which include the mental image of the intended action (movement project, leading engram, movement melody), gradually emerged during the first half of the 20th century. Concerted efforts addressing these difficult, complex aspects of motricity had to wait until recent years. Translations of Bernstein's work (unfortunately, only little of it so far), evidently played a key role in this development.

Communication among the researchers at an international level was much reduced during the difficult time of the two world wars, Nazism, and Stalinism. In the Soviet Union, the ideas of Pavlov were transformed into a political dogma; this meant that all research in the neurosciences had to conform with the dogma. As a consequence, Bernstein's work could not exert the influence it deserved (compare the most recent commentaries in Bernstein (1996)). Very little, if anything, penetrated to the Western scientific community until 1967. In Germany, the spirit of the pioneers was also hard hit by Nazism; there were no surviving schools left to follow masters like Bethe and Wachholder. Notable exceptions were von Holst, mentioned above, and Jung (for an autobiographical sketch see Jung 1973). Under Jung's guidance and with his participation, the *Bereitschaftspotentiale* (readiness potentials) (Kornhuber and Deecke 1965) and the *Zielpotentiale* (goal-related potentials) (Grünewald-Zuberbier and Grünewald 1978) were discovered. Before World War II, Jung had been working for some time at the Physiology Institute in Zurich headed by Hess, who had a lasting influence on Jung. Hess well recognized that posture is an integral part of volitional actions and that postural adjustment is a physical precondition for moving. He deduced from this that posture has to occur in anticipation of the movement, not as a reflex response. He illustrated this for his students by means of a delightful three-men model (Jung and Hassler 1960)(figure 5.10). Hess was also led to this concept by his pioneering work on freely moving cats for which he was rewarded with the Nobel prize in 1949. His integrative ideas about motricity were nearer to those of Bernstein (whose work Hess did not know) than to those of Sherrington. Like the work of the German pioneers, that of Hess was published in German and is also not well known in the English-speaking world. Some of the key articles have been translated into English and edited by Akert (1981).

The work before us is enormous—aiming at an understanding of the admirable but incredibly complex nature of human motor skills. This includes the mechanics of performance and their anticipatory mental representations, motor preparation, and motor memory. However, powerful tools are now available, as never before, for gradually investigating some of the underlying mechanisms. A number of plausible propositions and models also need to be tested experimentally.

Acknowledgments

The experiments described in the second part of this essay were performed at the Institute of Physiology, University of Fribourg, in collaboration with Brian Hyland, Oleg Kazennikov, Stephen Perrig, Urs Wicki, and Eric Rouiller. To all I extend my warm thanks for the excellent cooperation and for the many fruitful discussions. I am also grateful for the technical help received by Véronique Moret, Bernard Aebischer, and André Gaillard. Financial support was provided by the Swiss National Science Foundation (Grants 31-27569.89, 31-36183.92, and 4038-044053).

Figure 5.10 Three-men model of Hess for a goal-directed action with anticipatory postural control. Student 1 performs the goal-directed jump, student 2 is responsible for postural support, and student 3 is the tone setter. In *a–c*, the goal-directed jump is successful because of an adequate proactive (i.e., anticipatory) postural support. In *d–f*, however, the jumper miserably misses the goal because the tone setter was unprepared. This is a clear demonstration that a reactive adjustment of postural support (initiated after onset of the intended jump) is too late and inefficient to reach the goal correctly. Motion pictures were taken by Hess in the early forties (see Akert 1981). Drawings from these motion pictures were used by Jung and Hassler (1960) for their Handbook's chapter about the extrapyramidal system, published by the American Physiological Society. Reprinted from Jung and Hassler 1960.

References

Abbs, J.H., Cole, K.J. (1987) Neural mechanisms of motor equivalence and goal achievement. In *Higher Brain Functions*. S.P. Wise (Ed.), New York: Wiley, 15–43.

Akert, K. (1981) *Biological Order and Brain Organization; Selected Works of W.R. Hess*. Berlin: Springer, 1–347.

Bernstein, N.A. (1935) The co-ordination and regulation of movements. *Archiv biologiceskich nauk,* 38: 1–34 (In Russian).

Bernstein, N.A. (1967) *The Co-ordination and Regulation of Movements.* Oxford: Pergamon Press.

Bernstein, N.A. (1984) The Co-ordination and Regulation of Movements. In *Human Motor Actions; Bernstein Reassessed.* H.T.A. Whiting (Ed.), Amsterdam: North Holland, 1–633.

Bernstein, N.A. (1988) *Bewegungsphysiologie.* 2d ed. Leipzig: J.A. Barth, 1–272.

Bernstein, N.A. (1996) *Dexterity and its Development.* M.L. Latash, Turvey (Eds.), Mahwah, NJ: Erlbaum, 1–460.

Bethe, A. (1917) Beiträge zum Problem der willkürlich beweglichen Armprothesen. III. Die Konstruktionsprinzipien willkürlich beweglicher Armprothesen. *Münchener Medizinische Wochenschrift.* 64: 1625–1629.

Bethe, A. (1931) Plastizität und Zentrenlehre. In *Handbuch der normalen und pathologischen Physiologie. Arbeitsphysiologie II: Orientierung, Plastizität, Stimme und Sprache.* A. Bethe, G. von Bergmann, G. Embden, A. Ellinger (Eds.), Berlin: Springer, 1175–1221.

Bethe, A., Fischer, E. (1931) Die Anpassungsfähigkeit (Plastizität) des Nervensystems. In *Handbuch der normalen und pathologischen Physiologie. Arbeitsphysiologie II: Orientierung, Plastizität, Stimme und Sprache.* A. Bethe, G. von Bergmann, G. Embden, A. Ellinger (Eds.), Berlin: Springer, 1045–1129.

Brinkman, C. (1984) Supplementary motor area of the monkey's cerebral cortex: Short- and long-term deficits after unilateral ablation and the effects of subsequent callosal section. *J Neurosci,* 4: 918–929.

Bucy, P.C. (1957) Is there a pyramidal tract? *Brain,* 80: 376–392.

Bullock, D., Grossberg, S., Guenther, H. (1993) A self-organizing neural model of motor equivalent reaching and tool use by a multijoint arm. *Journal of Cognitive Neuroscience,* 5: 408–435.

Cajal, S.R. (1955) *Histologie du système nerveux de l'homme et des vertèbrès.* Vol. II Madrid: Instituto Ramon y Cajal (French translation of the original Spanish publication, 1909–1911).

Fentress, J.C. (1989) Comparative coordination (a story of three little P's in behavior). In *Perspectives on the Coordination of Movement.* S.A. Wallace (Ed.), Amsterdam: North Holland, 185–219.

Foerster, O. (1902) *Die Physiologie und Pathologie der Koordination.* Jena: Gustav Fischer.

Goldstein, K. (1931) Die Lokalisation in der Grosshirnrinde nach den Erfahrungen am kranken Menschen. In *Handbuch der normalen und pathologischen Physiologie, Arbeitsphysiologie II: Orientierung, Plastizität, Stimme und Sprache.* A. Bethe, G. von Bergmann, G. Embden, A. Ellinger, A. (Eds.), Berlin: Springer, 600–842.

Granit, R. (1980) Reflections on motricity. *Perspectives in Biology and Medicine,* 23: 171–178.

Grünewald-Zuberbier, E., and Grünewald, G. (1978). Goal-directed movement potentials of human cerebral cortex. *Exp. Brain Res.* 33: 135-138.

Hebb, D.O. (1949) *Organization of Behavior.* New York: Wiley, 153–157.

Henn, V. (1971) The history of cybernetics in the 19th century. In *Pattern Recognition in Biological and Technical Systems.* Berlin: Springer, 1–7.

Jeannerod, M., Marteniuk, R.G. (1992) Functional characteristics of prehension: From data to artificial networks. In *Vision and Motor Control. Advances in Psychology Series.* Vol. 85. L. Proteau, D. Elliott (Eds.), Amsterdam: North Holland (Elsevier), 197–232.

Jordan, M.I. (1990) Motor learning and the degrees of freedom problem. In *Attention and Performance XIII Motor Representation and Control.* M. Jeannerod (Ed.), Hillsdale, NJ: Erlbaum Ass. Publ., 796–836.

Jung, R. (1973) Some European neuroscientists: A personal tribute. In *The Neurosciences: Paths of Discovery.* F.G. Worden, J.P. Swazey, G. Adelman (Eds.), Cambridge, MA: MIT Press, 477–517.

Jung, R., Hassler, R. (1960) The extra-pyramidal motor system. In *Handbook of Physiology Neurophysiology.* Sect. 1, \ Vol. II. J. Field, H.W. Magoun, V.W. Hall, V.E. (Eds.), Washington, DC: Am. Physiol. Soc., 863–927.

Kawashima, R., Roland, P.E., O'Sullivan, B.T. (1994) Fields in human motor areas involved in preparation for reaching, actual reaching, and visuomotor learning: A positron emission tomography study. *Journal of Neuroscience,* 14: 3462–3474.

Kazennikov, O., Wicki, U., Corboz, M., Hyland, B., Palmeri, A., Rouiller, E.M., Wiesendanger, M. (1994) Temporal structure of a bimanual goal-directed movement sequence in monkeys. *European Journal of Neuroscience,* 6: 203–210.

Kohnstamm, O. (1901) Über Koordination, Tonus und Hemmung. *Zeitschr f diät u phys Therapie,* 4: 112–122.

Kornhuber, H.H., Deecke, L. (1965) Hirnpotentialänderungen bei Willkürbewegungen und passiven Bewegungen des Menschen: Bereitschaftspotential und reafferent Potentiale. *Pflügers Arch.ges.Physiol.* 284: 1-17.

Lashley, K.S. (1924) Studies in cerebral function in learning. V. The retention of motor habits after destruction of the so-called motor areas in primates. *Arch. Neurol. Psychiat. (Chic.)* 12: 249-276.

Lashley, K.S. (1930) Basic neural mechanisms in behavior. *Psych Rev.* 37: 1–24.

Lashley, K.S. (1933) Integrative functions of the cerebral cortex. *Physiol. Rev.,* 13: 1–42.

Lashley, K.S. (1942) The problem of cerebral organization in vision. In: *Biological Symposia, Vol.VII: Visual Mechanisms.* Kluever, H. (Ed.), Lancaster: Cattel Press, 301–322.

Macpherson, J.M., Marangoz, C., Miles, T.S., Wiesendanger, M. (1982a). Microstimulation of the supplementary motor area (SMA) in the awake monkey. *Exp.Brain Res.,* 45: 410–416.

Macpherson, J.M., Wiesendanger, M., Marangoz, C., Miles, T.S. (1982b). Corticospinal neurones of the supplementary motor area of monkeys: a single unit study. *Exp. Brain Res.,* 48: 81–88.

Paulignan, Y., MacKenzie, C., Marteniuk, R., Jeannerod, M. (1990) The coupling of arm and finger movements during prehension. *Exp Brain Res,* 79: 43–435.

Pavloski, R.P. (1989) The physiological stress of thwarted intentions. In *Volitional Action.* W.A. Hershberger (Ed.), Amsterdam: North Holland (Elsevier), 215–232.

Penfield, W., Jasper, H. (1954) *Epilepsy and the Functional Anatomy of the Human Brain.* Boston: Little, Brown.

Phillips, C.G. (1986) *Movements of the Hand.* Liverpool: Liverpool University Press.

Rouiller, E.M., Moret, V., Tanné, J., Boussaoud, D. (1996) Evidence for direct connections between the hand region of the supplementary motor area and cervical motoneurons in the Macaque monkey. *European Journal of Neuroscience,* 8: 1055–1059.

Schilder, P. (1923) *Das Körperschema.* Berlin: Springer.

Travis, A.M. (1955) Neurological deficiencies following supplementary motor area lesions in Macaca Mulatta. *Brain,* 78:174–198.

von Holst, E., Mittelstaedt, H. (1950) Das Reafferenzprinzip (Wechselwirkungen zwischen Zentralnervensystem und Peripherie). *Naturwissenschaften,* 37: 464–476.

von Monakow, C., (1911). Aufbau und Lokalisation der Bewegungen beim Menschen. In *Bericht IV Kongress Exp. Psychol. Insbruck 1910.* Leipzig, Germany: Johan Ambrosius Barth, 1–28.

Wachholder, K. (1928). Willkürliche Haltung und Bewegung. *Ergebn Physiol,* 26: 568–775.

Weiss, P.A. (1922). Die Funktion transplantierter Amphibienextremitaeten. *Anz Akad Wiss Wien,* 59: 199–201.

Weiss, P.A. (1951) Central versus peripheral factors in the development of coordination. In *Patterns of Organization in the Central Nervous System.* P. Bard (Ed.), Baltimore: Williams & Wilkins, 3–23.

Wiesendanger, M. (1997) Paths of discovery in human motor control; A short historical perspective. In *Perspectives of Motor Behavior and its Neural Basis.* M.C. Hepp-Reymond, G. Marini (Eds.), Basel, Switzerland: Karger, 103-134.

Wiesendanger, M., Wicki, U., Rouiller, E. (1994) Are there unifying structures in the brain responsible for interlimb coordination? In *Interlimb Coordinaton: Neural, Dynamical and Cognitive Constraints.* S.P. Swinnen, H. Heuer, J. Massion, P. Casaer (Eds.), San Diego: Academic Press, 179–207.

Wright, C.E. (1990) Generalized motor programs: Reexamining claims of effector independence in writing. In *Attention and Performance XIII: Motor Representation and Control.* M. Jeannerod (Ed.), Hillsdale, NJ: Erlbaum, 294–320.

6

Coordinated Control of Posture and Movement: Respective Role of Motor Memory and External Constraints

Jean Massion

Laboratoire de Neurobiologie et Mouvements,
CNRS, Marseille, France

Alexey Alexandrov

Institute of Higher Nervous Activity and Neurophysiology,
Academy of Sciences of Russia, Moscow, Russia

Sylvie Vernazza

Laboratoire de Neurobiologie et Mouvements,
CNRS, Marseille, France

During most motor acts, multiple goals are simultaneously achieved during the same general action. For example, when reaching for an object located in the surrounding space, one has to orient the gaze toward the object, transport the hand before grasping, and shape the fingers to the width of the object (Arbib 1981; Jeannerod 1988). Coordination between the various aspects of the task occurs via timing signals that synchronize the main steps of the various, parallel controls. This also occurs in other tasks such as the bimanual grasping task described by Wiesendanger et al. (1994). Coordinate structures are supposed to provide the appropriate timing signals and spatial coordination. In contrast, during other coordinate tasks such as gaze control, where the head and eye controls are integrated to achieve the same goal, the interactions between head and eye control occur through a combination of feedforward and feedback circuits. These compensate for the different inertias of the two controlled segments (Bizzi 1974).

Coordination between posture and movement is another example of a coordinated task where multiple goals have to be controlled simultaneously during the same motor act. The pioneer work by Belenkiy et al. (1967) showed that arm raising by a standing subject was preceded by postural adjustment aimed at minimizing the postural and balance disturbance due to the movement. This indicated that balance and/or body segment orientation had to be preserved during arm movement. It also showed that a postural change starting before the onset of the movement was initiated in order to preserve the postural functions that are continually exerted during the movement.

Anticipatory Postural Adjustments: Unanswered Questions

Many investigations have been devoted to the analysis of the coordination between posture and movement (Horak and Macpherson 1996; Massion 1992, 1994). However, several basic questions still remain concerning why and how the coordination is organized. These questions were raised by Bernstein (1984) in his illuminating work. He did propose a theoretical framework that is still the most appropriate for discussing these questions.

Goal

A prerequisite for understanding what is controlled by the anticipatory postural adjustments is to define which goals are achieved by these anticipations. In the field of posturo-kinetic coordination, the identification of goals has not always been clearly made. Therefore, a short analysis will be made of the main postural functions and how they have to be exerted during movement.

Control

Concerning control of the anticipatory postural adjustments, two main questions have to be raised.

First, how are the movement and postural control coordinated in the task? Nashner and McCollum (1985) proposed that for postural control, a repertoire of synergies does exist, mainly based on the biomechanical characteristics of the multijoint chain. When an external perturbation does occur, one of the postural synergies is triggered in order to restore the initial posture or to preserve the antero-posterior position of the center of mass (CM). When considering the anticipatory postural adjustments, they suggested (see also Gahéry and Massion 1981; Lee 1984) that the same repertoire of postural adjustments could be used in a feedforward manner in association with the voluntary movement. However, interactions between the control systems for the anticipatory postural adjustment and for the movement had to take place (Cordo and Nashner 1982). In this point of view, the postural adjustments were based on activation of a set of muscles of the trunk and legs, which represented a "synergy." The synergy was fixed. It depended on the activation of specific neural networks, either inborn or built through learning. The concept of fixed synergies was later challenged by Macpherson (1991), who proposed the concept of "flexible synergies". She noticed that by changing the direction of an imposed disturbance, the muscle pattern also changed. The idea that the same synergies were activated for the postural reactions and the anticipatory postural adjustments was also questioned (Massion 1992).

Whatever the neural basis for the anticipatory postural adjustments, the literature generally admitted that two parallel and coordinated control systems did exist, one for the postural adjustment, the other for the movement (Massion 1992).Contrasting this view are considerations previously formulated by Bernstein (1984). These led to the thought that a single coordinated control exists, achieving simultaneously the movement and the posture adjustment. In this view, the central control would be simultaneously addressed to the various joints involved in the task, whether they mainly contribute to the movement or to the associated postural adjustment (Aruin and Latash 1995a,b). This single, coordinated control would result from experience and practice, and is aimed at mastering the redundant number of degrees of freedom. At the end of the acquisition process, a "motor skill" would have emerged from the interaction between the learned central command, the assisting feedback loops and the dynamic interactions between segments. Thus, the controversial views on whether or not the associated postural adjustments depend on a separate control coordinated with that of the movement will be at the center of the discussions.

Second, the multijoint kinematic chain that must be controlled for postural maintenance is characterized by intersegmental dynamic interactions. Due to the external and internal constraints, the control exerted by the CNS cannot be directly addressed to the muscles, as noticed by Bernstein (1984). He proposed a hierarchical model of control in which a higher level of control exerted by the CNS is the "motor field," which he supposed to be mainly topological and metrical. Implementation by muscle force occurs at a lower level. In order to identify the higher level of control, he proposed to use the principle of "equal simplicity," by which using constraints force the controller invariant characteristics to emerge. Are we able to identify the motor field under control in the anticipatory postural adjustments?

Motor Memory and Adaptation to Constraints

At the end of the acquisition of the anticipatory postural adjusments, a motor memory has been built. How far does this motor memory become independent from the external constraints and how far does it adapt to these constraints are questions to be solved. Bernstein mentioned that anticipation implied a process of representation of the external world and a prediction of the interactions between the external world and the internal constraints during movement. Are these representations used for adapting the motor memory to the constraints?

In this paper, the discussions will focus on why the anticipatory adjustments occur and how they are controlled. First, their main functions will be recalled. A short survey of the various categories of anticipatory postural adjustment will be examined in order to identify their goals and their control. Finally, some conclusions will be drawn in light of Bernstein's concepts.

Why Anticipation in the Postural Domain?

In order to understand why anticipation occurs in the postural domain, one should ask which functions posture serves during activities of daily living. Two main functions can be identified. A primary one is an antigravity function that has two aspects. The first aspect consists of building the actual posture against the force of gravity. Regulating joint stiffness by using muscle and postural tone is the main tool to achieve this. The second aspect of antigravity posture is to ensure balance. This means that under static conditions, the vertical projection of the center of mass (CM) onto the support base should remain inside that base. Another function of posture is to interface with the external world for perception and action. The orientation of various segments, such as the head and the trunk, is often aligned with the vertical. This serves as a reference frame (Berthoz 1991; Paillard 1991). The head axis reference frame aids perception of the body movements relative to the external world through retinal receptors reacting to visual moving stimuli. These provoke body postural reorientation with respect to this reference frame. Head-trunk orientation to the vertical axis is also used to estimate the location of objects within the environment with respect to the body's position. Additionally, it organizes the trajectory for reaching and grasping (Jeannerod 1988; Soechting and Flanders 1991). The trunk posture is also used to evaluate foot trajectory in the external space; leg position is calculated using the trunk axis as a reference value (Mouchnino et al. 1993). When the trunk axis is aligned with the vertical axis, for example, during locomotion (Winter 1990) or during leg raising while dancing (Mouchnino et al. 1992), the position of the leg and thereby the position of the foot are directly expressed in external coordinates.

Categories of Anticipatory Postural Adjustments

When considering the examples of postural adjustments reported in the literature, the large diversity of these adjustments with respect to their function and their organization becomes impressive.

Set Control

A first level of anticipation, which is not usually quoted among the anticipatory processes in the postural domain, is the presetting of the postural reaction gains. These aim to restore the initial posture when an external disturbance of known amplitude is predicted (figure 6.1). The correcting response amplitude is then adjusted in advance by the CNS (Horak and Macpherson 1996; Johansson and Westling 1990; Prochazka 1989). The following example is related to the forearm posture maintenance after releasing the load supported by the arm. As seen in figure 6.1, the monosynaptic unloading reflex gain increases (increased inhibition) with repeated unloading. This increase in the reflex response is associated with a reduction of the forearm flexion amplitude due to the unloading. This reduced forearm postural displacement is due to an increase of the reflex correction. It is not due to a postural change preceding the disturbance, because the maximal velocity measured during the first 100 ms after unloading, which reflects the passive effect of the load release on the forearm posture, remains unchanged (Dufossé et al. 1985).

In contrast with the phasic anticipatory postural adjustments, the set control is not time related with each postural disturbance. Instead, it is gated for the whole session where postural disturbances are expected (see figure 6.2). The setting concerns not only the postural reactions but also the anticipatory postural adjustments. The amplitude of the anticipatory postural adjustments is also reported to be preset as a function of the external constraints or of the specific instructions related to the task (figure 6.2). Thus, by providing an additional support during stance, such as touching a wall, the anticipatory postural adjustments aimed at minimizing the postural and equilibrium disturbance during arm raising are markedly reduced and may even disappear (Belenkiy et al. 1967). Conversely, increasing the balance constraint such as unipodal stance is associated with an increased postural anticipation amplitude during arm raising (Vernazza et al. 1996b).

Anticipation to External Disturbances

A second level of anticipation is represented by postural adjustments where the intensity of the postural disturbance and its timing have to be predicted. This occurs, for example, when a load is released from various heights onto the subject's hand (Lacquaniti and Maioli 1989). Here the impact of the load is predicted in terms of timing and impact intensity. Two characteristics of the anticipatory adjustment are

IMPOSED UNLOADING
effect of repetition

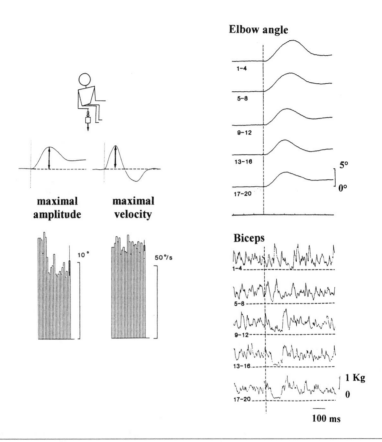

Figure 6.1   Setting of the postural reaction. The subject is instructed to maintain the forearm in a horizontal position. The forearm is loaded with a 1 kg weight suspended by an electromagnet. The load release is triggered by the experimenter. The left part of the figure shows maximal amplitude and maximal velocity of the forearm movement after unloading in each trial in the sequence. In the right part of the figure, a series of four traces averaged in a serial order represent the elbow angle before and after unloading. The time of unloading is indicated by a vertical line. Note the reduction of maximal amplitude and the maintenance of maximal velocity, indicating that no feedforward control occurred. At bottom right an EMG trace (leaky integrator with a time constant of 10 ms) of the biceps averaged as for the angle traces is shown. Note the increasing unloading reflex during repetition of the trials. The vertical bar indicates the background level integrated activity for a 1 kg load (upper level) and a 0 load (lower level) used for calibration.

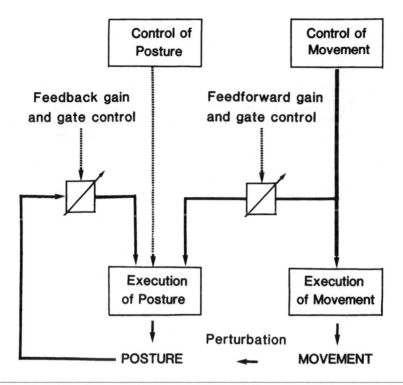

Figure 6.2   Feedforward and feedback adjustment of posture. The diagram represents the two mechanisms involved in compensating for a postural perturbation. The central control of posture is indicated by a striped line. Two phasic mechanisms minimize the postural disturbance. They operate through a feedback loop and a feedforward control. The feedforward control acts through internal collaterals from the movement control pathways on an adaptative network involved in postural control. Both mechanisms are under adaptative gate and gain control.

noticed. A first phase is characterized by a forearm movement starting just after the load release whose amplitude is related to the estimated impact of the load. A second phase is where an EMG change occurs starting 100 ms before the impact, indicating that a very precise evaluation of the time of impact has been made. Anticipatory postural adjustments belonging to the same category are observed in other tasks: self-initiated forearm perturbation with a hammer (Struppler et al. 1993), grip load force adjustment when an external disturbance is imposed (Johansson 1991), or landing from various heights (McKinley and Smith 1983). In all these tasks, a precise prediction of the impact time and the impact force exists based on either visual or proprioceptive cues. In order to understand how these predictions occur, one has to hypothesize that some representation of the load characteristics and of their interactions with the body biomechanics has been built. These serve as a base for evaluating the postural disturbance to occur and for organizing the spatiotemporal characteristics of the anticipatory control. These representations are built slowly during ontogenesis and become fully effective rather late in infancy (Eliasson et al. 1995).

Anticipatory postural adjustments, where the intensity and timing of the disturbance are predicted, have been also reported when the disturbance recurs rhythmically. For example, when subjects standing on a treadmill were subjected to sinusoidal movements of the platform (Dietz et al. 1994), anticipatory rhythmic changes of posture were observed. They occurred in such a way that the inertial disturbance caused by reversal of the imposed movement was minimized. In the same way, imposed rhythmic changes in the depth of a structured visual field were imposed on quiet, standing subjects (Dijkstra et al. 1994a, b). Here, dynamic coupling between postural orientation and the visual stimulus was noticed. The coupling minimized the distance changes between the head and the visual frame. These examples illustrate the close coupling between perception and action in the postural domain.

Anticipation to Movement Related Disturbances

In the previously reported examples of anticipation, the occurrence of anticipatory postural adjustments were related to a disturbance initiated by an external force. In most cases reported in the literature, however, the anticipatory postural adjustments are closely related to disturbances due to voluntary movements. In brief, two types of disturbances result from movement performance. The first one is due to the reaction forces and intersegmental interactions caused by movement. The reaction forces act on the supporting segments and as a result, tend to displace these segments. For example, during arm raising, the trunk is submitted to the reaction forces associated with the movement. These forces will disturb both the trunk position and balance. The second type of disturbance is related to the change in body configuration provoked by the movement. In this instance, a displacement of the center of gravity (CG) in the horizontal plane is observed toward the moving segment. Depending on how much the CG is displaced, disequilibrium or an imbalance occurs. Among the many examples of anticipatory postural adjustments described in the literature, most minimize the postural or balance disturbance due to the movement. This occurs, for example, with arm-raising, trunk-bending, or bimanual load-lifting tasks (see Massion 1992 for review).

Most anticipatory postural adjustment goals are aimed at minimizing the postural or equilibrium disturbance provoked by the voluntary movement. Two other types of goals have also been identified.

The first one is associated with leg movement. It is aimed at shifting the CG toward the other leg in order to let the limb freely move as shown by Brenière and Do (1991), Crenna and Frigo (1991), Mouchnino et al. (1992), and Rogers and Pai (1990). This occurs with leg flexion, leg raising, or gait initiation. Although this kind of activity precedes the movement onset, it differs from the usual anticipatory adjustments because it does not minimize a postural disturbance. By contrast, this type of activity shifts the CG prior to the next step of the action, which is the leg movement. In this sense, this type of posturo-kinetic coordination is closer to a preparatory postural adjustment for the forthcoming movement than an associated postural adjustment minimizing the posture and balance disturbance.

A second type of anticipatory postural adjustment is aimed at assisting the movement performance. An example of this type of anticipation is seen when the subject is asked to exert increasing pulling forces on a handle. A whole body movement is associated with the task of increasing the pulling forces (Lee et al. 1990). The spatiotemporal characteristics of the whole body movement aimed at increasing the pulling forces are constrained by the support condition. The maximal shift of the center of pressure, which provides the acceleration of the center of mass, should remain beyond the extent of the base of support. Therefore, increased pulling forces are associated with increased durations of the center of pressure shift. This observation is in line with the concept of posturo-kinetic capacity, which is an evaluation of how much the postural system is able to assist the movement in terms of velocity or force according to the constraints (Bouisset et al. 1992).

Rules of Anticipatory Postural Control

In this section, two models of posturo-kinetic coordination will be analyzed in detail. One is a bimanual load-lifting task where the forearm position is preserved after unloading. The other is an upper trunk movement that endangers balance. For each model, several aspects of the anticipatory postural adjustments will be analyzed. The rules that prevail for coordinating posture and movement in each task and those that are common for both tasks will be identified.

The Bimanual Load-Lifting Task

The load-lifting task represents an interesting model for the study of posturo-kinetic coordination because one arm is the moving arm and the other is the postural arm. In this scenario, clear-cut, anatomical separation exists between the postural and moving side, even if they interact during the task.

During this task, the load is supported by a force platform fixed to the postural forearm, which the subject is instructed to maintain in a horizontal position. The subject is asked to lift the load with the other hand in response to a tone. In the control series, the load is suspended by an electromagnet supported by the force platform fixed to the forearm. The load release is triggered by the experimenter. Thus, when the load lifting is performed by the subject's moving arm, it is associated with an anticipatory adjustment of the postural forearm that precedes the onset of the unloading and is time related to the onset of the movement (Dufossé et al. 1985; Hugon et al. 1982; Paulignan et al. 1989). The anticipatory adjustments are graded as a function of the load to be lifted. This indicates, at first sight, that both aspects of the task have a common control. It also indicates that an increase in the movement-lifting force is accompanied by an increase in the anticipatory postural adjustment. This observation agreed with similar observations in the literature about other tasks, indicating that the intensity of the postural adjustment was graded as a function of the postural disturbance to be expected from the movement. For example, Bouisset and Zattara (1987) came to that conclusion for standing subjects when adding a load to the arm to be raised. Lee et al. (1987) indicated that the anticipatory adjustment intensity depended on the movement velocity and thus on the disturbing action of the movement. Aruin and Latash (1995b) also came to the same conclusion with the release of various weights by the arms. All these observations would have been in line with a schema of a central control where the control pathway for the movement would also grade the anticipatory postural adjustment intensity. However, as will be mentioned later, the link between movement and posture is not as simple as this first observation seems to indicate.

Role of the Voluntary Movement

Does the anticipation-related, postural forearm unloading occur in absence of voluntary movement? The answer to this question was negative in the authors' experimental conditions (Dufossé et al. 1985). When a tone preceded the forearm unloading provoked by the experimenter by a fixed interval, no anticipation occurred. Thus, the anticipatory adjustment only occurred when the forearm unloading was provoked by a voluntary movement. This contrasts the situation where a load falls onto the hand (Lacquaniti and Maioli 1989). There, anticipatory adjustments are observed when the load falling does not result from the subject's own voluntary movement.

Role of the Movement Parameters

Is the anticipation aimed at maintaining the postural forearm position specifically related to the lifting movement of the other arm? Can it instead be produced by any voluntary movement regardless of the joint involved and the movement parameters, provided that they are associated with a disturbance of the forearm posture? Using an artificial mechanical link between the movement performance and the postural forearm unloading, anticipatory postural adjustments were observed after short-term learning, not only with elbow flexion but also with elbow extension and even with leg movements (Forget and Lamarre 1990, 1995). In addition, the anticipatory adjustments minimizing the postural forearm disturbance were observed when the load release of the postural forearm was triggered by an electronic switch during various types of elbow movements: isometric contraction, elbow joint rotation, and lifting a load with the hand (Paulignan et al. 1989). Therefore, a large flexibility between the location of the segment to be moved, the movement parameters, and the anticipatory postural adjustments was found to exist. However, surprisingly, a distal movement such as pressing a switch with a finger did not provoke anticipation. Other experiments did show that a finger-lifting movement was able to cause anticipatory postural adjustment. During a load-lifting task where the load was suspended by a postural finger and lifted by a finger of the other hand, anticipation was observed (Kaluzny and Wiesendanger 1992). The main difference between the previous paradigm and the last one was that in daily life, the forearm position is not disturbed by a finger movement. As stressed by Konorski (1967), acquiring a task is made easier when it is close to naturally existing tasks. Aruin and Latash (1995a) came to the same conclusions in a paradigm where the subject supported a weighted balloon between the hands and was asked to pop the balloon by a small movement of the index finger to which a small pin was attached. In this case, too, very little anticipatory adjustment of the trunk was observed.

Role of the Postural Disturbance

In this paradigm and within given limits, voluntary movement is needed to build an anticipatory postural adjustment. However, the link between the voluntary movement and the postural adjustment is flexible in terms of the joint involved and movement parameters. Evaluation of the disturbance liable to occur on the forearm posture when performing the movement determines the anticipatory postural adjustment intensity and timing. For example, in a learning session where the postural forearm unloading (1 kg load) was triggered by an electronic switch activated when the other arm was lifting a load, the anticipatory postural adjustment remained unchanged whatever the load (0.5 kg, 1 kg, 1.5 kg) lifted by the moving arm (figure 6.3, right). In this paradigm, the voluntary movement does not provide the anticipatory adjustment parameters. It delivers a timing signal that triggers the appropriate adjustment as a function of the disturbance that will occur with the load release (Paulignan et al. 1989). These conclusions are close to those reported by Kazennikov et al. (1994) for another bimanual task.

Characteristics of the "Motor Memory"

What are the rules for acquiring this type of anticipatory postural adjustment? A first remark concerns the role of sensory afferences. As shown by Forget and Lamarre (1990, 1995), the anticipatory postural adjustments remained when postural forearm unloading was performed in a deafferented patient, thus in absence of sensory kinesthetic feedback. This means that in an already trained task, a stable motor memory has been built. The network responsible for the anticipatory postural adjustment is still functioning in a feedforward manner when the voluntary movement responsible for lifting a load is performed. This type of anticipatory adjustment may be compared to an overtrained motor skill that does not need to be reinforced anymore by the knowledge of the result on the basis of proprioceptive feedback. However, in contrast, acquiring a new coordination, for example, between foot movement and forearm unloading, is not possible in the absence of sensory feedback.

A second remark concerns the anticipatory control that is built while acquiring the task. Acquiring anticipatory postural adjustment can be initiated in a modified paradigm where the load release is triggered through an electronic switch in relation to the voluntary movement of the other arm. Usually, 40 to 60 trials where the unloading is provoked artificially by the lifting movement are needed to build a stable anticipatory

LOAD LIFTING TASK
Acquisition of the coordination

Effect of training   **Effect of change in the lifted load**

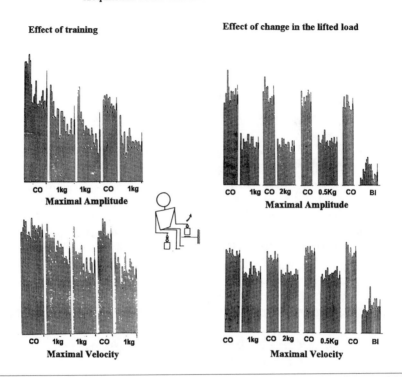

Figure 6.3   Acquisition of the anticipatory postural adjustment. On the left is acquisition of the anticipatory postural adjustment of the postural forearm position. The subject is instructed to maintain horizontal the right forearm position. The forearm supports a 1 kg load suspended by an electromagnet. The unloading is triggered by the experimenter (control, CO) or by the subject lifting a 1 kg load with the other hand (1 kg) through an electronic switch. Note the reduction of both forearm maximal amplitude and velocity after unloading due to repetition of the trials. The right shows the same conditions as the left. The subject lifts various loads, 0.5 kg, 1 kg, 2 kg, with the left hand. Note that the amplitude of the maximal amplitude and velocity of the forearm rotation are not influenced by the load lifted by the hand. Instead, they depend only on the load released from the postural forearm.

postural adjustment (figure 6.3, left). An analysis based on dynamic equations using a linear approximation (Biryukova et al. 1995) suggests that the acquisition is based on changing the elbow extensor's equilibrium point and increasing the joint stiffness before the onset of unloading. This observation agrees with the equilibrium point hypothesis (Feldman 1986; Latash 1993) since the main variables controlled by the CNS in this paradigm are the equilibrium point and joint stiffness.

Is the network that is built to acquire the anticipatory adjustment with the postural forearm on one side transferable after changing the postural and the moving arm?

The authors expected that a transfer would occur and that acquiring the anticipatory postural adjustment after the exchange would at least be greatly facilitated after the first learning. The results obtained in eight subjects showed no evidence of transfer (Ioffe et al. 1996). Moreover, the new learning that occurred after changing the postural and moving arm (transfer learning) was not facilitated by the previous acquisition. This result agrees with the recent observation by Jeka and Lackner (1995). They showed that while adapting to Coriolis forces during a reaching movement, no transfer from one arm to the other concerning the trajectory happened. However, a transfer did occur concerning the estimation of the target location, which involves a higher level of control.

Insight Into Central Organization

Finally, how could the central organization of such a coordinated bimanual load-lifting task be conceived? Results obtained from patients have provided some insight into the central organization of the task (Viallet et al. 1992). Observations first showed that the anticipatory postural adjustments were still present after corpus callosum section. This indicated that the time coordinating signal was not due to direct connections between both cortices and, thus, was possibly subcortical. Moreover, patients with supplementary motor area region (SMA) lesions did show impaired anticipatory postural adjustments only when the lesion was contralateral to the postural arm (figure 6.4). In order to interpret these data, the following functional scheme was proposed. It was first hypothesized that the main aspect of the task is not to lift the load but to preserve the postural forearm position. This is a prerequisite for bimanual activities where the forearm position serves as a reference value for the task. Since the control of that position is impaired after a SMA lesion contralateral to the postural forearm, one may propose that this area is crucial for selecting and maintaining the forearm position as a reference value for the future task. Is this area directly responsible for the phasic anticipatory postural adjustments that occur during load lifting? Since the anticipatory adjustment was still present in a patient with callosal section, the coordination between the two sides may have occurred at a subcortical level. If so, the role of the SMA would be to gate the adaptive network responsible for the anticipatory adjustment and to adjust its gain according to the expected disturbance. In this scheme, two levels of control are represented. The highest includes the SMA (Gurfinkel and Elner 1988) and basal ganglia that are in charge of selecting and stabilizing the postural reference frame for the task (in the present case, the forearm position). The lower level is in charge of executing the task and timing of the anticipatory postural adjustment.

To conclude, in the load-lifting task model, most questions raised in the introduction could be answered. The *goal* of the anticipatory postural adjustment is to provide a stable postural reference frame during the task. Concerning the *control*, the experimental results indicate that two parallel controls exist, one for the postural anticipation and the other for the load-lifting movement. The main role of the voluntary movement is to provide a timing signal for the disturbance to occur. A

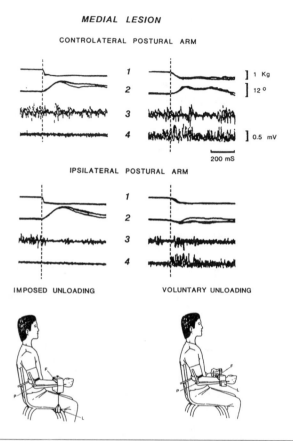

Figure 6.4   Central organization of the coordination between posture and movement in the load-lifting task. On this page, recordings of the postural forearm unloading (1), of the postural forearm movement (2), of the postural forearm biceps EMG (3), and of the moving arm biceps EMG (4) in a patient with a unilateral medial lesion predominating in the SMA area. Superposition of four traces. Imposed and voluntary unloadings are compared. When the postural forearm is contralateral to the lesion, the anticipatory postural adjustment is weak or absent. The forearm position after unloading is moving as in the imposed unloading situation. By contrast, when the postural forearm is ipsilateral to the lesion, the anticipatory postural adjustment is preserved. A scheme is proposed on the following page to account for the central organization of coordination between posture and movement in this bimanual task. (M1) primary motor cortex. (SMA) supplementary motor area region. In this scheme, the motor cortex on one side (M1) controls the load-lifting movement (continuous line), whereas the motor cortex on the other side controls the postural maintenance of the postural forearm (dashed line). The coordination between the two controls is not performed through the corpus callosum (CC). The control pathway for movement sends collaterals at a subcortical level toward the postural arm, which are triggering the anticipatory postural adjustment. The possibility of using this collateral pathway (gate) depends on a supraspinal control from the SMA contralateral to the postural arm and also from the contralateral basal ganglia (BG).

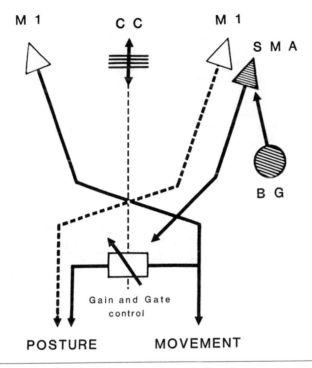

Figure 6.4 *(continued)*

hierarchical organization is suggested in the task. The highest control level is related to the selection of the postural reference position for the task to be performed by the moving hand. A lower level of control is in charge of executing the movement and the anticipatory postural adjustment as well. The motor field under control during the anticipatory postural adjustments is compatible with a change in equilibrium point. Finally, acquisition of the task reveals that the sensory afferents are needed during the acquisition process, not for the performance once the acquisition process has been achieved. The memorized motor skill is lateralized and not transferable.

Control of Balance During Movement

In the bimanual load-lifting task, two goals are simultaneously achieved. One is the forearm postural maintenance, the other is lifting a load. The two goals are achieved by distinct neural networks that are coordinated in terms of timing.

Coordination within another task where the two goals are also achieved simultaneously will now be examined. One is to perform a trunk movement with a given velocity and amplitude (trunk bending), the other is to maintain balance. Remember that the sole change in body geometry that the prescribed movement will produce should provoke imbalance by shifting the CG horizontal projection outside the support area or close to its border. Moreover, the only way to preserve balance during upper trunk bending is to change the position of the other body segments in such a way that the CG shift due to the movement is minimized. This was noticed by Babinski (1899) who described what he called the synergies between the upper and lower segment displacements during trunk bending. He claimed that they were related to preserving equilibrium.

When considering the control of this task, two preliminary remarks should be formulated. First, the upper and lower segments that are moved during upper trunk movement belong to the same postural kinematic chain. Movement of any joint in the chain provokes dynamic interactions with the other segments involved in the task (Ramos and Stark 1990). It is thus difficult to conceive that two parallel controls would exist, one for the balance control, the other for the movement control. More likely, a single control for both tasks does exist. The second remark concerns the redundancy of the number of degrees of freedom. In this task, at least three main joints are involved, the ankle, knee, and hip joints. Three degrees of freedom have to be controlled. The trunk movement and balance control represent two degrees of freedom. Since these two constraints are identified, there is a redundancy in the number of degrees of freedom to be mastered, at least for backward movements where no mechanical limitation of knee joint movement exists as for forward movement (limited knee extension).

Evidence for a single central control.   In the present task, several experimental data suggest that a single central control does exist. This rules out both the upper trunk movement and the balance. Data also suggest that a coupling between the movement of the three joints takes place in such a way that a single degree of freedom is controlled. First, preceding the onset of movement, a burst of activity occurs in a set of muscles of the shank, thigh, and trunk segments, indicating that a central control initiates movement of the various joints involved in the task (Crenna et al. 1987; Oddsson and Thorstensson 1986, 1987).

Second, the peak velocity of the markers placed on the shoulder, hip, and knee synchronize. This would not occur if the interaction between segments were the sole source of the kinematic synergy (Ramos and Stark 1990). It suggests that a synchronized central control is addressed to all joints participating in the task.

Third, a principal component analysis indicates that the first component (PC1) represents 96–99% of the movement of the three joints for both forward and backward movements. This indicates that only one degree of freedom is controlled and that the three joint movements are linked together by fixed ratios (Alexandrov et al. 1997). Intraindividual variations of these ratios were less than 5%.

Are the links between joint movements of central origin or are they due to dynamic interactions between segments? During fast trunk bending, the forces related to trunk acceleration are much higher than during slow movements, and the dynamic interactions between trunk and leg markedly increase. However, in both fast and slow movements, PC1 represents the same proportion of the joint's movement (more than 99% for forward and from 96 to more than 99% for backward). This suggests that PC1 represents the action of a central control which takes into account in a predictive way the interaction between segments.

Influence of external constraints.   Is the central automatic control closely tied to the balance constraint due to gravity? In order to answer this question, upper trunk bending was performed in microgravity, during long-term space flights and parabolic flights. The kinematic synergies associated with upper trunk bending still appeared to be present in microgravity in absence of balance constraint (Massion et al. 1993), and minimization of the CM shift was still observed (figure 6.5). However, implementation of these synchronized joint changes in terms of muscle force were markedly changed. For example, for forward bending, the pattern of EMG changes before and during the movement were markedly changed in microgravity as compared with normal gravity. The invariant aspect of the central control seems to be coded in terms of joint angle control and not in muscle control.

If a central automatic control of the CM position is still present in the absence of gravity constraints, the question remains as to whether this control can be influenced by changes in the balance constraints in normal gravity. In other words, are the kinematic synergies under normal gravity conditions stable and invariant whatever the changes in the body mass or the support area? Do they instead adapt to these changes in order to preserve the CM projection onto the ground and thereby balance? A first set of experiments was performed with the subjects loaded with 10 kg placed on the shoulders (Vernazza et al. 1996a). By performing a forward bending movement, inclination of the trunk with the additional load appears to provoke a further forward shift of the CM by about 4 cm. If during the performance of the upper trunk movement the control does not take into account the effect of the additional load on the CM position, the final CM position at the end of the movement should be 4 cm further forward than without load, and the kinematic synergies should be unchanged. If, on the other hand, the additional load is taken into account in the control, the final CM position should remain the same, and the kinematic synergy should be consequently changed. The experiments performed in five subjects did show that three out of five subjects preserved the same final CM position with and without load and that, consequently, the kinematic synergy was modified in the presence of a load (Vernazza et al. 1996a). This indicated that these subjects did predict the effect of the load on the CM position at the end of the movement and changed the kinematic synergy in anticipation of the additional CM shift that would occur during the movement. This prediction was necessarily based on sensory inputs provoked by the additional load. However, in two other subjects, the CG shift was changed when the additional load was present and the kinematic synergy remained the same.

In order to further understand why the adaptation to the additional load did not occur in every subject, another set of experiments was made with subjects performing arm raising in the frontal plane. The CM mechanical shift due to the movement is much weaker there (1 cm in place of 8 cm), even when the hand lifts a load of 3 kg (2 cm). The disturbing effect on balance due to arm raising is therefore reduced with normal bipedal stance. The experiments did show that during quiet bipedal stance,

PARABOLIC FLIGHT

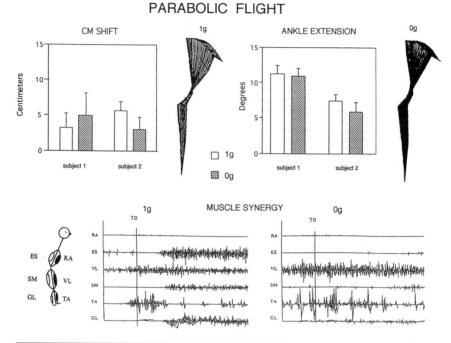

Figure 6.5 Forward, upper trunk movement. Subject stands with feet fixed to the floor. Comparison between normal gravity and microgravity during parabolic flight. Upper left, center of mass (CM) shift at the end of the upper trunk movement, reconstructed from the body kinematics and the anthropometric model by Winter (1990). Note the similarity between the CM displacements observed under normal gravity and microgravity. Upper right—ankle extensions during upper trunk movement. They are comparable in both conditions as are the stick diagrams indicating the opposite upper and lower body segment shifts in normal gravity and microgravity. Lower traces, muscle synergy during upper trunk forward bending. On the left, *1 g*. Note that the movement onset was preceded by a reduced activity of the erector spinae (ES) and a burst in the vastus lateralis (VL) and tibialis anterior (TA). After movement onset, a braking activity occurred in the antagonistic muscles (VL), semimembranosus (SM), and gastrocnemius lateralis (GL). On the right, *0 g*. VL and TA were activated prior to movement onset. No clear-cut braking phase was observed. Superimposed recordings from five trials.

the CM shift due to arm raising both without and with load was poorly compensated. However, as soon as the balance constraints were increased, such as when standing on one foot, the CG shift both with and without load was largely minimized (Vernazza et al. 1996b). This result indicates that the balance constraints are taken into consideration when organizing central control of the kinematic synergies associated with movement performance.

Conclusions. To conclude, the central organization of upper trunk movement and of balance rely on the following items (figure 6.6).

BALANCE CONTROL DURING TRUNK BENDING

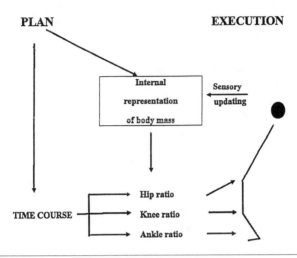

Figure 6.6   Schema indicating how balance is controlled during upper trunk movement. For explanation, see text.

1. A memorized representation exists of the ratios between angles appropriate for maintaining the CM position when a movement is performed. This sensorimotor memory was probably learned in childhood. It remains in microgravity for upper trunk movements in absence of balance constraints. This memory involves a representation of body segment geometry and of the respective center of inertia of the segments and their position with respect to the vertical gravity vector. On the basis of this representation, a set of ratios between the various angles is selected for the movement to be performed. The memorized set of angle ratios can be updated in normal gravity as a function of balance constraints such as additional loads or a reduced support area.

2. A central time course signal is addressed to the various joints. It defines the amplitude and duration of the movement and of the associated synergies.

The central nervous system therefore seems to code the axial synergies associated with upper trunk bending in terms of motor fields (Bernstein 1984). These consist of movement topology, which reflects the invariant features of the movement (PC1 loadings, describing the relative changes in the joint angle), and of movement metrical properties. This last aspect is illustrated by the time course of PC1 score, which changes the amplitude and the duration in accordance with the movement instruction. The topological and metrical aspects of the control are also in line with the concept of the generalized motor program proposed by Schmidt (1988).

Concluding Remarks

Three main questions were raised in the present survey on the anticipatory postural adjustments. They concern their goals, their control, and the adaptation of the built motor memory to constraints.

The first conclusion refers to the *goals* which reveals a large diversity. The first goal category is directly related to preserving the "postural" functions that are still to be achieved during the movement, such as maintaining the orientation of given body segments (head, trunk, forearm) or preserving balance. Both examples analyzed in this paper, the bimanual load lifting task and the trunk bending task, belong to this category of anticipatory postural adjustments. The anticipatory postural adjustments are then aimed at minimizing the disturbance to occur when the movement is performed. A second category of goals involves preparing the postural configuration as a function of the forthcoming movement in order to preserve balance during the movement. This occurs with leg movements, which are preceded by a CG shift toward the supporting leg. A third category of anticipatory adjustments is directly assisting the movement performance in terms of force or velocity while preserving balance. This diversity of goals permits predicting that the central organization of the anticipatory postural adjustments will not be uniform.

Concerning the *organization of the central control* of the anticipatory postural adjustments, they were analyzed in detail in the bimanual load lifting task and the trunk bending paradigm. A primary question was related to the coordination between the control of the anticipatory postural adjustment and that of the movement. Do two parallel controls exist, one for posture and the other for movement, or only a single one which is common for both aspects of the task? In the load-lifting task, two parallel controls appear to exist, one for the anticipatory postural adjustment, the other for movement. The first one is mainly dependent on the disturbance to be expected, independent of the location of the moving segment or of the movement parameters that are at the origin of the disturbance. The main function of the movement seems to provide a timing signal as in other bimanual coordinated tasks (Wiesendanger et al. 1994). By contrast, in the upper trunk-bending paradigm when movement and balance have to be controlled simultaneously, a single control appears to exist, addressed to the various joints involved in the task. This integration between postural and movement control was predicted by Bernstein (1984) and proposed by Aruin and Latash (1995a,b).

Another question related to the control of the anticipatory postural adjustments was the identification of the "motor field" controlled in these tasks. In agreement with Bernstein (1984), the control was not primarily organized in terms of muscle control. In the load-lifting task, the reference value that was preserved is the position of the forearm with respect to space. What is actually controlled in anticipation is the joint stiffness and the forearm extensor muscles' equilibrium point, both being explained in the frame of the equilibrium point theory by Bizzi et al. (1992) and Feldman and Levin (1995). In the upper trunk-bending task, the CM position is preserved at the end of the movement due to the kinematic synergies; the motor control seems coded in terms of topology and metrics as proposed by Bernstein (1984). The topology is reflected by the fixed ratios between changes in the joint angles and metrics is represented by the time course (amplitude, velocity) signal (PC1 score) addressed simultaneously to the various joints involved in the kinematic synergy.

The third question was related to the *motor memory and its adaptation to constraints*. In both examples, evidences have been given that the anticipatory postural adjustments are based on a motor memory built during the acquisition of the task. Sensori-motor memories have been reported in other tasks such as during lifting objects, where the ratio between grip and load forces is very stable (Johansson and Westling 1990). An interesting aspect of the motor memory in specific tasks that were analyzed here is that they persist unchanged independently from the constraints. The anticipatory adjustment of the postural forearm is still present after years in absence of any afferent feedback from the postural forearm (Forget and Lamarre 1990). In the example of trunk bending, the kimematic synergies are preserved in microgravity and still stablilize the antero-posterior position of the center of mass in absence of balance constraints.

A stable motor memory might prevent any adaptation to changes in constraints. Actually, adaptation was observed in each of the models that were analyzed in the present report. A first mechanism for adaptation was related to an internal representation of the external and internal constraints. This concept of representation was already proposed by Bernstein. It seems largely supported by the two models reported in the present paper and by data from the literature. It includes the body segment geometry, the body segment weight and inertia, and their relation with the vertical gravity vector, as previously proposed by Clement et al. (1984), Gurfinkel et al. (1988), and Lestienne and Gurfinkel (1988). This representation is actually used in the load lifting task to adapt the intensity of the anticipatory postural adjustments to the load to be lifted and in the trunk bending paradigm in adjusting the kinematic synergy as a function of an additional load supported by the trunk. An internal representation should also explain other examples of adaptation such as the selection of "motor equivalents" representing a new combination of joint angle changes for achieving the same goal in presence of a constraint (Abbs and Gracco 1984; Abbs and Cole 1987; Jeannerod 1991). A second mechanism for adaptation is related to short-term learning. This occurs for example during the load lifting task when changing the relations between the movement's performance and the postural perturbation.

To conclude, understanding how and why the anticipatory adjustments are controlled is still enlightened by the framework of the conceptual construction of Bernstein, which remains the most extensive and prospective approach on how the brain, the biomechanical system, and the external forces interact in an integrated motor task.

Acknowledgments

The authors are very grateful to Patricia McKinley for her critical reading of the manuscript and to the Centre National de la Recherche Spatiale for its financial support.

References

Abbs, J.H., Cole, K.J. (1987) Neural mechanisms of motor equivalence and goal achievement. In *Higher Brain Functions: Recent Explorations Of The Brain's Emergent Properties*. S.P. Wise (Ed.), New York: Wiley, 15–43.

Abbs, J.H., Gracco, V.L. (1984) Control of complex motor gestures: Orofacial muscle responses to load perturbations of lip during speech. *Journal of Neurophysiology*, 51: 705–723.

Alexandrov, A., Frolov, A., Massion, J. (1997) Axial synergies during human upper trunk bending. *Exp Brain Res* (In press).

Arbib, M.A. (1981) Perceptual structures and distributed motor control. In *Handbook of Physiology, Section 1, The Nervous System*. Vol II, Part 2. V.B. Brooks (Ed.) Bethesda, MD: American Physiological Society, 1449–1480.

Aruin, A.S., Latash, M.L. (1995a) The role of motor action in anticipatory postural adjustments studied with self-induced and externally triggered perturbations. *Exp Brain Res*, 106: 291–300.

Aruin, A.S., Latash, M.L. (1995b) Directional specificity of postural muscles in feed-forward postural reactions during fast voluntary arm movements. *Exp Brain Res*, 103: 323–332.

Babinski, J. (1899) De l'asynergie cérébelleuse. *Rev Neurol*, 7: 806–816.

Belenkiy, V.E., Gurfinkel, V.S., Paltsev, E.I. (1967) On elements of control of voluntary movements. *Biofizica*, 12: 135–141 (In Russian).

Bernstein, N.A. (1984) In *Human Motor Actions. Bernstein Reassessed*. H.T.A. Whiting (Ed.), Amsterdam: North—Holland.

Berthoz, A. (1991) Reference frames for the perception and control of movement. In *Brain and Space*. J. Paillard (Ed.), Oxford: Oxford University Press, 81–111.

Biryukova, E.V., Roschin, V.Y., Frolov, A.A., Ioffé, M.E., Massion, J., Dufossé, M. (1995) Elbow joint stiffness dynamics during learning of posture maintenance in the process of arm unloading. *XVth Congress of the International Society of Biomechanics. Book of Abstracts*. Finland, July 2–6: University of Jyvaskyla, 106–107.

Bizzi, E. (1974) The coordination of eye-head movement. *Scientific American*, 231: 100–106.

Bizzi, E., Hogan, N., Mussa-Ivaldi, F.A., Giszter, S. (1992) Does the nervous system use equilibrium-point control to guide single and multiple joint movements? *Behav Brain Sci*, 15: 603–613.

Bouisset, S., Do, M.C., Zattara, M. (1992) Posturo-kinetic capacity assessed in paraplegics and parkinsonians. In *Posture and Gait: Control Mechanisms*. Vol. II. M. Woollacott, F. Horak (Eds.), Eugene, OR: University of Oregon Books, 19–22.

Bouisset, S., Zattara, M. (1987) Biomechanical study of the programming of anticipatory postural adjustments associated with voluntary movement. *Journal of Biomechanics*, 20: 735–742.

Brenière, Y., Do, M.C. (1991) Control of gait initiation. *Journal of Motor Behavior*, 23: 235–240.

Clément, G., Gurfinkel, V.S., Lestienne, F., Lipshits, M.I., Popov, K.E. (1984) Adaptation of postural control to weightlessness. *Exp Brain Res*, 57: 61–72.

Cordo, P.J., Nashner, L.M. (1982) Properties of postural adjustments associated with rapid arm movements. *Journal of Neurophysiology*, 47: 287–302.

Crenna, P., Frigo, C. (1991) A motor program for the initiation of forward oriented movement in man. *Journal of Physiology (London)*, 437: 635–653.

Crenna, P., Frigo, C., Massion, J., Pedotti, A. (1987) Forward and backward axial synergies in man. *Exp Brain Res*, 65: 538–548.

Dietz, V., Trippel, M., Ibrahim, I.K., Berger, W. (1994) Human stance on sinusoidally translating platform: Balance control by feed-forward and feedback mechanisms. *Exp Brain Res*, 93: 352–362.

Dijkstra, T.M.H., Schoner, G., Gielen, C.C.A.M. (1994a) Temporal stability of the action-perception cycle for postural control in a moving visual environment. *Exp Brain Res,* 97: 477–486.

Dijkstra, T.M.H., Schoner, G., Giese, M.A., Gielen, C.C.A.M. (1994b) Frequency dependence of the action-perception cycle for postural control in a moving visual environment: Relative phase dynamics. *Biological Cybernetics,* 71: 489–501.

Dufossé, M., Hugon, M., Massion, J. (1985) Postural forearm changes induced by predictable in time or voluntary triggered unloading in man. *Exp Brain Res,* 60: 330–334.

Eliasson, A.C., Forssberg, H., Ikuta, K., Apel, I., Westling, G., Johansson, R. (1995) Development of human precision grip. V. Anticipatory and triggered grip actions during sudden loading. *Exp Brain Res,* 106: 425–433.

Feldman, A.G., Levin, M.F. (1995) The origin and use of positional frames of reference in motor control. *Behavioral Brain Sci,* 18: 723–806.

Forget, R., Lamarre, Y. (1990) Anticipatory postural adjustment in the absence of normal peripheral feedback. *Brain Res,* 508: 176–179.

Forget, R., Lamarre, Y. (1995) Postural adjustments associated with different unloading of the forearm: Effects of proprioceptive and cutaneous afferent deprivation. *Can J Physiol Pharmacol,* 73: 285–294.

Gahéry, Y., Massion, J. (1981) Coordination between posture and movement. *TINS,* 4: 199–202.

Gurfinkel, V.S., Elner, A.M. (1988) Participation of secondary motor area of the frontal lobe in organization of postural components of voluntary movements in man. *Neurophysiology,* 20: 7–14.

Gurfinkel, V.S., Levik, Y.S., Popov, K.E., Smetanin, B.N. (1988) Body scheme in the control of postural activity. In *Stance and Motion: Facts and Concepts.* V.S. Gurfinkel, M.E. Ioffe, J. Massion, J.P. Roll (Eds.), New York: Plenum Press, 185–193.

Horak, F.B., Macpherson, J.M. (1996) Postural orientation and equilibrium. In *Handbook on integration of motor circulatory, respiratory, and metabolic control during exercise.* L.B. Rowell, J.T. Sheperd (Ed.s), Bethesda: Amer. Physiol. Soc., 255-292.

Hugon, M., Massion, J., Wiesendanger, M. (1982) Anticipatory postural changes induced by active unloading and comparison with passive unloading in man. *Pflügers Arch,* 393: 292–296.

Ioffé, M., Massion, J., Gantchev, N., Dufossé, M., Kulikov, M.A. (1996) Coordination between posture and movement in a bimanual load-lifting task: Is there a transfer? *Exp Brain Res,* 109: 450–456.

Jeannerod, M. (1988) *The neural and behavioral organization of goal-directed movements.* Oxford: Clarendon Press.

Jeannerod, M. (1991) The interaction of visual and proprioceptive cues in controlling reaching movements. In *Motor Control: Concepts and Issues.* D.R. Humphrey, H.-J. Freund (Eds.), New York, NY: Wiley, 277–291.

Jeka, J.J., Lackner, J.R. (1995) The role of heptic cues from rough and slippery surfaces in human postural control. *Exp Brain Res,* 103: 267–276.

Johansson, R.S. (1991) How is grasping modified by somatosensory input? In *Motor Control: Concepts and Issues.* D.R. Humphrey, H.-J. Freund (Eds.) New York, NY: Wiley, 331–355.

Johansson, R.S., Westling, G. (1990) Tactile afferent signals in the control of precision grip. In *Attention and Performance.* Vol. XIII. M. Jeannerod (Ed.) Hilldale, NJ: Erlbaum, 677–713.

Kaluzny, P., Wiesendanger, M. (1992) Feedforward postural stabilization in a distal bimanual unloading task. *Exp Brain Res,* 92: 173–182.

Kazennikov, O., Wicki, U., Corboz, M., Hyland, B., Palmeri, A., Rouiller, E.M., Wiesendanger, M. (1994) Temporal and spatial structure of a bimanual goal-directed movement sequence in monkeys. *European Journal of Neuroscience,* 6: 203–210.

Konorski, J. (1967) *Integrative activity of the brain. An interdisciplinary approach.* Chicago: The University of Chicago Press.

Lacquaniti, F., Maioli, C. (1989) The role of preparation in tuning anticipatory and reflex responses during catching. *Journal of Neuroscience,* 9: 134–148.

Latash, M.L. (1993) *Control of Human Movement.* Champaign, IL: Human Kinetics.

Lee, W.A. (1984) Neuromotor synergies as a basis for co-ordinated intentional action. *J Motor Behav,* 16: 135–170.

Lee, W.A., Buchanan, T.S., Rogers, M.W. (1987) Effects of arm acceleration and behavioural conditions on the organization of postural adjustments during arm flexion. *Exp Brain Res,* 66: 257–270.

Lee, W.A., Michaels, C.F., Pai, Y.C. (1990) The organization of torque and EMG activity during bilateral handle pulls by standing humans. *Exp Brain Res,* 82: 304–314.

Lestienne, F., Gurfinkel, V.S. (1988) Postural control in weightlessness: A dual process underlying adaptation to an unusual environment. *TINS,* 11: 359–363.

Macpherson, J.M. (1991) How flexible are muscle synergies? In *Motor Control: Concepts and Issues.* D.R. Humphrey, H.-J. Freund (Eds.), New York, NY: Wiley, 33–47.

Massion, J. (1992) Movement, posture and equilibrium: Interaction and coordination. *Progress in Neurobiology,* 38: 35–56.

Massion, J. (1994) Postural control system. *Current Opinion in Neurobiology,* 4: 877–887.

Massion, J., Gurfinkel, V., Lipshits, M., Obadia, A., Popov K. (1993) Axial synergies under microgravity conditions. *Journal of Vestibular Research,* 3: 275–287.

McKinley, P.A., Smith, J.L. (1983) Visual and vestibular contributions to prelanding EMG during jump-down in cats. *Exp Brain Res,* 52: 439–448.

Mouchnino, L., Aurenty, R., Massion, J., Pedotti, A. (1992) Coordination between equilibrium and head-trunk orientation during leg movement: A new strategy built up by training. *Journal of Neurophysiology,* 67: 1587–1598.

Mouchnino, L., Aurenty, R., Massion, J., Pedotti, A. (1993) Is the trunk a reference frame for calculating leg position? *NeuroReport,* 4: 125–127.

Nashner, L.M., McCollum, G. (1985) The organization of human postural movements: A formal basis and experimental synthesis. *Behav Brain Sci,* 8: 135–172.

Oddsson, L., Thorstensson, A. (1986) Fast voluntary trunk flexion movements in standing: Primary movements and associated postural adjustments. *Acta Physiol Scand,* 128: 341–349.

Oddsson, L., Thorstensson, A. (1987) Fast voluntary trunk flexion movements in standing: motor patterns. *Acta Physiol Scand,* 129: 93–106.

Paillard, J. (1991) Motor and representational framing of space. In *Brain and Space.* J. Paillard (Ed.), Oxford: Oxford University Press, 163–182.

Paulignan, Y., Dufossé, M., Hugon, M., Massion, J. (1989) Acquisition of co-ordination between posture and movement in a bimanual task. *Exp Brain Res,* 77: 337–348.

Prochazka, A. (1989) Sensorimotor gain control: A basic strategy of motor systems? *Progress in Neurobiology,* 33: 281–307.

Ramos, C.F., Stark, L.W. (1990) Simulation experiments can shed light on the functional aspects of postural adjustments related to voluntary movements. In *Multiple Muscle Systems: Biomechanics and Movement Organization.* J.M. Winters, L.-Y. Woo (Eds.) New York: Springer-Verlag, 507–517.

Rogers, M.W., Pai, Y.C. (1990) Dynamic transitions in stance support accompanying leg flexion movements in man. *Exp Brain Res,* 81: 398–402.

Schmidt, R.A. (1988) *Motor Control and Learning.* Champaign, IL: Human Kinetics.

Soechting, J.F., Flanders, M. (1991) Deducing central algorithms of arm movement control from kinematics. In *Motor Control: Concepts and Issues.* D.R. Humphrey, H.-J. Freund (Eds.), New York, NY: Wiley, 293–306.

Struppler, A., Gerilovsky, L., Jakob, C. (1993) Self-generated rapid taps directed to opposite forearm in man: Anticipatory reduction in the muscle activity of the target arm. *Neurosci Lett,* 159: 115–118.

Vernazza, S., Alexandrov, A., Massion, J. (1996a) Is the center of gravity controlled during upper trunk movement? *Neurosci Lett,* 206: 77–80.

Vernazza, S., Cincera, M., Pedotti, A., Massion, J. (1996b) Balance control during lateral arm raising in humans. *NeuroReport,* (In Press).

Viallet, F., Massion, J., Massarino, R., Khalil, R. (1992) Coordination between posture and movement in a bimanual load lifting task: Putative role of a medial frontal region including the supplementary motor area. *Exp Brain Res,* 88: 674–684.

Wiesendanger, M., Wicki, U., Rouiller, E. (1994) Are there unifying structures in the brain responsible for interlimb coordination? In *Interlimb Coordination: Neural, Dynamical, and Cognitive Constraints.* S. Swinnen, H. Heuer, J. Massion, P. Casaer (Eds.) San Diego, Academic Press, 179–207.

Winter, D.A. (1990) *Biomechanics and Motor Control of Human Movement.* New York: Wiley.

CHAPTER

Mechanical, Neural, and Perceptual Effects of Tendon Vibration

Paul J. Cordo

Robert S. Dow Neurological Sciences Institute,
Portland, OR, U.S.A.

*David Burke, Simon C. Gandevia,
and John-Paul Hales*

Prince of Wales Medical Research Institute, Sydney, Australia

Tendon vibration, a method to stimulate the proprioceptive system, has been used to investigate various aspects of motor control over the last 70–80 years. Among the earliest uses of tendon vibration was eliciting the tonic vibration reflex, a manifestation of the stretch reflex. Tonic vibration reflexes have been studied in both humans (DeGail et al. 1975; Eklund and Hagbarth 1966; Hansen and Hoffman 1922) and animals (Matthews 1966; Rymer and Hasan 1981). Like the tendon tap, they provide a quantitative measure of the level of spinal excitability and tonus.

Tendon Vibrator

Tendon vibration affects the sensorimotor system via excitation of muscle spindle Ia afferents. Other muscle afferents such as muscle spindle group II and Golgi tendon organ Ib afferents are relatively insensitive to this form of mechanical stimulation (Brown et al. 1967; Burke et al. 1976). Muscle spindle afferents can be activated directly by vibrating the area of muscle overlying the receptor (DeGail et al. 1975; Eklund and Hagbarth 1966) or indirectly by vibrating the tendon (Brown et al. 1967; Burke et al. 1976; Cordo et al. 1995b; Goodwin et al. 1972).

During artificial proprioceptive stimulation by tendon vibration, conscious proprioceptive sensations are evoked (Goodwin et al. 1972). In contrast, natural proprioceptive stimulation, such as joint rotation, results in subconscious processing of kinematic information (Cole 1991). Tendon vibration simultaneously evokes two conscious proprioceptive sensations—an illusory displacement of the relevant joint to another static position and an illusion of continuous joint rotation without reference to positions within the joint space (Goodwin et al. 1972; Sittig et al. 1985, 1987). When the limb is not moved (i.e., isometric) during the application of tendon vibration, these illusions are consistent with perceived lengthening of the vibrated muscle (Goodwin et al. 1972; Sittig et al. 1985). For example, if the isometric biceps brachii is vibrated, the elbow is perceived to be positioned at a more extended angle and to be rotating in the extension direction.

In addition to the study of reflexes and perception, tendon vibration has been used to study movement coordination. So far, most movement coordination studies have investigated whether proprioception is used to coordinate motor behavior, and more specifically, whether a particular movement is influenced by tendon vibration. In typical studies examining the effect of tendon vibration on movement coordination, a human subject is asked to rotate a joint to a target actively while one of the tendons crossing the joint is vibrated. Tendon vibration produces errors at the movement end point consistent with perceived lengthening of the vibrated muscle. In all of these studies, the effect of vibration was examined when the limb was at rest, either while maintaining a constant posture (Goodwin et al. 1972; Sittig et al. 1985) or at the end point of a targeted movement (Bullen and Brunt, 1986; Capaday and Cooke 1983; Cody et al. 1990; Inglis and Frank 1990; Kasai et al. 1994). At rest, muscle spindle afferents are either inactive or fire at low rates compared with those attained during movement. Consequently, examining the effects of tendon vibration when muscle spindles are relatively inactive can lead to a misinterpretation of how proprioceptive input is interpreted by the central nervous system (CNS). Despite these limitations, proprioception clearly influences the coordination of a variety of motor behaviors.

Given the proprioceptive coordination of any given movement, a number of more detailed questions can be addressed with tendon vibration. For instance, tendon vibration can be used to investigate the characteristics of proprioceptive coordination. For example, how significant is the proprioceptive influence, as indicated by the size of the errors produced? When during the movement does the CNS use proprioceptive input, as indicated by the time when transient vibration produces its maximal effect? Other questions that can be addressed with tendon vibration include: What is the mechanism of proprioceptive coordination? What specific kinematic information is extracted and used by the CNS to coordinate movements? On an even more fundamental level, one can ask how kinematic information is encoded in proprioceptive input at the peripheral and more central levels of the nervous system.

Servo-Controlled Tendon Vibrator

To address the relatively simple question of whether a movement is qualitatively influenced by proprioceptive input, controlling the specific mechanical characteristics of vibration may not be very critical. However, to address more detailed questions, such as identifying the characteristics and mechanisms of proprioceptive control as well as the coding of kinematic information in proprioceptive input, the effect of vibration on the firing patterns of muscle spindle afferents must be predictable. Therefore, a tendon vibrator was designed and constructed with which the frequency, amplitude, background force, and pulse waveform could be servo controlled (Cordo et al. 1993).

This servo-controlled tendon vibrator (figure 7.1) was used in the four experiments described in this chapter. In the first experiment, the vibrator was used to investigate whether changing the frequency, amplitude, and background force of tendon compression produced consistent and predictable changes in intramuscular length in localized regions of a muscle. In the second experiment, the vibrator was used to determine whether changing the frequency, amplitude, and background force of tendon vibration produced consistent and predictable changes in the firing patterns of muscle spindle afferents. In the third experiment, the vibrator was used to investigate the characteristics and mechanisms of proprioceptive coordination—specifically, how the frequency and timing of tendon vibration influenced a proprioceptively coordinated movement (Cordo et al. 1995b). In the fourth experiment, the vibrator was used to investigate the coding of kinematic information by muscle spindles—specifically, how random noise can improve the coding and transmission of proprioceptive information during movement (Cordo et al. 1996).

Effect of Dynamic Tendon Indentation on Intramuscular Length

The purpose of this experiment was to determine if local intramuscular displacements could be reliably controlled by the servo-controlled tendon vibrator. Pulse stimuli with a duration of 6 ms were applied to the tibialis anterior (TA) tendon of normal human subjects. Square pulses were applied with different amplitudes (0.25, 0.5, 1.0, and 2.0 mm) upon a background force of 1–25 N, either singly or in trains

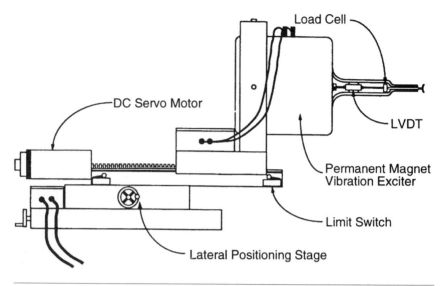

Figure 7.1   Servo-controlled tendon vibrator. Background force is controlled by a DC servo motor with input from a load cell and LVDT mounted on the shaft between the exciter and probe. Amplitude is controlled by the LVDT input, independent of background force.

at frequencies of 10, 20, 40, 80, and 100 pps. The amplitude of local intramuscular displacement was measured at the center of the muscle, relative to the proximal-to-distal and medial-to-lateral directions. The measurement device consisted of a miniature accelerometer imbedded in the hub of a 25-gauge hypodermic needle inserted into the belly of the TA to a depth of 20 mm. The acceleration signal was integrated twice to produce a measure of displacement.

Increasing the background force of the vibrator probe on the tendon increased the static length of the TA over a limited range of background forces. Between 1–6 N, the static length of the TA increased at a rate of approximately 50 μm/N background force. In terms of ankle angle, this corresponds to about 0.2 deg/N. Above a background force of 6 N, further increases in background force produced only modest increases in the static length of the TA. This saturation-like effect was presumed to reflect the compression of the tendon against underlying hard tissues such as bone.

Increasing the pulse amplitude produced a correspondingly larger level of intramuscular displacement. An increase of 1 mm in the pulse amplitude of tendon indentation produced approximately 165 μm of intramuscular displacement. This relationship was relatively independent of the background force level, except at higher force levels where saturation occurred.

When repetitive trains of pulses (i.e., tendon vibration) were applied at different frequencies, the amplitude of intramuscular displacement decreased by about 40% as the vibration frequency increased from 10–100 pps. This relationship suggests that the mechanical properties of muscle effectively act as a low-pass filter (Partridge 1965). The pulse amplitude produced by the tendon vibrator was constant over this range of frequencies.

In summary, the amount of intramuscular stretch produced by indentation of the TA tendon by the servo-controlled tendon vibrator was predictable over a range of background forces, pulse amplitudes, and frequencies of repetitive trains of pulses. Increasing background force within limits increased the static length of the muscle. Increasing the pulse amplitude increased the amplitude of the dynamic muscle stretch. Increasing vibration frequency decreased the amount of dynamic muscle stretch.

Effect of Different Vibration Parameters on Muscle Spindle Firing Patterns

The study described above demonstrated that local intramuscular changes in muscle length could be reliably predicted based on the specific mechanical parameters of tendon indentation. One hypothetical consequence is that the firing patterns of muscle spindle afferents might also be predicted from the mechanical parameters of tendon vibration. This would be useful for designing experiments to examine proprioceptive coordination of targeted movement. Transverse stimulation of tendons had previously been shown to influence the firing patterns of muscle spindle Ia afferents. In some cases, it caused these afferents to fire one action potential for each vibration pulse (Bianconi and Van der Meulen 1963; Burke et al. 1976; Roll and Vedel 1982; Roll et al. 1989).

The activity of individual muscle spindle afferents originating in the TA were recorded from normal human subjects using a tungsten microelectrode (i.e., microneurography) during vibration of the TA tendon. Vibration was delivered transversely to the tendon with the servo-controlled tendon vibrator in 1 s trains of pulses separated by 1 s periods without vibration. The amplitude of the pulses varied (0.25, 0.5, 1.0, and 2.0 mm) at frequencies of 20, 40, 80, and 110 pps upon a background force of 1–25 N. The responses of afferents were quantified based on the level of entrainment, which was defined as the percentage of vibration pulses that produced a time-locked action potential in an afferent. Thus, an afferent that fired in response to every pulse stimulus was said to be 100% entrained.[1]

Of the 14 TA muscle spindle afferents tested with tendon vibration, 13 responded with action potentials time locked to the stimulus. The average level of entrainment for these 13 afferents was just 41%. Only four of these afferents responded to tendon vibration with 100% entrainment under any conditions of stimulation. The degree of entrainment observed under the authors' experimental conditions—when the mechanical parameters of vibration could be optimized—were considerably lower than those previously reported (Bianconi and Van der Meulen 1963; Burke et al. 1976; Roll and Vedel 1982; Roll et al. 1989).

---

[1]No attempt was made to distinguish between muscle spindles' primary and secondary afferents.

The responses of the 13 afferents to changes in background force, and the amplitude and frequency of tendon vibration corresponded closely to expectations based on our earlier observations of intramuscular stretch. The response of an individual muscle spindle afferent (N1) to tendon vibration (40 pps) is shown in figure 7.2A for two different pulse amplitudes and background forces. In A, the afferent fired at a 27.5% entrainment level with a background force of 0.75 N and an amplitude of 0.5 mm. In figure 7.2B, the background force was raised to 1.5 N while maintaining a 0.5 mm amplitude, and the afferent fired at 100% entrainment, demonstrating a positive effect of background force on the entrainment level (e.g., Cordo et al. 1993). In C, the pulse amplitude was raised to 1 mm, while maintaining a background force of 0.75 N, and the afferent fired at a 100% entrainment level, demonstrating a positive effect of amplitude on the entrainment level. The inset in figure 7.2A shows the vibrator displacement for a single pulse stimulus to the tendon and the corresponding response of the afferent.

To show the overall effect of tendon vibration on the sample of muscle spindle afferents examined in this study, the entrainment levels were pooled irrespective of vibration frequency, amplitude, and background force. In figure 7.3, the entrainment characteristics are shown for two individual afferents, both of which had no background discharge, as well as the entire sample of 13 afferents responding to vibration. The afferent designated "S1" (top) was weakly activated by tendon vibration, and in 68% of trials responded only at the 0-10% entrainment level. The average entrainment level of afferent S1 was 8% (vertical dashed line). In contrast, the afferent designated "N3" (middle) was more responsive to tendon vibration, and in 35% of trials responded above the 90% entrainment level. The average entrainment level of afferent N3 was 64%. In a summary of the responses of all 13 muscle spindle afferents (bottom), the average entrainment level was 37%. This distribution includes both trials in which optimal and non-optimal vibration parameters were used; nevertheless, it shows that for any given set of vibration parameters, most muscle afferents are only partly entrained by the stimulus. On the other hand, given that 70% of the spindle afferents had no background discharge, vibration markedly increased the overall level of afferent firing, with a maximal achievable discharge of, on average, 69 pps per ending against a background rate of 2.3 pps. There was no correlation between background discharge or dynamic index and the entrainment level for the sample of spindle afferents in this study.

The frequency of tendon vibration had a pronounced effect on the response of muscle spindle afferents, higher frequencies being less effective at entraining the firing of these afferents. This negative effect of vibration frequency is shown in figure 7.4 for an individual afferent. In figure 7.4A, percent entrainment is plotted against vibration frequency (20, 40, 80 and 110 pps), each symbol representing the average level of entrainment for all trials at a particular frequency and amplitude of vibration. Each plot (i.e., symbol type and linear regression) shows, therefore, the effect of vibration frequency independently of vibration amplitude; however, a range of background forces are represented in each data point. For vibration amplitudes of 0.25, 0.5 and 1 mm (i.e., open squares, filled squares, and open circles,

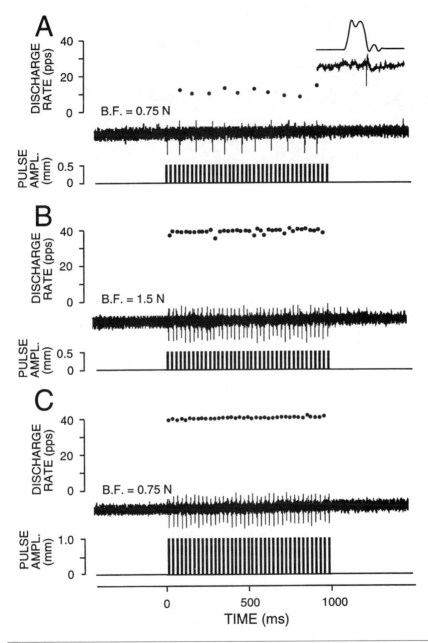

Figure 7.2   Tendon vibration responses of an individual muscle spindle afferent. The vibrator displacement amplitude, raw nerve recording and instantaneous frequency (bottom to top) are shown in each panel for the same TA muscle spindle afferent during tendon vibration at 40 pps. In A and B, the vibration amplitude was 0.5 mm, and in C, it was 1 mm. In A and C, the background force was half of that in B.

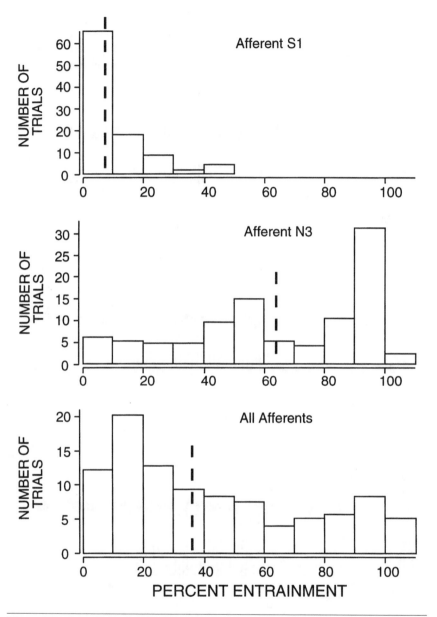

Figure 7.3 Overall entrainment level of afferents to tendon vibration. The percent of all trials—irrespective of background force, vibration amplitude, and frequency—in which a particular level of entrainment was obtained is plotted against the level (i.e., "percent") of entrainment. Percent entrainment is shown in bin widths of 10%. Top and middle, the overall entrainment levels are shown for two individual afferents, and bottom, levels are shown for the average of all 13 afferents. The vertical dashed lines show the mean entrainment level for all trials in each bar graph.

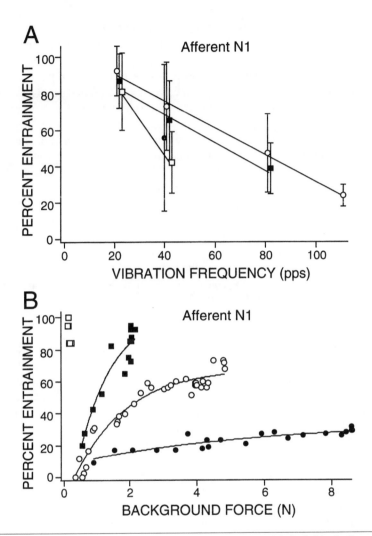

Figure 7.4   Effect of vibration frequency on entrainment level in individual afferents. In A, percent entrainment is plotted against vibration frequency at each of four vibration amplitudes (open squares-0.25 mm; filled squares-0.5 mm; open circles-1 mm; filled circles-2 mm). Each symbol and error bar represents the mean and standard deviation of all trials at a particular vibration frequency and amplitude (≥ 5). Linear regressions were calculated based on individual points rather than averages—for 0.25 mm (slope = -1.97 pps^{-1}), 0.5 mm (slope = -0.79 pps^{-1}), 1.0 mm (slope = -0.74 pps^{-1}), and 2.0 mm (only one frequency value). In B, percent entrainment is plotted against background force at each of 4 vibration frequencies (open squares-20 pps; filled squares-40 pps; open circles-80 pps; filled circles-110 pps). Each symbol represents the average entrainment level for one trial (i.e., a 1-s train of vibration pulses at a given frequency). Exponential fits for 40, 80 and 110 pps plots were statistically significant ($p \leq 0.05$) with exponents of -0.010 (40 pps), -0.006 (80 pps) and -0.002 (110 pps).

respectively), increasing vibration frequency decreased the entrainment level. Trials at 40 pps were obtained only at 2.0 mm amplitude in this afferent, resulting in a single data point (i.e., filled circle) in figure 7.4A.

The plot in figure 7.4B shows the negative effect of vibration frequency as a function of background force. The level of entrainment is plotted against the background force at vibration frequencies of 20, 40, 80 and 110 pps (i.e., open squares, filled squares, open circles, and filled circles, respectively), and each data point represents a measurement from an individual trial (i.e., 1 s long vibration train). The overall entrainment level decreased as a function of vibration frequency, with an entrainment level consistently > 80% at 20 pps vibration declining to < 30% at 110 pps vibration. At a given background force level (i.e., any arbitrary vertical line), the entrainment level decreased as the vibration frequency increased, and a given entrainment level (i.e., any arbitrary horizontal line) was achieved at a lower background force levels for lower frequencies.

Figure 7.4B also shows a facilitatory effect of background force. At each frequency of tendon vibration, increasing the background force over a range of 0-8.5 N had the effect of increasing the entrainment level in the muscle spindle afferent. At higher background forces (i.e., >10 N) this facilitatory effect saturated, or even decreased, presumably because the tendon vibrator was compressing the tendon against the underlying bone.

Increasing the amplitude of tendon vibration also enhanced the entrainment level of afferents. In most, the effect was more modest than might be expected from our previous observations on intramuscular stretch. Nevertheless, 7 of the 10 afferents tested for the effects of vibration amplitude exhibited a significant positive correlation with entrainment level. Enhancement of entrainment level was primarily observed at the lower vibration amplitudes, however, and most afferents were unaffected by an increase in pulse amplitude from 1.0 mm to 2.0 mm.

The two experiments described above demonstrated that, when the mechanical parameters of tendon indentation are controlled, the amount of intramuscular stretch and the timing patterns of muscle spindle afferents produced by the indentation can be predicted. However, the most significant finding was that only a small fraction of these afferents become 100% entrained to the vibration stimulus. Some afferents do not respond at all. Others are entrained at an average level of less than one action potential for every other vibration pulse. Those afferents not sensitive to tendon vibration or entrained at levels <100% are presumably free to respond to other types of stimuli, such as muscle stretch due to active or passive joint rotation.

Effect of Vibration Frequency and Timing on Coordination of Targeted Movements

Tendon vibration has been used to investigate the coordination of a variety of different movement tasks. In most, however, proprioception has been examined at the end point of movement (Bullen and Brunt 1986; Capaday and Cooke 1983; Cody et al. 1990; Inglis and Frank 1990; Kasai et al. 1994). Other studies have examined the perceptions of static joint angle and movement velocity in the isometric limb (Gilhodes 1993; Goodwin et al. 1972; Sittig et al. 1985). However, the effect of tendon vibration during movement—when muscle spindles fire at much higher rates than at rest—has not been studied in detail (however, see Sittig et al. 1987). In the experiment described in this section, tendon vibration was used to examine how the frequency and timing of vibration influenced a targeted limb movement.

Normal human subjects placed the right arm into a hydraulic manipulandum designed to rotate the elbow passively. Two electrical contacts were attached to the index finger and thumb of the right hand to detect hand opening. Visual contact with the right arm was prevented by an opaque screen. The elbow was passively extended from a starting angle of 120 deg (i.e., 60 deg flexion) through and beyond a target angle of either 145 deg or 157 deg. Elbow extensions occurred at one of seven constant velocities (15–70 deg/s), but the velocity in any given trial was randomized. The subject's task was to open the right index finger and thumb briskly when the elbow passed through the target angle. The target angle and the actual elbow angle at which the subject opened could be displayed on a video screen to provide the subject with knowledge of results about accuracy.[2]

In 20% of the trials, vibration was applied to the biceps brachii tendon, which lengthens during elbow extension. The background force of the vibrator was set to the level at which the subject reported experiencing the most potent illusion of movement. Vibration was applied at three different frequencies (20, 30, and 40 pps) and at three different times with respect to the elbow rotation:

- beginning 5 s before and ending at the onset of elbow rotation *(before)*
- beginning at the onset and continuing through the elbow rotation *(during)*
- beginning 5 s before and continuing through the elbow rotation *(before and during)*

No knowledge of results was provided in trials with tendon vibration.

In the absence of tendon vibration, performance was extremely accurate—usually within ±1.0–1.5 deg of the target angle (Cordo 1990, Cordo et al. 1994). When the tendon was vibrated, the subjects produced errors with a direction and amplitude that depended on the frequency of vibration and the velocity of elbow rotation. First, the effect of vibration frequency was investigated with the "during timing." When vibration was applied at 40 pps, the subjects opened the hand before the elbow reached the target angle (figure 7.5, circles). This result is consistent with the illusion that the vibrated muscle was longer or elongating faster than it really was as previously observed for the isometric limb (Goodwin et al. 1972; Sittig et al. 1985) or at the end point of movement (Bullen and Brunt, 1986; Capaday and Cooke 1983; Cody et al. 1990; Inglis and Frank 1990; Kasai et al. 1994). The amount of under-shoot depended on the velocity of joint rotation. The error decreased as velocity

---

[2]See Cordo et al. (1995a, b) for a more detailed description of the methods.

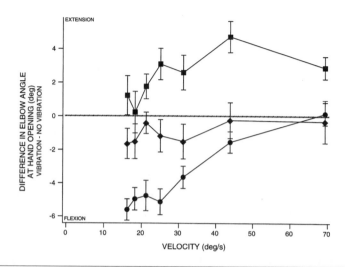

Figure 7.5 Effect of vibration frequency on elbow angle at hand opening. The difference in elbow angle at hand opening between trials with and without tendon vibration is shown for vibration at 20 pps (squares), 30 pps (diamonds) and 40 pps (circles) during elbow rotations through the 157° target angle. Each data point and error bar represents the grand mean and standard deviation, respectively, for six subjects.
Reprinted from *Journal of Neurophysiology* 1995.

increased and the movement time decreased. This diminishing influence was an effect of velocity rather than movement time (Cordo et al. 1995b).

Our analysis of low frequency tendon vibration during passive joint rotation reveals illusions not previously reported. When 20 pps vibration was applied to the biceps during the elbow rotation (figure 7.5, squares), the effect of tendon vibration switched to an overshoot of the target angle. This oppositely directed effect is consistent with the illusion that the vibrated muscle is *shorter* or *lengthening more slowly* than it actually is.

When vibration was applied at 30 pps (figure 7.5, diamonds), the result was intermediate between those evoked at 20 pps and 40 pps; no significant error occurred. Thus, when muscle spindles are activated by tendon vibration during movement, the perception of joint angle and velocity (Cordo et al. 1995b) depends on the vibration frequency. The difference between these results and those previously reported might relate to the higher firing rates of muscle spindles during movement compared to firing rates in the stationary limb.

The timing of tendon vibration with respect to the onset of elbow rotation also influenced the direction of performance errors. In figure 7.6, the time of tendon vibration at 40 pps is plotted against the performance accuracy. When vibration was applied *during* elbow rotation (figure 7.6, squares), the subject undershot the target by an amount that depended on the velocity of elbow rotation. When vibration was applied *before and during* elbow rotation (figure 7.6, circles), the undershoot errors increased slightly. Thus, undershoot errors produced by tendon vibration during

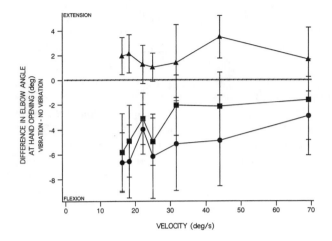

Figure 7.6   Effect of vibration timing. The difference in elbow angle at hand opening between trials with and without tendon vibration is plotted against elbow velocity when tendon vibration occurred *before* (triangles), *during* (squares), and *before and during* (circles) the elbow rotation. Each data point and error bar represents the grand mean and standard deviation, respectively, of 12 subjects.
Reprinted from *Journal of Neurophysiology* 1995.

elbow rotation were enhanced by the addition of vibration before elbow rotation. In contrast, vibration restricted to the 5 s period *before* elbow rotation (figure 7.6, triangles) resulted in an overshoot of the target angle. Thus, when muscle spindles are activated by tendon vibration at different times with respect to movement, the perception evoked depends dramatically on the vibration timing.

The effects of tendon vibration illustrated in figures 7.5 and 7.6 suggest that conventional interpretations of the effect of vibration might be simplistic. The conventional interpretation is that, on the level of individual receptors, vibration is occlusive—receptors sensitive to vibration tend to respond only to vibration. A historical focus on those muscle spindle afferents 100% entrained by vibration has produced the impression that tendon vibration occludes the output of most muscle spindle afferents. Moreover, most previous studies of muscle spindle firing patterns during tendon vibration have been restricted either to isometric conditions (Burke et al. 1976; Roll and Vedel 1982) or to movements with very slow velocities (Al-Falahe et al. 1990; Roll et al. 1989; Sittig et al 1987)—when these receptors fire at relatively low levels.

The studies of muscle spindle responses to tendon vibration described above (see also, Cordo et al. 1993) reveal that only a small proportion of muscle spindle afferents can be 100% entrained and, as a consequence, completely occluded by vibration. At least as many receptors are insensitive to vibration as those that respond with 100% entrainment; most receptors that do respond to vibration are entrained below the 50% level. Thus, during vibration, most muscle afferents are capable of also responding to more natural forms of mechanical stimulation such as muscle stretch resulting from active or passive joint rotation. Because muscle spindle afferents fire at much higher levels during movement compared with isometric conditions, muscle stretch can be as effective a stimulus to muscle spindle afferents as vibration. This creates the potential for two relatively independent channels of proprioceptive input to exist. One channel could transmit proprioceptive activity evoked by joint rotation in which the activity of responding afferents is unsynchronized. The other channel could transmit proprioceptive activity evoked by tendon vibration in which the activity of responding afferents is synchronized to the action of the vibrator probe.

The illusions of increased or decreased muscle length and velocity evoked by vibration during movement have been hypothesized as the difference or average— not the summation—of the hypothesized vibration- and movement-evoked channels (Cordo et al. 1995b). If elbow rotation produced a higher firing rate in biceps muscle spindle afferents than the vibration frequency (figure 7.5, squares) at the moment the elbow passed through the target angle, the difference between movement- and vibration-evoked activity would be negative. The subject would perceive that the elbow had not yet reached the target angle. Alternatively, if elbow rotation produced a lower firing rate in biceps muscle spindle afferents than the vibration frequency (figure 7.5, circles) at the moment the elbow passed through the target angle, the difference would be positive. The subject would perceive that the elbow had already passed through the target angle. A similar argument can be made to explain the relationships of movement velocity and vibration timing on performance errors (Cordo et al. 1995b). Thus, the effects of tendon vibration require a more complex explanation than previously presumed. For the effects of tendon vibration on movement coordination to be fully revealed, they must be studied during movement while the mechanical parameters of vibration are carefully controlled.

Effect of Random Noise Input on Muscle Spindle Responses to Small Amplitude Movements

Tendon vibration is a potentially powerful technique for investigating the properties of muscle spindle firing patterns and the coding of proprioceptive information within these patterns. In no other sensory system is it possible to activate receptors such that they can be made to fire artificially with any arbitrary pattern of activity. Tendon vibration can activate a subset of the population of muscle spindle afferents on a one-to-one basis, thereby distorting the kinematic information transmitted to the CNS. Using behavioral paradigms such as that described above (Cordo et al. 1995b; Gilhodes et al. 1993), different perceptions of joint kinematics can be evoked by changing the pattern of vibration. Thus, the combination of tendon vibration and motor behavioral paradigms can potentially lead to a better understanding of sensory processing.

In the final experiment described in this chapter, the servo-controlled vibrator was used to investigate the effect of a random noise input to muscle spindle receptors during small amplitude movements (see also Matthews and Watson 1981). (The tendon vibrator can produce mechanical stimuli with arbitrary waveforms, not just pulse stimuli.) The specific purpose of this experiment was to determine whether a random noise input to muscle spindles would enhance the coding and transmission of proprioceptive information related to small amplitude joint rotations. Such enhancement, termed stochastic resonance, has been previously identified in a variety of biological (Douglas et al. 1993; Levin and Miller 1996) and nonbiological systems (Hibbs et al. 1995). Prior to the experiment described below, stochastic resonance had not been demonstrated in humans. Demonstration of stochastic resonance in muscle spindle afferents has important implications with respect to the mechanisms of fusimotor activation of muscle spindles (Cordo et al. 1996).

Normal human subjects sat at a table with a hydraulic manipulandum designed to rotate the wrist passively. With the subject relaxed, recordings were obtained from single muscle spindle afferents in the radial nerve using standard microneurographic techniques. Muscle spindle afferents in the radial nerve originate primarily from the extensor muscles of the wrist, thumb, and fingers. The vibrator probe was placed in contact with the muscle from which the muscle spindle afferent originated.

The wrist was rotated with a sinusoidal waveform (0.5 Hz or 0.8 Hz) with a peak-to-peak amplitude sufficient to activate the muscle spindle afferent weakly, without evoking a strong burst of firing synchronized to the wrist rotation. Typical peak-to-peak amplitudes employed in this experiment were 1–3 deg. Approximately 2 min recordings were obtained from the afferent during wrist rotation with a particular level of random noise stimulation applied to the tendon. Several control trials (i.e., noise level = 0) were run throughout the recording session, which lasted from 30–105 min. Two subjects received a complete anesthetic block of the brachial plexus prior to the recording session, effectively deafferenting and deefferenting the arm.[3]

Selected recordings from an extensor carpi radialis (ECR) afferent are shown in figure 7.7 during a 2 deg peak-to-peak wrist rotation at 0.5 Hz. In the top panel, the afferent fired weakly during the flexion phase of rotation, during which the ECR was lengthening. The tendon was unstimulated in this control trial. When a small amplitude (10 μm rms) random noise input was applied to the ECR tendon (middle panel), the afferent fired strongly during the flexion phase of wrist rotation. When a larger amplitude (60 μm rms) random noise input was applied to the tendon (bottom panel), the afferent began to fire during both the flexion and extension phases of wrist rotation.

The signal-to-noise ratio was calculated from the power spectrum of the spike train. The signal was defined as the integral of the peak in the power spectrum at the wrist rotation frequency. In figure 7.8, plots of the signal-to-noise ratio are shown for two afferents as a function of the amplitude of the applied random noise input to the tendon. The relationship shown in the upper panel is typical of stochastic resonance behavior. The signal-to-noise ratio increases to a peak at low noise

---

[3]See Cordo et al. (1996) for a more detailed description of the methods.

Cordo et al.

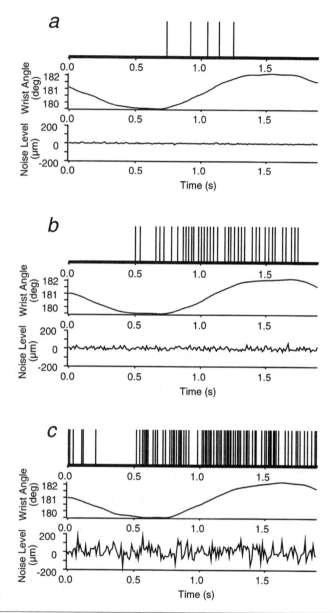

Figure 7.7 Representative 2 s samples of the afferent responses and input signals for three experimental conditions: (*a*) no input noise, (*b*) input noise of moderate intensity, and (*c*) input noise of high intensity. Top: Spike trains from a single muscle spindle afferent. Middle: Sinusoidal wrist angle imposed by the manipulandum. Bottom: Random noise signal applied by the actuated vibrator probe to the distal tendon of the parent muscle. Reprinted from *Nature* 1996.

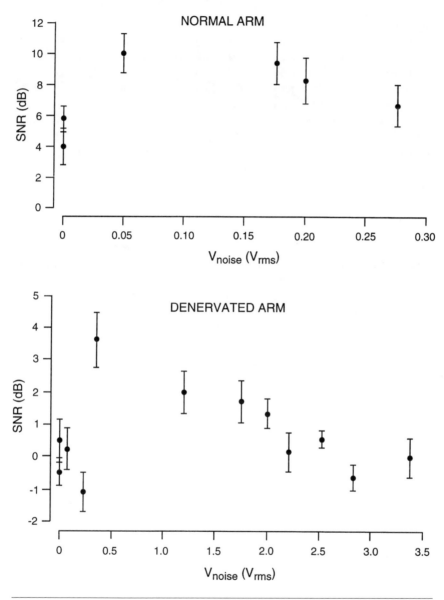

Figure 7.8   Values of the output signal-to-noise ratio (SNR) versus the input noise rms voltage for single spindle afferents in two different subjects: (*top*) a healthy subject, and (*bottom*) a subject who was administered an anesthetic block of the brachial plexus. The output SNR for each noise intensity level was computed from the spike-train power spectra in 0.15 Hz increments. The error bars on the SNR values represent the standard deviation of the noise value of the center frequency of each 0.15 Hz bin.
Reprinted from *Nature* 1996.

amplitudes and declines at larger noise amplitudes. The afferent shown below, which also demonstrated stochastic resonance-like behavior, was obtained in a subject with an anesthetized and paralyzed arm. Thus, the enhancement of signal coding and transmission by noise input was produced by a direct effect on the receptor, rather than indirectly via fusimotor reflexes (Gandevia et al. 1994; Johansson and Sojka 1985; Murphy and Martin 1995).

Enhancement of muscle spindle signal coding and transmission by a noise input suggests that the fusimotor system might operate via a stochastic resonance type of mechanism. During the performance of precision and novel movements, fusimotor input to muscle spindles is enhanced (Prochazka 1989; Prochazka et al. 1985). When gamma motoneurons are activated, the resulting contraction of intrafusal muscle could evoke a fused contraction, thereby producing steady tension on the sensory region of muscle spindle. Steady tension could prevent the sensory region from being unloaded during shortening of extrafusal muscle. It could also make the receptor more sensitive to changes in the length during lengthening of extrafusal muscle. Steady stretch of the muscle spindle capsule should evoke an increased discharge rate without producing a significant increase in interspike variability. However, during gamma activation produced by either electrical stimulation in animals (Matthews and Stein 1969) or active contraction in humans (Burke et al. 1979), the interspike interval distribution of muscle spindle afferents becomes more variable, suggesting the action of a stochastic or random process.

An alternative mechanism for fusimotor action is that the activation of gamma motoneurons, especially dynamic gamma motoneurons, produces a noise input to the muscle spindle receptor analogous to a biological dither signal. This noise input could place the receptor close to threshold more frequently but also randomly and could make it more responsive to small movements, such as while performing precision tasks. This hypothesis awaits more careful testing at the receptor level. A recent report about cutaneous perception in humans showed that introduction of a noise input enhances discrimination of tactile stimuli (Collins et al. 1996), suggesting that noise might also be capable of enhancing the perception of proprioceptive input.

Potential Impact of Tendon Vibration on the Understanding of Motor Coordination

Experiments involving tendon vibration as well as a host of other techniques have demonstrated the involvement of proprioception in a wide variety of movements (Andersson and Grillner 1983; Bässler 1986; Conway et al. 1987; Wolf and Pearson 1988). Studies of deafferented humans (Cole and Sedgwick 1992; Rothwell et al. 1982; Saling et al. 1992; Sanes et al. 1985) support this contention.

While the question of whether movements are proprioceptively influenced has become less and less of an issue in motor-control research, other more important questions have risen to the surface. What are the characteristics (e.g., timing, gain, and situational relevance) of proprioceptive influences on movement coordination? What are the central mechanisms of proprioceptive coordination? How is kinematic information in proprioceptive input encoded, decoded, and interpreted by the peripheral and central nervous systems?

To answer these more fundamental questions concerning movement coordination, more sophisticated experimental probes will be needed. Tendon vibration could potentially become such a probe. However, it is necessary to apply vibration in a way that both the qualitative and quantitative effects on muscle afferent firing patterns can be predicted. A clearer understanding of the perceptual interactions between afferent input evoked by vibration and movement must also be achieved. Not all studies will require the use of a tendon vibrator as sophisticated as that described in this chapter. In all cases, however, the level of stimulus control should be carefully matched to the requirements of the hypothesis being tested.

Acknowledgments

The work described in this chapter was performed in collaboration with a number of other researchers. The first two experiments were carried out in Sydney, Australia by Simon Gandevia, David Burke, and John-Paul Hales. The behavioral experiment was carried out in collaboration with Victor Gurfinkel, Leslie Bevan, and Graham Kerr. The stochastic resonance experiment involved contributions by Timothy Inglis, Sabine Verschueren, Jim Collins, Dan Merfeld, Scott Buckley, Stuart Rosenblum, and Frank Moss. This work was supported by the U.S. National Institutes of Health (P.C., L.B., V.G.), Medical Research Council of Australia (S.G., D.B.), National Fund for Scientific Research of Belgium (S.V.), National Science Foundation (J.C.), Ivey Research Fund (T.I.), and Office of Naval Research (F.M.). The author also wishes to express his appreciation to the organizers of the *Bernstein Conference,* in particular to Mark Latash.

References

Al-Falahe, N.A., Nagoaka, M., Vallbo, A.B. (1990) Response profiles of human muscle afferents during active finger movements. *Brain,* 113: 325–346.

Andersson, O., Grillner, S. (1983) Peripheral control of the cat's step cycle. II. Entrainment of the central pattern generator by sinusoidal hip movements during "fictive locomotion". *Acta Physiol. Scand.,* 118: 229–239.

Bässler, U. (1986) Afferent control of walking movements in the stick insect *Cuniculina impigra.* II. Reflex reversal and the release of the swing phase in the restrained foreleg. *J. Comp. Physiol.,* 150: 351–362.

Bianconi, R., Van der Meulen, J.P. (1963) The response to vibration of the end organs of mammalian muscle spindles. *Journal of Neurophysiology,* 26: 177–190.

Brown, M.C., Engberg, I., Matthews, P.B.C. (1967) The relative sensitivity to vibration of muscle receptors of the cat. *Journal of Physiology (London),* 192: 773–800.

Bullen, A.R., Brunt, D. (1986) Effect of tendon vibration on unimanual and bimanual movement accuracy. *Exper. Neurol.,* 93: 311–319.

Burke, D., Hagbarth, K., Löfstedt, L., Wallin, B.G. (1976) The responses of human muscle spindle endings to vibration of non-contracting muscles. *Journal of Physiology (London)*, 261: 673–693.

Burke, D., Skuse, N.F., Stuart, D.G. (1979) The regularity of muscle spindle discharge in man. *Journal of Physiology (London)*, 291: 277–290.

Capaday, C., Cooke, J.D. (1983) Vibration-induced changes in movement-related EMG activity in humans. *Exp. Brain Res.,*52: 139–146.

Cody, F.M., Schwartz, M.P., Smit, G.P. (1990) Proprioceptive guidance of human voluntary wrist movements studied using muscle vibration. *Journal of Physiology (London)*, 427: 455–470.

Cole, J.D. (1991) *Pride and a Daily Marathon*. London: Duckworth.

Cole, J.D., Sedgwick, E.M. (1992) The perceptions of force and of movement in a man without large myelinated sensory afferents below the neck. *Journal of Physiology (London)*, 449: 503–515.

Collins, J.J., Imhoff, T.T., Grigg, P. (1996) Psychophysical stochastic resonance: Noise-enhanced tactile sensation in humans. *Nature*, 383: 770.

Conway, B.A., Hultborn, H., Kiehn. O. (1987) Proprioceptive input resets central locomotor rhythm in the spinal cat. *Exp. Brain Res.,*68:643–656.

Cordo, P.J. (1990) Kinesthetic control of a multijoint movement sequence. *Journal of Neurophysiology*, 63: 161–172.

Cordo, P.J., Gandevia, S.C., Hales, J.P., Burke, D., Laird, G. (1993) Force and displacement-controlled tendon vibration in humans. *EEG Clin. Neurophysiol.,*89: 45–53.

Cordo, P.J., Bevan, L., Carlton, L., Carlton, M.J., Kerr, G. (1994) Kinesthetic coordination of movement sequences: Role of velocity and position information. *Journal of Neurophysiology*, 71: 1848–1861.

Cordo, P.J., Bevan, L., Gurfinkel, V.S., Carlton, L., Carlton, M.J., Kerr, G. (1995a) Proprioceptive coordination of discrete movement sequences: Mechanisms and generality. *Canad. J. Physiol. Pharm.*, 73: 305–315.

Cordo, P.J., Gurfinkel, V.S., Bevan, L., Kerr, G. (1995b) Proprioceptive consequences of tendon vibration during movement. *Journal of Neurophysiology*, 74: 1675–1688.

Cordo, P.J., Inglis, J.T., Verschueren, S.M.P., Collins, J.J., Merfeld, D., Buckley, S., Rosenblum, S., Moss, F. (1996) Stochastic resonance in human muscle spindle afferents. *Nature*, 383: 769-770.

DeGail, P., Lance, J.W., Neilson, P. (1975) Differential effects of tonic and phasic reflex mechanisms produced by vibration of muscles in man. *J. Neurol. Neurosurg. Psychiat.*, 29: 1–11.

Douglass, J.K., Wilkens, L., Pantazelou, E., Moss, F. (1993) Noise enhancement of information transfer in crayfish mechanoreceptors by stochastic resonance. *Nature*, 365: 337–340.

Eklund, G., Hagbarth, K.E. (1966) Normal variability of tonic vibration reflexes in man. *Exp. Neurol.*, 16: 80–92.

Gandevia, S.C., Wilson, L., Cordo, P.J., Burke, D. (1994) Fusimotor reflexes in relaxed forearm muscles produced by cutaneous afferents from the human hand. *Journal of Physiology*. 479: 499–508.

Gilhodes, J.C., Coiton, Y., Roll, J.P., Ans, B. (1993) Propriomuscular coding of kinaesthetic sensation. Experimental approach and mathematical modeling. *Exp. Brain Res.*, 68: 509–517.

Goodwin, G.M., McCloskey, D.I., Matthews, P.B.C. (1972) The contribution of muscle afferents to kinaesthesia shown by vibration induced illusions of movement and by the effects of paralysing joint afferents. *Brain*, 95: 705–748.

Hansen, K., Hoffman, P. (1922) Electrische methoden. *Zeitshr. f. Biol.*, 74: H. 5/6.

Hibbs A.D., Singsaas A.L., Jacobs E.W., Bulsara A.R., Bekkedahl, J.J., Moss, F. (1995) Stochastic resonance in a superconducting loop with a Josephson junction. *J. Appl. Phys.*, 77: 2582–2590.

Inglis, J.T., Frank, J.S. (1990) The effect of agonist/antagonist muscle vibration on human position sense. *Exp. Brain Res.*, 81: 573–580.

Johansson, H., Sojka, P. (1985) Actions on gamma-motoneurones elicited by electrical stimulation of cutaneous afferent fibres in the hind limb of the cat. *Journal of Physiology (London)*, 366: 343–363.

Kasai, T., Kawanishi, M., Yahagi, S. (1994) Effects of upper limb muscle vibration on voluntary wrist flexion-extension movements. *Percept. Motor Skills.*, 78: 43–47.

Levin, J.E., Miller, J.P. (1996) Broadband neural encoding in the cricket cercal sensory system enhanced by stochastic resonance. *Nature*, 380: 165–168.

Matthews, P.B.C. (1966) The reflex excitation of the soleus muscle of the decerebrate cat caused by vibration applied to its tendon. *Journal of Physiology (London)*, 184: 450–472.

Matthews, P.B.C., Stein, R.B. (1969) The regularity of primary and secondary muscle spindle afferent discharges. *Journal of Physiology (London),* 202: 59–82.

Matthews, P.B.C., Watson, J.D.G. (1981) Action of vibration on the response of cat muscle spindle Ia afferents to low frequency sinusoidal stretching. *Journal of Physiology (London),* 317: 365–381.

Murphy, P.R., Martin, H.A. (1995) Fusimotor neurone responses to medial plantar nerve stimulation in the decerebrate cat. *Journal of Physiology.* 482: 167–177.

Partridge, L.D. (1965) Modifications of neural output signals by muscles: A frequency response study. *Journal of Applied Physiology,* 20: 150–156.

Prochazka, A. (1989) Sensorimotor gain control: A basic strategy of motor systems? *Prog. Neurobiol.,* 33: 281–307.

Prochazka, A., Hulliger, M., Zangger, P., Appentung, K. (1985) 'Fusimotor set': New evidence for alpha-independent control of gamma-motoneurones during movement in the awake cat. *Brain Res.,* 339: 136–140.

Roll, J.P., Vedel, J.P. (1982) Kinaesthetic role of muscle afferents in man, studied by tendon vibration and microneurography. *Exp. Brain Res.,* 47: 177–190.

Roll, J.P., Vedel, J.P., Ribot, E. (1989) Alteration of proprioceptive messages induced by tendon vibration in man: A microneurographic study. *Exp. Brain Res.,* 76: 213–222.

Rothwell, J.C., Traub, M.M., Day, B.L., Obeso, J.A., Thomas, P.K., Marsden, C.D. (1982) Manual motor performance in a deafferented man. *Brain,* 105: 515–542.

Rymer, W.Z., Hasan, Z. (1981) Prolonged time course for vibratory suppression of stretch reflex in the decerebrate cat. *Exp. Brain Res.,* 44: 101–112.

Saling, M., Sitarova, T., Vejsada, R., Hnik, P. (1992) Reaching behavior in the rat: Absence of forelimb peripheral input. *Physiol. Behav.,* 51: 185–191.

Sanes, J.N., Mauritz, K.-H., Dalakas, M.C., Evarts, E.V. (1985) Motor control in humans with large-fiber sensory neuropathy. *Human Neurobiology,* 4: 101–114.

Sittig, A.C., Denier van der Gon, J.J., Gielen, C.C.A.M. (1985) Separate control of arm position and velocity demonstrated by vibration of muscle tendon in man. *Exp. Brain Res.,* 60: 445–453.

Sittig, A.C., Denier van der Gon, J.J., Gielen, C.C.A.M. (1987) The contribution of afferent information on position and velocity to the control of slow and fast human forearm movements. *Exp. Brain Res.,* 67: 33–40.

Wolf, H., Pearson, K.G. (1988) Proprioceptive input patterns elevator activity in the locust flight system. *Journal of Neurophysiology,* 59: 1831–1853.

8

CHAPTER

On the Number of Degrees of Freedom in Biological Limbs

*Stan Gielen, Bauke van Bolhuis,
and Erik Vrijenhoek*
Department of Medical Physics and Biophysics,
University of Nijmegen, Nijmegen, The Netherlands

Many studies about motor control have revealed a consistent and reproducible pattern of motor behavior across subjects (Flanders et al. 1992; Theeuwen et al. 1994a). Considering the large number of joints and muscles, this is a remarkable result. The large number of available joints and muscles provides many degrees of freedom, with respect both to the kinematics as well as to the dynamics of movements. Since subjects reveal a reproducible pattern of motor behavior, they have learned to use either a particular muscle activation pattern or use an algorithm to reduce the number of degrees of freedom. A more or less unique motor repertoire evolves. The problem, that various muscle activation patterns may lead to the same desired behavior, was referred to as the motor-equivalence problem by Bernstein (1967). He had already provided general hypotheses about dealing with this problem. This chapter discusses how reducing the number of degrees of freedom of the motor system may reduce the number of potentially feasible muscle activation patterns. It also discusses what this may teach about underlying motor-control strategies.

The available degrees of freedom in the human arm are related to a broad range of different aspects. First, each limb has multiple joints. A joint can have one degree of freedom, such as the phalangeal finger joints. Other joints may have more rotational degrees of freedom. The elbow has two degrees of freedom (supination/

pronation and flexion/extension), the shoulder has three degrees of freedom. For joints with three degrees of freedom, the limb position does not depend only on the angles of rotations along each of the three axes but also on the order of the rotations. Rotating an object and changing the order of any two, noncolinear rotations of that object demonstrates this. The final orientation of the object will differ depending on the order of rotations. This has large implications for human arm movements. It implies that both the position and orientation of the fingers in space depend on the order of shoulder joint rotations.

Since most limbs have multiple joints, the total number of degrees of freedom may be larger than six. Six is the minimum number of degrees of freedom necessary to reach any point in space with any arbitrary orientation of the hand. How to deal with multiple degrees of freedom is frequently referred to as the kinematic redundancy problem.

Another set of degrees of freedom is related to the number of muscles acting across a joint. In general, the number of muscles acting across a joint exceeds the number of rotational degrees of freedom in that joint. As a result, the same joint torque can be obtained by various muscle activation patterns. For example, the same flexion torque in the elbow can be obtained by activating m. brachialis or m. brachioradialis (both monoarticular elbow flexors) or by any linear combination of activations. Several studies (Buchanan et al. 1986, 1989; ter Haar Romeny et al. 1982, 1984; van Zuylen et al. 1988) have revealed a consistent muscle activation pattern for elbow flexion torques. It consists of activating not only m. brachialis and m. brachioradialis but also m. biceps brachii and m. pronator teres. It is not yet known why this complicated activation pattern occurs instead of the many, much simpler muscle activation patterns. Because biceps brachii is a biarticular muscle, the problem related to the large number of muscles acting across a joint is enlarged because many muscles act across two or more joints. As a consequence, elbow flexion requires shoulder muscle involvement as well.

Later studies have demonstrated that the relative activation of muscles acting across the elbow is different in isometric contractions, in concentric contractions, and in eccentric contractions (Tax et al. 1989, 1990a, b; Theeuwen et al. 1994a). This illustrates that subjects do have the flexibility to use various muscles in different combinations depending on the motor task.

The last aspect of degrees of freedom pertains to individual muscles. Each muscle has a large number of motor units with various properties. The best-known difference involves the isometric recruitment threshold. It differs for all motor units and is thought to be related to the size of the soma of the motoneuron belonging to each motor unit (Cope and Pinter 1995; Henneman 1981). All motoneurons in a muscle are thought to receive the same synaptic input (also called homogeneous activation). Motor units are supposedly recruited in an orderly fashion corresponding to the motoneuronal soma. Although some evidence shows that not all motoneurons receive the same input (Kernell and Hultborn 1990), the total synaptic input is thought to be distributed across the motoneurons. This causes motoneurons with the smallest soma to be recruited first. This idea has become well known as the size principle (Henneman 1981). Previous studies (ter Haar Romeny et al. 1982, 1984; van Zuylen et al. 1988) have demonstrated that within a muscle, recruitment reversal may occur. This reversal is because many muscles contain several populations of motor units. Each motor unit seems to receive a different, task-dependent, but homogeneous input. As a consequence, two motor units from different subpopulations may become active in any order depending on the task (e.g., some motor units are recruited for elbow flexion and others for supination). The existence of a few groups of subpopulations of motor units within a muscle (also called task groups, see Loeb 1985) is no reason to abandon the concept of the size principle. All evidence suggests that within each subpopulation, the size principle holds. However, the existence of subpopulations each with a specific activation pattern enlarges the number of degrees of freedom above the number of muscles.

Motor units differ in many more aspects than only their recruitment order. A covariation of the following properties exists: recruitment threshold, fatigability, twitch amplitude, twitch contraction time, and axonal conduction velocity (Bodine et al. 1987; Botterman and Cope 1988a, b; Thomas et al. 1990; Zajac and Faden 1985; Zengel et al. 1985). The homogeneous activation of motor units has been studied mainly during static and/or isometric conditions. However, if it can be generalized to dynamic conditions, this would imply that the properties of motor units with a high isometric recruitment threshold, which seem to be ideally suited for fast cyclic contractions (e.g., for running at full speed), cannot be selectively used without a slowdown by the slow twitches (with a long duration) of motor units with low isometric recruitment thresholds. Recently, some evidence has been presented in favor of a selective activation of motor units with high isometric recruitment thresholds (Howell et al. 1995; Nardone and Schiepatti 1988; Nardone et al. 1989). If selective activation of motor units with different functional properties were really possible, this would tremendously increase the number of degrees of freedom available to the motor system.

This chapter will survey the various aspects of degrees of freedom available to the motor apparatus. It will review how the motor system uses or does not use these degrees of freedom in various motor tasks.

Methods

This chapter reviews several experimental observations related to reducing degrees of freedom. The most important methodological techniques will be described only briefly, allowing the reader to appreciate the results. For more detailed information, see van Bolhuis et al. (1997), Miller et al. (1992), and Theeuwen et al. (1994a, b).

Experimental Setup

In order to study joint rotation in multijoint movements, the position and orientation of the upper arm and forearm were measured with an OPTOTRAK system (Northern Digital), which measures the position of infrared, light-emitting diodes (IREDs).

Crosses with IREDs on each of the four tips were attached to the upper arm just proximal to the elbow joint and at the backside of the hand. The arm lengths of the crosses were 5 cm for the cross on the forearm and 10 cm for the one on the upper arm. The wrist was fixated with a bracelet. The subjects were strapped with their shoulders to the chair, fixating their shoulders. These precautions were taken to ensure that movements only in the elbow and shoulder joints remained. The position of each IRED was sampled at a frequency of 100 Hz with a resolution of 0.1 mm in a range of approximately 2 m³. The position of the upper arm and hand were calculated as the average of the positions of the four IREDs on the cross. The orientation of the upper arm and forearm was calculated from the orientation of the four IREDs in three-dimensional space (Miller et al. 1992). This setup allowed relatively unrestricted movements to be made within most of the natural space.

The positions of the four IREDs on a cross were measured in a coordinate system fixed in space. From the position of the IREDs on the cross, the position of the cross (including the orientation) was expressed as a three-dimensional rotation vector (Haustein 1989). This vector described the position and orientation of the cross as the result of a single-axis rotation relative to a reference position. The rotation vector was parallel to the axis of rotation from the reference position to the position of the cross. The vector magnitude was equal to the tangent of half the rotation angle. For small rotations, the magnitude was approximately equal to the angle in degrees divided by 100 (Haslwanter 1995).

The data describing the position of the arm segments were fitted to a flat plane. Although rotation vectors are better described by a slightly curved plane rather than by a flat plane (Hore et al. 1992; Miller et al. 1992; Theeuwen et al. 1993), a local linear approximation was justified and appeared to be well suited. It provided a comparison of the arm segment positions in various experimental paradigms.

Studies about muscle activation during isometric contractions and during movements were performed with each subject seated on a chair with the right arm in a horizontal plane through the shoulder joint. The forearm was supported by a cloth sling attached to the ceiling, requiring no activation of shoulder muscles to keep the arm in the horizontal plane. The right shoulder was securely strapped to the chair using straps. The shoulder joint angle was always the same (0 deg anteflexion), and the hand was always in the middle between full supination and pronation. A lightweight, aluminum bracelet was fixed around the subject's wrist and was connected via a cable to a torque motor. The cable was guided over several pulleys to let the torque motor pull in various directions.

For isometric contractions, each subject's wrist was fixated. Subjects were instructed to exert a prescribed force of 30 N at the wrist in 13 different equidistant directions in the range between 180–360 deg. Force measured by strain gauges near the wrist was displayed on an oscilloscope in front of a subject to provide feedback to the subject. This enabled the subject to exert the prescribed force. In conditions where the subject had to vary isometric force gradually (e.g. sinusoidally), the oscilloscope presented a target signal that the subject had to track during the experimental trial.

For eccentric (lengthening) contractions at a constant velocity, the subject's wrist was pulled back by the torque motor with a speed of 1.5 cm/s, which corresponds to an angular velocity in the elbow of about 3.7 deg/s. For shortening contractions at a constant velocity, the cable was released by the torque motor at a speed of 1.5 cm/s. Subjects had to keep the magnitude of the force at 30 N while the hand moved in one of the 13 different directions. The movement was made over a range of about 20 cm (about 46 deg) centered at the position in which the isometric contraction was performed. EMG activity was recorded, however, only in a range of 6 cm (about 14.8 deg) centered around the isometric test position to prevent any effects of the force-length relation. The duration of the 6 cm movement was about 4 s. Each trial was repeated three times. For each direction, task and elbow angle EMG activity were averaged over the three signals obtained in the three trials.

In other trials, subjects were instructed to track a sinusoidally moving signal on the oscilloscope. In this condition, wrist position, instead of force at the wrist, was shown to the subject.

In order to prevent cross talk, EMG activity was recorded using intramuscular, fine-wire, nylon-coated electrodes with a diameter of 25 μm. The wires were inserted by means of a hypodermic needle into the middle of the muscle belly. The insertion depth in the muscle could be determined by referring to the visible length of the hypodermic needle.

After amplification, rectification, and filtering (with a fourth-order Bessel filter; 3 dB high-pass frequency at 3 Hz and 3 dB low-pass frequency at 150 Hz), the EMG signals were sampled at a rate of 500 Hz. The mean amplitude of the EMG activity at the isometric force or in the 6 cm range (for concentric or eccentric contractions), corrected for the mean EMG activity at rest, was considered the amplitude of EMG activity in each trial.

Results

Reduction of Rotational Degrees of Freedom in a Single Joint

For a single joint with only one degree of freedom, a unique joint torque exists for each external load in order to maintain that position or joint angle. The relation between joint torque and joint angle has been elegantly described by the equilibrium-point hypothesis by Feldman (1966, 1986). According to this hypothesis, the position of a limb corresponds to the joint angle where external forces (e.g., due to gravity) balance the forces generated by muscles. This hypothesis relates joint angle to muscle force by the stiffness, which is determined by muscle mechanical properties and by reflex actions. However, how this equilibrium-point hypothesis can explain the kinematics of multijoint limbs is not yet clear. Following the ideas underlying the equilibrium-point hypothesis, the posture of a multijoint limb is determined by the stiffness, where posture corresponds to the minimum potential energy. However, since stiffness results from muscle mechanical properties, this hypothesis shifts the problem of finding the correct posture in the redundant joint space into solving the correct muscle activation pattern in the (even higher dimensional) redundant muscle space.

When joints with more than one degree of freedom are involved, the situation is even more complex. It is well known that rotations in three-dimensions do not commute. The final result of two subsequent, noncolinear rotations depends on the order of the rotations. How the motor system deals with this problem has been investigated extensively for the eye (Donders 1875; von Helmholtz 1925; Tweed and Vilis 1987, 1990; Straumann et al. 1991). During the last decade, these studies about eye movements have been extended to the head and shoulder (Hore et al. 1992; Miller et al. 1992; Theeuwen et al. 1993). For eye movements, head movements, and pointing with the extended arm, only those angular positions are selected that can be obtained by single-axis rotations from a specific reference position, called the primary position. All rotation vectors describing the position of the eye, head, or arm are contained in a two-dimensional surface. As a consequence, the number of degrees of freedom is effectively reduced from three to two. This is illustrated in figure 8.1. The upper panels show the position of the hand (figure 8.1) in a three-dimensional workspace. The lower panels provide a front view, side view, and top view, respectively, of the plane with the rotation vectors, which correspond to the positions shown in the upper panels. The data in the lower panels show that the plane is not infinitely thin. The thickness of the plane, defined as the standard deviation of the data relative to the plane, is typically 1 deg for the eye and somewhat larger (3–4 deg) for the arm, head, and shoulder. Actually, the rotation vectors are not contained in a flat plane but in a slightly curved plane (Hore et al. 1992; Miller et al., 1992). However, the degree of freedom is reduced from three to two. The reduction of the number of degrees of freedom for eye, head, and shoulder mean that the orientation of the eye, head, and upper arm is uniquely determined by the gaze direction of the eye or head and by the direction the arm points, respectively.

Controlling the Orientation of the Hand in Multijoint Arm Movements

The previous section showed that the orientation of the arm while pointing with the arm extended can be described by rotation vectors in a two-dimensional surface. Similar results have been obtained both for the upper arm and forearm (Miller et al. 1992; Theeuwen et al. 1993). This result illustrates that both the upper arm and forearm obey Donders' law, which says that the orientation is uniquely determined by the pointing direction and does not depend on how the arm reached that pointing direction. The fact that hand positions can also be described by rotation vectors in a two-dimensional surface is not trivial since supination/pronation in the elbow can affect the orientation of the hand in space. The study by Theeuwen et al. (1993) showed that the position of the upper arm and forearm during pointing cannot be described by the same curved plane. This demonstrates that supination/pronation

frontal view   side view   top view

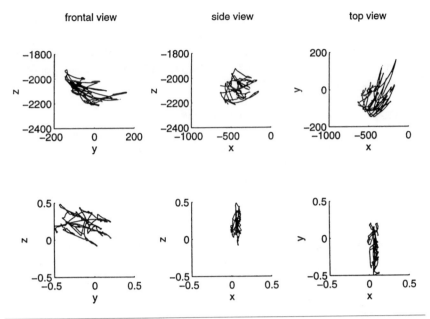

Figure 8.1 Position of the hand during pointing movements with a fully extended arm. The upper panels show the position of the hand in x-, y-, and z-coordinates (in mm) in a three-dimensional workspace. The lower panels show the rotation vectors (Haustein 1989) describing the position of the hand in a front view, side view, and top view of the plane with rotation vectors. In order to obtain the amplitude of the rotation vectors in degrees, the units in the lower panels have to be multiplied by a factor of about 100.

does contribute to the orientation of the forearm but in a consistent way for each pointing direction.

A more detailed analysis (Gielen et al. 1997; Soechting et al. 1995) has shown that Donders' law is not strictly obeyed. However, the deviations from Donders' law are small. Presumably, they explain why the thickness of the surface containing the rotation vectors describing the orientation of the upper arm and forearm (expressed as the standard deviation of the deviation of the rotation vectors relative to the surface) is larger for the upper arm and forearm than for the eye. For the upper arm and forearm, it is typically about 3 deg (Hore et al. 1992; Miller et al. 1992; Theeuwen et al. 1993). For the eye, it is typically 1 deg (Tweed and Vilis 1987, 1990). This chapter will not discuss these small deviations in detail.

When the arm is not fully extended, the situation becomes more complicated. The position and orientation of the hand do not depend only on shoulder rotations but also on elbow flexion. Moreover, torsion at the shoulder affects not only the orientation of the hand but also the position of the hand when the elbow is not fully extended. Yet, the result is the same. The orientation of the hand can always be described by a single-axis rotation vector, which transforms the orientation of the hand in the primary position to the actual orientation, containing all rotation vectors in a plane. This is illustrated in figure 8.2. The upper panels show the position of the hand in Cartesian coordinates. The lower three panels show the corresponding rotation vectors for pointing movements to targets at a distance of 25, 45, or 65 cm from the shoulder in a horizontal and vertical range of 45 deg. A full mathematical explanation of this result can be found elsewhere (Gielen et al. 1997). The thickness of the plane with rotation vectors for the hand, averaged over many subjects, is approximately 3 deg. This does not differ significantly from the thickness of the plane with rotation vectors for the upper arm, approximately 3 deg (Gielen et al. 1997; Theeuwen et al. 1993).

The number of degrees of freedom reduce both in the shoulder and in the elbow. In the shoulder, the reduction is from three to two. Since the elbow adds two degrees of freedom to the available three degrees of freedom in the shoulder, the total number of available degrees of freedom for the wrist is at least five. However, the orientation of the hand appears to be reduced to two degrees of freedom. Together with the amount of elbow flexion/extension, which determines the distance of the hand

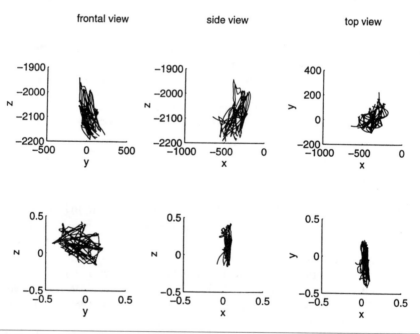

Figure 8.2 Position of the hand during pointing movements to targets at a distance of 25, 45, or 65 cm from the shoulder in a horizontal and vertical range of 45 deg. The upper panel shows the position of the hand in x-, y-, and z-coordinates (in mm) in a three-dimensional workspace. The lower panels show the rotation vectors (Haustein 1989) describing the position of the hand in a front view, side view, and top view of the plane with rotation vectors. In order to obtain the amplitude of the rotation vectors in degrees, the units in the lower panels have to be multiplied by a factor of about 100.

relative to the shoulder, three degrees of freedom describe the position and orientation of the hand in three-dimensional space. These results appear to be consistent between subjects with only small variations in the position of the primary position between subjects.

Based on the assumption that shoulder movements from an initial to a final position are single-axis rotations, the trajectory of the shoulder during the movement is given by rotation vector

$$r_{21} = \frac{r_2 - r_1 + r_1 \times r_2}{1 + r_1 \times r_2}$$

where r_1 and r_2 represent the rotation vectors describing the initial and final position of the shoulder, respectively (Hepp 1990).

When the shoulder has only two degrees of freedom left and only flexion/extension is added as an additional degree of freedom to adjust the position of the hand relative to the shoulder, the hand trajectory while moving between two positions should be determined completely. This can be understood from the following. For each position of the hand in space, the values of at least three degrees of freedom have to be specified. The initial and final elbow joint angles are completely determined by the distance of the initial and final target position relative to the shoulder. Given the elbow joint angle, a unique upper arm position brings the hand to the target. This unique upper arm position provides the additional two degrees of freedom, which specify the position of the upper arm for each position of the hand in space. Assuming that the ratio of elbow and shoulder joint angles is constant during each movement (Lacquaniti et al. 1986), the hand trajectory during aiming movements can be predicted. A comparison of measured hand trajectories and simulated trajectories with the model described above gave a good correspondence (Gielen et al. 1997).

Reduction of Degrees of Freedom in Muscle Activation Patterns

Previous studies have shown that the relative contribution of elbow flexors to elbow joint torque is reproducible and consistent across subjects (Jamison and Caldwell 1993; Jørgenson and Bankov 1971; Theeuwen et al. 1994a). This indicates that a large reduction of the number of degrees of freedom is imposed on the large number of degrees of freedom available through the number of muscles. However, no model exists to explain or to understand the observed muscle activation patterns.

The relative activation of muscles appears to differ for the same elbow torque depending on the type of contraction. Theeuwen et al. (1994a) found that the relative activation of flexor muscles differs for concentric/eccentric contractions as compared with isometric contractions. In another experiment, Tax et al. (1990a, b) instructed subjects to control either force or position while moving against a constant preload. Although force and position at the hand were physically the same in these tasks, the relative activation of muscles differed in these two conditions.

A model to explain the relative activation of monoarticular and biarticular muscles was proposed by van Ingen Schenau (1989). This model was based on the fact that using biarticular muscles can greatly improve movement efficiency (Gielen and van Ingen Schenau 1992). This improvement occurs because the joint torque and the change in joint angle may have an opposite sign for many movements. Due to the opposite sign, that joint delivers negative work. Since the total work delivered by a limb is the sum of the work delivered by each of the joints, the negative work in one joint requires that another joint produce an excess of work in order to make up the difference between external work and the negative work. Gielen and van Ingen Schenau (1992) have explained how dissipation of energy due to negative work can be prevented by using biarticular muscles. Recently, Doorenbosch and van Ingen Schenau (1995) reported that the EMG activity of monoarticular muscles is proportional to torque in the joint. In contrast, the EMG activity of the biarticular muscles is proportional to the difference in torque in the two joints. Based on these results, they concluded that the biarticular muscles function in a reciprocal way to regulate the distribution of the net torques in the joints.

Reducing Degrees of Freedom in a Muscle

Previous studies (ter Haar Romeny et al. 1982, 1984; Kandou and Kernell 1989; Segal 1992; Segal et al. 1991; van Zuylen et al. 1988) have shown that many muscles have several subpopulations of motor units, each with its own task-dependent activation pattern. This indicates that there is no homogeneous activation of all motor units. Rather, the results are compatible with the notion that within each subpopulation, a homogeneous activation exists. These results imply that the subpopulations of motor units should be considered the basic functional force generating elements in a muscle, not a single muscle itself. This section will not discuss this muscle partitioning any further. Instead, it will focus on the motor units within a single subpopulation.

Each muscle has different types of muscle fibers, each with different functional properties. These different properties become evident in the fatigability and in the twitches of various motor units. If all motoneurons belonging to a muscle receive a homogeneous activation, the selective use of motor units with properties especially suited for a particular motor task would be excluded. For example, motor units with long twitch contraction times might hamper fast repetitive movements. Using only motor units with short twitch contraction times, which, in addition, have large twitch amplitudes, would be preferable. Such a selective activation has been reported for reflex-induced, fast oscillatory paw shakes in hind limb muscles in cats (Smith et al. 1980). In humans, a selective activation of motor units with high isometric recruitment thresholds has been reported during lengthening contractions in the walking cycle in m. gastrocnemius (Nardone and Schieppati 1988; Nardone et al. 1989). However, it is not clear whether these selective activations aim to use the specific functional twitch properties of motor units with a high isometric recruitment threshold.

Recently, Howell et al. (1995) reported different recruitment patterns of motor units during isometric, concentric, and eccentric contractions in the first dorsal interosseus in humans. During repeated concentric/eccentric contractions, the majority of motor units followed an expected firing rate pattern. First, recruitment occurred during the concentric phase with increasing firing rates as the concentric movement progressed. Second, decreasing firing rates and decruitment occurred during the eccentric phase. However, some motor units, which had a relatively high isometric recruitment threshold, were recruited during the eccentric contraction phase of a movement at a time when other units were decreasing their discharge rate or being decruited.

In order to investigate the dynamic aspects of movements on the recruitment behavior of motor units, the motor-unit activity during sinusoidal modulations of isometric force and during sinusoidal movements in the elbow have been studied. Various frequencies were tested in the range between 0.2–1.5 Hz. In these studies, intramuscular, fine-wire electrodes were used. Into each muscle, several electrodes were simultaneously inserted, allowing the activity of several (generally two to four) motor units to be simultaneously recorded.

For sinusoidally modulated isometric contractions, very consistent motor-unit activity behavior was found. At all frequencies, motor units were recruited in the same order as during slow isometric ramp contractions. With increasing frequency, there is a progressive phase advance of motor-unit activity relative to the isometric force (figure 8.3).

The thin dashed lines in figure 8.3 represent the phase lead of the action potential bursts relative to the force at the wrist. The crosses show the frequency-dependent phase lead for a motor unit (A) with an isometric recruitment threshold of 15 N. The circles represent the phase lead for another motor unit (B), simultaneously recorded, with an isometric recruitment threshold exceeding 50 N. Figure 8.3 shows that motor-unit B fired only for isometric contractions with frequencies above 1.0 Hz. The thick dashed line shows the phase relation of a second-order, low-pass system fitted to the data. This phase relationship is compatible with the fact that muscle twitches can, in good approximation, be modeled by the impulse response of a second-order, low-pass system.

The thin solid lines represent the phase leads of the action potential bursts with respect to the position of the wrist for both motor-unit A (crosses) and motor-unit B (circles). The thick solid line shows the phase relation of a second-order, low-pass system with a time constant of 70 ms (thick dashed line) followed by a second-order, low-pass system with a time constant of 115 ms. It shows that the phase lead of motor-unit A, with a relatively low isometric recruitment threshold (crosses) and which is representative for the majority of motor units, can be well fit by this fourth-order system. Motor-unit B (circles) clearly shows a larger phase lead with respect to the position of the wrist than motor-unit A.

During sinusoidal movements, several differences in motor-unit activity were found relative to the behavior observed during sinusoidal isometric contractions. First, the increasing phase lead of motor-unit activity for higher frequencies is larger than during isometric contractions (solid lines in figure 8.3). Due to the

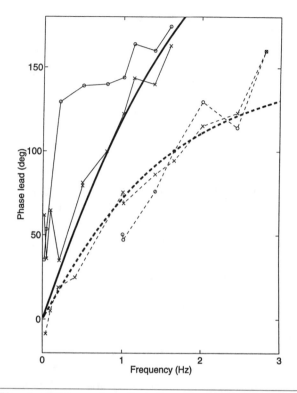

Figure 8.3   Phase lead of two motor units, indicated by crosses and circles, as a function of frequency for sinusoidally modulated isometric contractions with peak-to-peak amplitude of 30 N (dashed lines) and for sinusoidal movements with peak-to-peak amplitude of 20 cm (solid lines). The data represented by crosses are from a motor unit with a recruitment threshold, which was lower than that of the motor unit represented with circles. The thick dashed line represents the phase lead of a second-order, low-pass filter fitted to the motor unit data, obtained during the isometric contractions. The thick solid line represents the phase lead of a second-order, low-pass filter fitted to the motor unit data with a relatively low recruitment threshold in the movement condition.

second-order, mass-spring behavior of the arm mechanics, arm position progressively lags behind the force for higher frequencies (van Bolhuis et al. 1997).

In addition, motor units with a different isometric recruitment threshold may reveal a different phase relationship. The behavior of motor units with a relatively low isometric recruitment threshold is compatible with a model where the motor-unit firing rate (crosses in figure 8.3) is fed into a second-order, low-pass system modeling the twitch, followed by a second-order, low-pass system representing the arm biomechanics. Some motor units with a relatively high isometric recruitment threshold (circles in figure 8.3) tend to shift their activity period into the lengthening phase of the movement (figure 8.3). These motor units tend to have relatively high recruitment thresholds during isometric contractions and are sometimes hard to activate at all during isometric contractions.

Figure 8.4 shows the firing pattern of the same motor units, A and B, from figure 8.3 for a cyclic movement of the forearm at a frequency of 0.8 Hz. The top trace shows the position of the wrist in centimeters. Elbow flexion caused a movement of the wrist in the positive direction. The lower two panels show the firing rates of motor-units A and B, respectively. Vertical lines have been drawn at the position signal minimum (corresponding to the position of largest elbow extension) in order to illustrate the phase lead of the bursts in each cycle. The mean phase lead of the bursts with respect to the position signal of motor-unit A is 100 deg and that of motor-unit B is 140 deg.

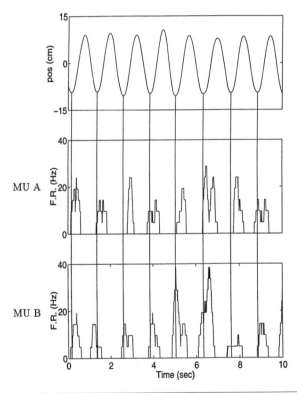

Figure 8.4   Position of the hand (upper panel) during sinusoidal flexion/extension movements of the arm at a frequency of 0.8 Hz. The top trace shows the position of the wrist in cm. Flexion of the elbow caused a movement of the wrist in the negative direction. The lower two panels show the firing rates of the motor units whose data were represented by crosses and circles, respectively, in figure 8.3. Vertical lines have been drawn at the minimal position signal (corresponding to largest flexion position of the elbow) in order to illustrate the phase lead of the bursts in each cycle. The mean phase lead of the bursts with respect to the position signal of motor unit MU A is 100 deg and that of motor unit MU B is 140 deg.

In summary, this section demonstrates two things. First, each muscle has several subpopulations of motor units, each with a different activation. Second, within the population of motor units, motor units with different isometric recruitment thresholds can be activated selectively. These results indicate that the number of degrees of freedom of all muscles acting across a joint is much larger than the number of muscles itself. Yet, the use of all these degrees of freedom is consistent and reproducible, both within and between subjects.

Conclusion

This chapter has given an overview of the various degrees of freedom available to the central nervous system to control the motor apparatus. However, when discussing any solutions for the redundancy problem, keep in mind that the genuine redundancy problem emerges when choosing how to make a movement. That is, the redundancy should be considered at the control level rather than at the peripheral mechanical level. The most conspicuous phenomenon is that the number of degrees of freedom is much larger than a superficial glance might suggest. Yet, the use of the large number of degrees of freedom is reproducible and consistent within and between subjects. What can be learned from this about motor control?

By studying skilled multijoint muscle coordination, several studies have suggested looking for constraints that may reduce the number of redundant degrees of freedom based on some sort of optimization. Historically, optimization approaches can be distinguished in two categories. The first approach is frequently referred to as inverse optimization. It tries to solve the load-sharing (redundancy) problems given skeletal kinetic data or muscle activation patterns (Pedotti et al. 1978; Theeuwen et al. 1996). It uses the data to extract the best solution, e.g., the best distribution of force contributed by all muscles, which is compatible with the kinetic or EMG data. The second approach is referred to as forward optimization. It tries to select the optimum (with respect to specific goals) coordination strategy out of a set of alternatives (Winters 1995; Zajac and Winters 1990). Usually, optimization is based on some criteria (e.g., minimizing total muscle force (Yeo 1976), optimizing force or position control, or optimizing movement efficiency (van Ingen Schenau 1989)). It then tries to predict the kinematic pattern of movement or the muscle activation pattern that follows from the optimization criteria. It has become clear during the last decade that optimizing one single criterion or set of criteria cannot explain the experimental material. This becomes evident from the observation that the muscle activation pattern differs depending on the instruction given to the subject (Tax et al. 1990a, b; Theeuwen et al. 1994a). This task-dependent muscle activation that depends on whether force or position is controlled provides additional support for the hypothesis that the redundancy problem should be considered and solved at the level of control variables available to the motor system. This implies that in looking for criteria that may impose constraints on the available degrees of freedom, at best, the criteria should be flexible and depend on the task. This was suggested by Loeb and Levine (1990).

Another aspect, which has been ignored in most studies, is that many so-called degrees of freedom are not independent. For example, the degrees of freedom related to joints in multijoint limbs are related to the degrees of freedom of a set of monoarticular and biarticular muscles. In particular, the specific action of biarticular muscles in multijoint movements might give rise to a specific activation pattern for monoarticular and biarticular muscles (as predicted by van Ingen Schenau 1989). This has consequences for the joint torques and, therefore, also for the limb kinematics and kinetics. In fact, the model proposed by Rosenbaum et al. (1995) hypothesizes that the problem of inverse kinematics is solved by selecting a weighted average of stored postures based on knowledge about the body and its interactions with the environment. It involves a combination of multiple degrees of freedom to come to a unique solution for postures.

Another model that should be discussed in the context of dealing with redundant motor systems is the λ-model (Feldman 1966, 1986). The original version of the λ-model dealt with the control of single-joint movements only. However, recently (Feldman and Levin 1995), an extension toward redundant multijoint systems has been proposed. According to this hypothesis, the redundancy could be solved by relating the threshold for muscle activation for each muscle to the actual joint angles. For example, in a situation where the actual joint angles correspond to the threshold lengths of the muscles, a movement could be made by activating the muscles. This would cause the change in threshold length to mirror the anatomical relationship between joint angles and the actual lengths of muscles spanning the joints. This would imply that muscles with large lever arms have larger changes in threshold during movements than muscles with small lever arms. This certainly provides a solution for the redundancy problem. However, it is not clear as to whether the predictions made by this model quantitatively agree with the observed muscle activation patterns in redundant muscle systems. It is also not clear how this hypothesis can explain the differences in muscle activation patterns in force and position control tasks, as observed by Tax et al. (1990a, b) and by Theeuwen et al. (1994a).

An important question, which cannot be answered yet, concerns which neuronal pathways are involved in the control of all available degrees of freedom. Evidently, the muscle activation pattern for isometric contractions and movements originates from commands from the central nervous system. However, it is not clear whether different task-dependent commands arise from motor cortex or whether different task-dependent muscle activations reflect variable contributions from different (e.g., motocortical or rubrospinal) pathways. Moreover, it is not clear whether the differences in the muscle activation during isometric contractions and during movements are exclusively due to different commands from the central nervous system or whether reflex actions may explain part of the differences. The activity of muscle spindles is quite different during isometric contractions and during concentric/eccentric contractions. Therefore, any differences in motor-unit activity in these conditions could be explained by additional input from muscle afferents, which do not project homogeneously to all motor units. In fact, some evidence indicates preferential synaptic projections to different types of motoneurons (Garnett and Stephens 1981; Heckman et al. 1994; Luscher and Vada 1989; Powers and Binder 1985). Tax et al. (1990b) measured motor-unit activity in the same physical conditions (i.e., force, position, and movement velocity were identical) but with different instructions to the subject. The differences in motor-unit activity in this experiment could not be attributed to additional input to the motoneuron pool only. By exclusion, they concluded that the observed differences in motor unit activity had to be attributed to changes in input from central stages. In summary, progress in understanding the control of the large number of available degrees of freedom is somewhat disappointing despite the large increase of experimental data since Bernstein.

References

Bernstein, N. (1967) The problem of co-ordination and localization. In *The Co-ordination and Regulation of Movements*. New York: Pergamon, 15–59.

Bodine, S.C., Roy, R.R., Eldred, E., Edgerton, V.R. (1987) Maximal force as a function of anatomical features of motor units in the cat tibialis anterior. *Journal of Neurophysiology*, 57: 1730–1745.

Botterman, B.R., Cope, T.C. (1988a) Maximum tension predicts relative endurance of fast-twitch motor units in the cat. *Journal of Neurophysiology*, 60: 1215–1225.

Botterman, B.R., Cope, T.C. (1988b) Motor-unit stimulation patterns during fatiguing contractions of constant tension. *Journal of Neurophysiology*, 60: 1198–1214.

Buchanan, T.S., Almdale, D.P.J., Lewis, J.L, Rymer, W.Z. (1986) Characteristics of synergic relations during isometric contractions of human elbow muscles. *Journal of Neurophysiology*, 56: 1225–1241.

Buchanan, T.S., Rovai, G.P., Rymer, W.Z. (1989) Strategies for muscle activation during isometric torque generation at the human elbow. *Journal of Neurophysiology*, 62: 1201–1212.

Cope, T.C., Pinter, M.J. (1995) The size principle: Still working after all these years. *News in Physiological Sciences*, 10: 280–286.

Donders, F.C. (1875) Ueber das Gesetz der Lage der Netzhaut in beziehung zu der Blickebene [On the law that describes the retinal position during gaze]. *Albrecht von Graefes Archieve fuer Ophthalmologie*, 21: 125–130.

Doorenbosch, C.A.M., van Ingen Schenau, G.J. (1995) The role of mono- and bi-articular muscles during contact control tasks in man. *Human Movement Science*, 14: 279–300.

Feldman, A.G. (1966) Functional tuning of the nervous system during control of movement or maintenance of a steady posture. II. Controllable parameters of the muscle. *Biophyzika*, 11: 565–578.

Feldman, A.G. (1986) Once more on the equilibrium-point hypothesis λ-model for motor control. *Journal of Motor Behavior*, 18: 17–54.

Feldman, A.G., Levin, M.F. (1995) The origin and use of positional frames of reference in motor control. *Behavioral and Brain Sciences*, 18: 723–806.

Flanders, M., Tillery, S.I.H., Soechting, J.F. (1992) Early stages in sensorimotor transformation. *Behavioral and Brain Sciences*, 15: 309–320.

Garnett, R., Stephens, J.A. (1981) Changes in the recruitment thresholds of motor units produced by cutaneous stimulation in man. *Journal of Physiology*, 311: 463–473.

Gielen, C.C.A.M., van Ingen Schenau, G.J. (1992) The constrained control of force and position in multi-link manipulators. *IEEE Transactions on Systems, Man and Cybernetics*, 22: 1214–1219.

Gielen, C.C.A.M., Neggers, B., Vrijenhoek, E., Flash, T. (1997) Arm position constraints during pointing and reaching in 3-D space. *Journal of Neurophysiology*, 78: 660–673.

Haslwanter, T. (1995) Mathematics of 3-dimensional eye rotations. *Vision Research*, 35: 1727-1739.

Haustein, W. (1989) Considerations of Listing's law and the primary position by means of a matrix description of eye position control. *Biol. Cybern.*, 60: 411–420.

Heckman, C.J., Miller, J.F., Munson, M., Paul, K.D., Rymer, W.Z. (1994) Reduction in postsynaptic inhibition during maintained electrical stimulation of different nerves in the cat hindlimb. *Journal of Neurophysiology*, 71: 2281–2293.

Henneman, E. (1981) Recruitment of motoneurons: The size principle. In *Motor Unit Types, Recruitment, and Plasticity in Health and Disease. Progress in Clinical Neurophysiology*. Vol. 9. Basel, Switzerland: Karger, 26–60.

Hepp, K. (1990) On Listing's law. *Communications in Mathematical Physics*, 132: 285–292.

Hore, J., Watts, S., Vilis, T. (1992) Constraints on arm position when pointing in three dimensions: Donders' law and the Fick gimbal strategy. *Journal of Neurophysiology*, 68: 374–383.

Howell, J.N., Fuglevand, A.J., Walsh, M.L., Bigland-Ritchie, B. (1995) Motor unit activity during isometric and concentric-eccentric contractions of the human first dorsal interosseus muscle. *Journal of Neurophysiology*, 74: 901–905.

Jørgenson, K., Bankov, S. (1971) Maximum strength of elbow flexors with pronated and supinated forearm. In *Medicine and Sport 6: Biomechanics*. Vol II. Basel, Switzerland: Karger, 174–180.

Jamison, J.C., Caldwell, G.E. (1993) Muscle synergies and isometric torque production: Influence of supination and pronation level on elbow flexion. *Journal of Neurophysiology*, 70: 947–960.

Kandou, T.W., Kernell, D. (1989) Distribution of activity within the cat's peroneus longus muscle when activated in different ways via the central nervous system. *Brain Research*, 486: 340–350.

Kernell, D., Hultborn, H. (1990) Synaptic effects on recruitment gain: A mechanism of importance for the input-output relations of motoneurone pools? *Brain Research*, 507: 176–179.

Lacquaniti, F., Soechting, J.F., Terzuolo, S.A. (1986) Path constraints on point-to-point arm movements in three dimensional space. *Neuroscience*, 17: 313–324.

Loeb, G.E. (1985) Motoneuron task groups: Coping with kinematic heterogeneity. *Journal of Experimental Biology*, 115: 137–146.

Loeb, G.E., Levine, W.S. (1990) Linking musculoskeletal mechanics to sensorimotor neurophysiology. In *Multiple Muscle Systems: Biomechanics and Movement Organization*. J.M. Winters, S.L.-Y. Woo (Eds.), New York: Springer Verlag, 165–181.

Luscher, H.-R., Vadar, U. (1989) A comparison of homogeneous and heterogeneous connectivity in the spinal monosynaptic reflex arc of the cat. *Experimental Brain Research*, 74: 480–492.

Miller, L.E., Theeuwen, M., Gielen, C.C.A.M. (1992) The control of arm pointing movements in three dimensions. *Experimental Brain Research*, 90: 415–426.

Nardone, A., Romano, C., Schieppati, M. (1989) Selective recruitment of high-threshold human motor units during voluntary isotonic lengthening of active muscles. *Journal of Physiology*, 409: 451–471.

Nardone, A., Schieppati, M. (1988) Shift of activity from slow to fast muscle during voluntary lengthening contractions of the triceps surae muscles in humans. *Journal of Physiology*, 395: 363–381.

Pedotti, A., Krishnan, V.V., Stark, L. (1978) Optimization of muscle force sequencing in human locomotion. *Mathematical Biosciences*, 38: 57–76.

Powers, R.K., Binder, M.D. (1985) Distribution of oligosynaptic group I input to the cat medial gastrocnemius motorneuron pool. *Journal of Neurophysiology*, 53: 497–517.

Rosenbaum, D.A., Loukopoulos, L.D., Meulenbroek, R.G.J., Vaughan, J., Engelbrecht, S.E. (1995) Planning reaches by evaluating stored postures. *Psychological Review*, 103: 28–67.

Segal, R.L. (1992) Neuromuscular compartments in the human biceps brachii muscle. *Neuroscience Letters*, 140: 98–102.

Segal, R.L., Wolf, S.L., de Camp, M.J., Chopp, M.T., English, A.W. (1991) Anatomical partitioning of three multi-articular human muscles. *Acta Anatomica*, 142: 261–266.

Smith, J.L., Betts, B., Edgerton, V.R., Zernicke, R.F. (1980) Rapid ankle extension during paw shakes: Selective recruitment of fast ankle extensors. *Journal of Neurophysiology*, 43: 612–620.

Soechting, J.F., Buneo, C.A., Herrmann, U., Flanders, M. (1995) Moving effortlessly on three dimensions: Does Donders' law apply to arm movements ? *Journal of Neuroscience*, 15: 6271–6280.

Straumann, D., Haslwanter, T., Hepp-Reymond, M.C., Hepp, K. (1991) Listing's law for eye, head and arm movements and their synergistic control. *Experimental Brain Research*, 86: 209–215.

Tax, A.A.M., Denier van der Gon, J.J., Erkelens, C.J. (1990a) Differences in coordination of elbow flexor muscles in force tasks and in movement tasks. *Experimental Brain Research*, 81: 567–572.

Tax, A.A.M., Denier van der Gon, J.J., Gielen, C.C.A.M., Kleyne, M. (1990b) Differences in central control of m. biceps brachii in movement tasks and force tasks. *Experimental Brain Research*, 79: 138–142.

Tax, A., Denier van der Gon, J.J., Gielen, C.C.A.M., Tempel, C.M.M. van den (1989) Differences in the activation of m. biceps brachii in the control of slow isotonic movements and isometric contractions *Experimental Brain Research*, 76: 55–63.

ter Haar Romeny, B.M., Denier van der Gon, J.J., Gielen, C.C.A.M. (1982) Changes in recruitment order of motor units in the human biceps muscle. *Experimental Neurology*, 78: 360–368.

ter Haar Romeny, B.M., Denier van der Gon, J.J., Gielen, C.C.A.M. (1984) Relation of the location of a motor unit in human biceps muscle and its critical firing levels for different tasks. *Experimental Neurology*, 85: 631–650.

Theeuwen, M., Gielen, C.C.A.M., van Bolhuis, B.M. (1996) Estimating the contribution of muscle to joint torque based on motor-unit data. *Journal of Biomechanics*, 29: 881–889.

Theeuwen, M., Gielen, C.C.A.M., Miller, L.E. (1994a) The relative activation of muscles during isometric contractions and low-velocity movements against a load. *Experimental Brain Research*, 101: 493–505.

Theeuwen, M., Gielen, C.C.A.M., Miller, L.E., Doorenbosch, C. (1994b) The relation between surface EMG and recruitment thresholds of motor units in human arm muscles during isometric contractions. *Experimental Brain Research*, 98: 488–500.

Theeuwen, M., Miller, L.E., Gielen, C.C.A.M. (1993) Are the orientations of the head and arm related during pointing movements? *Journal of Motor Behavior*, 25: 242–250.

Thomas, C.K., Bigland-Ritchie, B., Westling, G., Johansson, R.S. (1990) A comparison of human thenar motor-unit properties studied by intraneural motor-axon stimulation and spike-triggered averaging. *Journal of Neurophysiology*, 64: 1347–1351.

Tweed, D., Vilis, T. (1987) Implications of rotational kinematics for the oculomotor system in three dimensions. *Journal of Neurophysiology*, 58: 832–849.

Tweed, D., Vilis, T. (1990) Geometric relations of eye position and velocity vectors during saccades. *Vision Research*, 30: 111–127.

van Bolhuis, B.M., Medendorp, W.P., Gielen, C.C.A.M. (1997) Recruitment and firing rate of motor units during isometric contractions and movements in flexion/extension direction at various frequencies. *Experimental Brain Research*, 107: 120-130.

van Ingen Schenau, G.J. (1989) From rotation to translation: Constraints on multi-joint movements and the unique action of bi-articular muscles. *Human Movement Science*, 8: 865–882.

van Zuylen, E.J., Gielen, C.C.A.M., Denier van der Gon, J.J. (1988) Coordination and inhomogeneous activation of human arm muscles during isometric torques. *Journal of Neurophysiology*, 60: 1523–1548.

von Helmholtz, H. (1925, 1867) *Handbuch der Physiologischen Optik*. 1st ed., Vol 3. Hamburg, Germany: Voss (Third edition translated into English by J.P.C. Southall (1925) as *Treatise on Physiological Optics*. Rochester, NY: *Optical Society of America.*)

Winters, J.M. (1995) How detailed should muscle models be to understand multi-joint movement coordination? *Human Movement Science*, 14: 401–442.

Yeo, B.P. (1976) Investigations concerning the principle of minimal total muscular force. *Journal of Biomechanics*, 9: 413–416.

Zajac, F., Winters, J.M. (1990) Modeling musculoskeletal movement systems: Joint and body-segment dynamics, musculotendon actuation and neuromuscular control. In *Multiple Muscle Systems: Biomechanics and Movement Organization*. J.M. Winters, S.L.-Y. Woo (Eds.), New York: Springer Verlag, 121–148.

Zajac, F.E., Faden, J.S. (1985) Relationship among recruitment order, axonal conduction velocity, and muscle-unit properties of type-identified motor units in cat plantaris muscle. *Journal of Neurophysiology*, 53: 1303–1322.

Zengel, J.E., Reid, S.A., Sypert, G.W., Munson, J.B. (1985) Membrane electrical properties and prediction of motor-unit type of medial gastrocnemius motoneurons in the cat. *Journal of Neurophysiology*, 53: 1323–1344.

9

Abnormal Muscle Synergies in Hemiparetic Stroke: Origins and Implications for Movement Control

W. Zev Rymer, Jules Dewald,
P.T. Joseph Given, and Randall Beer
Rehabilitation Institute of Chicago, Chicago, IL, U.S.A.

The great Russian experimentalist and theorist Bernstein articulated a coherent vision of the relationship between neural mechanisms and the mechanics of human voluntary movement. He recognized the important interactions between the neural command signals and the eventual mechanical outcome of such commands. The ultimate mechanical effect of a given set of motor commands is not invariant but depends greatly on the state of the periphery at the time that these neural commands arrive. Bernstein was also one of the earliest to emphasize the potential advantages of neural strategies that stereotypically coupled movements at different joints as a way to diminish the computational burden associated with controlling voluntary movements in multisegmented limbs. These stereotypical movement patterns were called synergies. They must be distinguished from the simpler muscle synergies described by Sherrington (1910), which referred to actions of muscles with similar mechanical effects, often acting across a single joint.

While the origins and functional impact of such movement synergies are still actively debated in nondisabled human subjects, the findings in neurologically disabled subjects are potentially easier to characterize and to understand. The synergies that emerge in subjects with neurological injury, such as stroke or other brain injuries, are both more distinct in character and more limited in number than is evident in nondisabled subjects.

To explain further, when nondisabled subjects are asked to acquire a target in the visible workspace manually, they can use a variety of strategies with comparably effective results. For point to point movements of the hand, for example, a subject will usually follow a straight-line trajectory with the hand. Plots of the tangential velocity of end point motion against time describe bell-shaped, essentially symmetrical curves. Under these conditions, the angular velocities at the shoulder and elbow usually show a strong linear correlation with a slope near unity. However, when subjects are asked to deviate intentionally from such a straight-line path and transit through a specific "via" point, they can readily produce curvilinear but smooth trajectories. Under these conditions, the tangential velocity profiles no longer show a simple, bell-shaped form. The angle to angle plots of shoulder and angle velocities also deviate from simple linearity. It appears that subjects can use one of an almost infinite set of strategies when instructed. However, they will settle on a relatively standard approach when allowed to choose freely.

Furthermore, when nondisabled subjects are asked to generate forces isometrically in different directions in the upper extremity, for example in a plane through the wrist orthogonal to the long axis of the forearm, the magnitude of EMG activity in both elbow and shoulder muscles continuously varies as a function of force direction (Buchanan et al. 1986). When EMG magnitude in one muscle is plotted against EMG magnitude in another muscle coactivated in the same tasks, no orderly relation usually occurs between EMG magnitudes for any arbitrary muscle pair. In other words, there is no strong evidence that simpler activation rules exist for different muscle pairs under such conditions.

The situation is quite different in many types of brain injury, especially stroke. Presumably because stroke consistently affects cerebral cortex and subcortical structures supplied by the middle cerebral artery, a relatively consistent pattern often occurs in the clinical features of the motor deficit that arises in the majority of hemiplegic stroke subjects.

Abnormal Synergies in Subjects With Hemiparetic Stroke

A number of investigators have described the alterations in voluntary movement, gait, and posture in subjects with hemiparetic stroke in terms of these changed patterns. One of the most extensive descriptions was provided by Brunnstrom (1970), who divided the typical patterns of upper limb posture and movement into flexor and extensor synergies (see table 9.1). These synergies primarily describe the postural abnormalities in a quiescent limb. They also characterize, at least in part, the typical relationships between kinematics at adjacent joints, particularly joint angle displacements and angular velocities, during normal, voluntary movements. These movement pattern disturbances are not simply descriptions of common choices

Table 9.1   Upper Limb Synergies in Hemiparetic Stroke

Extension synergy	Flexion synergy
Shoulder girdle	*Shoulder girdle*
Protraction	Retraction
Adduction	Abduction to 90°
Internal rotation	External rotation
Elbow	*Elbow*
Extension	Flexion
Pronation	Supination
Wrist extension	*Wrist flexion*
Finger extension	*Finger flexion*

Adapted from Brunnstrom (1970).

made by hemiparetic stroke subjects. They appear to represent legitimate, rigid constraints that reduce the choices available for patients to complete a designated voluntary movement task.

While these synergies have provided an important target for rehabilitation interventions, the origins of these stereotypical muscle activation patterns remain largely unexplored. The authors believe that an experimental investigation of these synergies could potentially be very important. An investigation may increase the understanding of the neural mechanisms underlying the motor impairments in stroke and may also shed light on the origins of muscular synergies in unimpaired subjects.

Origins of Muscular Synergies in Brain Injury

Three hypotheses exist regarding the potential origins of the abnormal synergies that arise in hemiparetic stroke. The first hypothesis is that the synergies are an expression of altered sensitivity of segmental interneuronal pathways to peripheral sensory inputs. The second hypothesis is that synergies reflect anatomical constraint inherent in residual descending pathways. The third hypothesis is that synergies result from stroke-related control reorganization.

In relation to the first hypothesis, the activity and spatial organization of interneuronal pathways now dictate the muscle activation patterns in both posture and voluntary movement. The principal segmental reflex pathway thought to underlie these synergies is the flexion-withdrawal reflex or flexor reflex. This well-known reflex has been described extensively by Sherrington and investigated extensively by others (Eccles and Lundberg 1959; Holmqvist and Lundberg 1959; Holmqvist and Lundberg 1961). It includes characteristic widespread and sequential activation of limb flexors induced by either noxious or nonnoxious stimulation of cutaneous and other deep afferent pathways. The proposed causal

sequence inducing abnormal movement synergies is that segmental interneuronal pathways, such as those involved in flexion withdrawal, are released from
descending inhibitory control. They are now free to influence the excitability of
spinal motoneurons to an unprecedented degree. This release would result from
one of two events. First, the descending spinal inhibitory pathways (especially
reticulospinal pathways) traversing the dorsolateral funiculus of the spinal cord
could be physically interrupted. Second, a loss of descending excitation could
affect reticulospinal neurons in the brain stem, causing their excitability to drop.
As a consequence, the threshold for spinal interneuronal activation declines.
Relatively modest sensory stimuli, such as those arising from innocuous stimulation of skin or even tonic discharge arising from cutaneous thermal receptors or
mechanoreceptors, may promote interneuronal excitation and, consequently,
preferential excitation of flexor motoneurons.

Substantial evidence favors augmented excitability of flexion-reflex pathways
in many lesions of the neuraxis. Shahani and colleagues (Fisher et al. 1979) have
demonstrated preferential excitation of flexion-reflex pathways in stroke. Extensive descriptions in the literature document augmented flexion-reflex responsiveness in animal models of spinal cord injury. If the hypothesis about augmented
flexion-reflex sensitivity is correct, the expected threshold for activation of flexion-
reflex pathways would be reduced in subjects with brain or spinal cord injury. The
spatial and temporal patterns of muscle activation in the abnormal synergies of
hemiparetic stroke would be expected to be comparable to those displayed in the
flexion-withdrawal response. The authors' findings about this do not clearly support
such a major role for flexion-reflex pathways in promoting the abnormal postural
synergies in hemiparetic stroke.

To characterize spatial patterns of voluntary muscle activation, nondisabled human
subjects were asked to generate isometric forces against a multiaxial load cell placed
at the wrist (see figure 9.1). The load cell was attached at the wrist via a polymer cast.
The forearm was held horizontal, with the elbow set to 90 degrees flexion. The
shoulder angle is 45 degrees below the horizontal. Under these conditions, a typical
muscle activation pattern became visible. For any direction in which a given muscle
as activated, muscles became more active for larger isometric loads. The level of
electromyographic activity in any given direction was essentially in proportion to
the load magnitude.

The more interesting data emerged during an examination of the spatial tuning
of muscle activity (see figure 9.2). For elbow flexors, the maximum EMG activity
was recorded when subjects generated forces in the optimum mechanical direction.
For example, brachialis and biceps brachii, both elbow flexors, were maximally
activated in (or near) pure flexion, whereas the triceps brachii muscle was maximally activated near pure extension. (The electrical activity peak varied slightly for
the different muscle heads of both biceps and triceps. It also varied with the initial
position of the shoulder. This reflects the fact that each of these muscle groups
traverses both the elbow and shoulder joints.) Less predictably, substantial electromyographic activity remained even when subjects generated forces quite far
away from the path of angular motion of the joint.

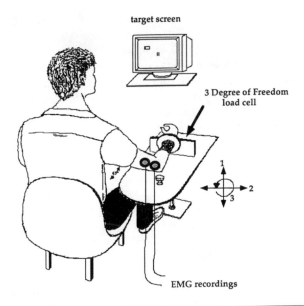

target screen

3 Degree of Freedom load cell

EMG recordings

Figure 9.1 Experimental setup. Subject sits with forearm in cast, attached to a multi-axial load cell. Subject views a display screen that indicates the magnitude and direction of voluntary force generation. Surface EMGs are recorded from biceps brachii, triceps brachii, brachioradialis, and shoulder muscles.

TRICEPS BRACHII MUSCLE POLAR PLOT

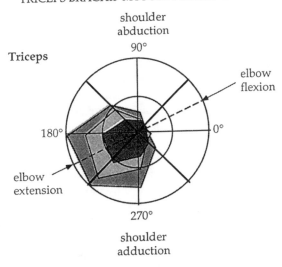

shoulder abduction
90°

Triceps

elbow flexion

180°   0°

elbow extension

270°

shoulder adduction

Figure 9.2 Polar plots of EMG activity in the triceps brachii as a function of torque magnitude and direction. EMG scales roughly with increasing load for multiple torque directions. Peak EMG corresponds with the plane of motion of the elbow joint. The outer circle reflects maximum EMG, and the inner circle reflects 0.5 of maximum.

As also illustrated in figure 9.2, significant electromyographic activity extended as far as 60 degrees on either side of the flexion-extension axis. Considerable EMG activity was recordable almost to 90 degrees in either direction (Buchanan et al. 1986). While these EMG Spatial tuning curves are not routinely symmetrical, they have been described quite accurately using cosine functions (Flanders and Soechting 1990). Similar patterns emerged for virtually all of the muscles that may contribute to force generation at the wrist. This occurred even when the muscles are located at a distance from the point of interaction with the load cell, such as those around the shoulder and neck (e.g., trapezius and deltoid).

These spatial tuning curves, often termed polar plots, have two uses. First, they provide a template of the spatial patterns of muscle activation in nondisabled limbs. Such templates are extremely helpful for characterizing and potentially understanding the patterns of muscle use observed in neurologically impaired subjects. Second, they allow systematic descriptions of the relations between different muscles, the muscular synergies, as a function of load magnitude and direction. (Here, the term synergies characterizes the quantitative relations between EMG activity in different muscles in the limb rather than the movements that result from such activities.) In most nondisabled subjects, the pattern of muscle coactivation recorded over all directions is quite scattered. There appears to be no simple rule by which different muscles are activated. In contrast, for subjects with hemiparetic stroke, the spatial profiles of muscle activation are often much broader. Consequently, the EMG-EMG plots, which are used as an index of muscle synergy, also become somewhat simpler (see the next section).

Figure 9.3 illustrates the altered spatial patterns of muscle activation in an impaired and contralateral extremity of a moderately impaired subject with hemiparetic stroke. Three characteristic differences distinguish these plots from those of the nondisabled subject or even those recorded from the contralateral limb. They can be summarized as differences in the angle or orientation of the peak EMG, in the angular range over which EMG is active, and in the angular locus of the mean EMG activity (Bourbonnais et al. 1989; Dewald et al. 1995).

In moderately severe hemiplegic stroke, peak elbow flexor EMG activity (biceps brachii, brachialis, and brachioradialis) routinely shifts upward toward humeral elevation (abduction) rather than toward the flexion and extension plane, the only plane in which these muscles can exert functionally useful moments. This means that when a subject is asked to generate voluntary flexion force at the elbow, the subject's efforts are relatively unsuccessful. However, when asked to abduct the paretic arm and shoulder, concurrent substantial excitation of elbow flexor muscles may occur. Conversely, when a subject is asked to extend the elbow, relying primarily on triceps brachii, there may be little activation in the direction of pure extension but significant extensor excitation when the limb is adducted. This results in a downward rotation of peak EMG activity in the triceps toward adduction or humeral depression. In addition, an increased angular range of muscle

POLAR PLOTS
UNIMPAIRED UPPER LIMB

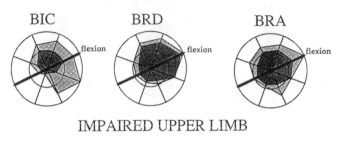

IMPAIRED UPPER LIMB

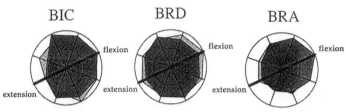

Figure 9.3   Polar plots are compared for impaired (contralesional) and unimpaired, ipsilesional limbs during controlled isometric loading for 8 different directions. EMGs were recorded from biceps (BIC), brachialis (BRA), and brachioradialis (BRD) at two load levels.

excitation and a shift in the center of mass of the polar plot often occur. This also results in redirection of the spatial locus of activation, referenced to the plane of the polar plot.

Figure 9.4 illustrates one approach toward characterizing the spatial patterns of muscle activation in paretic and unimpaired limbs. Here, the EMG activity in one muscle is plotted against the mean activity in a neighboring muscle for all directions in which both muscles are active. In the unimpaired limb, the scatter plot relating activity in biceps brachii and anterior deltoid shows no discernible relation, and the correlation coefficient is predictably very low. For the paretic side, a near linear relation exists between the two muscles, with a correlation coefficient of 0.81.

These differences indicate a substantial repertoire of choices on the unimpaired side, so that EMG-EMG plots are based an array of descending commands. In contrast, on the paretic side, the available choices for descending activation are sharply reduced, causing rather simpler patterns of coactivation to emerge. The authors believe these findings to be the functional expression of the abnormal synergies reported in stroke.

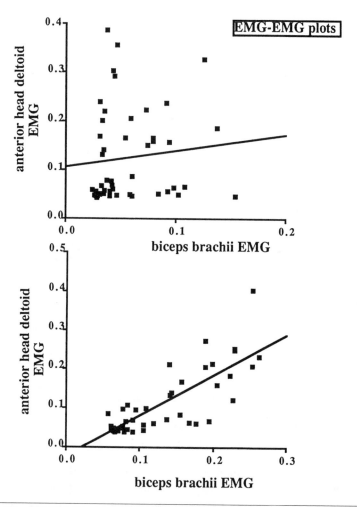

Figure 9.4 EMG-EMG plots for two muscle pairs (biceps and deltoid) drawn from both of the upper extremities of a stroke patient. Plots are based on EMGs collected for all directions tested.

Hypothesis I: Abnormal Flexor Synergies in Hemiparetic Stroke are Mediated via the Flexion-Withdrawal Reflex Response

These spatial patterns in stroke provide an interesting and valuable framework against which the spatial patterns of flexion-withdrawal responses may be compared. In order to derive quantitative comparisons between the spatial and temporal profiles of muscle activation elicited in the course of flexion withdrawal, the authors generated flexion-withdrawal responses in unimpaired limbs in either nondisabled controls or on the contralateral side of hemiparetic stroke subjects. We examined the spatial patterns of muscle coactivation that were evoked and the timing relations. These EMG patterns are difficult to compare fully with those elicited voluntarily. It is not possible to generate systematic tuning in elbow force over the full range of directions used in the course of voluntary isometric contractions. Nonetheless, the patterns of electromyographic activity in key arm and shoulder muscles for both tasks (i.e., voluntary contraction and flexion withdrawal) were compared.

Flexion-withdrawal responses were elicited by electrical stimulation of the palmar surface of the fingers using ring electrodes. Stimuli were delivered as a pulse train of duration and of a stimulus intensity sufficient to induce involuntary muscle activation. Stimulus intensity was increased progressively until a flexion-withdrawal response was induced. Induction of this response was manifested by the generation of flexion torque at the load cell coupled with systematic and widespread activation of elbow and shoulder flexor muscles.

In brief, findings for the flexion-withdrawal experiments were threefold. First, although large scale activation of elbow and shoulder flexors occurred during both voluntary activation and flexion withdrawal, the spatial patterns of activity significantly differed between the two. This indicates that subjects were probably not relying on comparable interneuronal and motoneuronal circuits on the two sides. Second, the threshold for activation of flexion reflexes in the paretic limb was not systematically reduced. If the flexion posture and preferred movement were a consequence of heightened responsiveness of flexion-withdrawal pathways, activation of flexion withdrawal responses would be expected to take place at lower stimulus intensities than were required in unimpaired limbs. This hypotheses was not sustained by experimental observations. Specifically, the stimulus intensity required to elicit the flexion-withdrawal response was no lower than and was often somewhat higher than that required in nondisabled individuals. Third, if the voluntary muscle activation pattern and the flexion reflex were similarly mediated, the temporal sequence of muscle activation from shoulder to hand should be similar in normal and paretic limbs. Instead, the typical pattern of muscle activation in the flexion-withdrawal response, in which proximal muscles at the shoulder are activated before distal arm muscles, was not evident. All active muscles appeared to be excited almost simultaneously.

Apparently, it is not possible to equate the abnormal synergy in hemiparetic stroke directly with an exaggerated flexion-withdrawal response. Since flexion-withdrawal responses also changed in the impaired limb, however, comparisons should be made cautiously. The very nature of the flexion-withdrawal response appears to be modified as a result of the stroke-induced brain injury.

To date, no similar studies about potential segmental origins of extensor synergies have been attempted. Perhaps this is because such synergies are much more difficult to elicit. They may be more readily induced by excitation of the lower extremities where weight-bearing extensor reflexes, such as the extensor thrust, are well known.

Hypothesis II: Abnormal Synergies Arise Because of Anatomical Constraints Inherent in Descending Brain Stem Pathways

In essence, this hypothesis argues that abnormal postures and movements arising as a result of stroke-induced brain injury, for example, result from forced reliance on simpler "ventromedial" spinal pathways. These become the primary means for distributing motor commands from cortex to the spinal cord.

To amplify this idea further, ventromedial spinal pathways, such as the vestibulospinal, reticulospinal, and tectospinal pathways, relay information originating in intact or partially intact cerebral locations. Many of these pathways show few, direct connections to spinal motoneurons. They are largely distributed to interneurons innervating motoneurons driving axial proximal limb muscles. Furthermore, anatomical studies indicate that the pattern of terminal fiber branching is often very diffuse, extending over multiple spinal cord segments. This would mean that the excitatory effects at any given spinal level would be relatively modest, promoting weakness. However, they would also be associated with obligatory coactivation of muscles innervated from quite separate parts of the spinal cord.

At the present time, relatively little evidence is available to support the direct involvement of these various brain stem pathways. Recent studies have examined the patterns of muscle activation induced as part of a startle response generated by a loud noise. Rothwell et al. (1994) have argued that such a startle response reflects the distribution patterns of reticulospinal pathways to brain stem-innervated structures and potentially beyond. While such effects are difficult to study because the response cannot be repeated frequently without significant adaptation, the spatial patterns elicited by loud sounds do not appear to resemble the synergies observed in brain-injured subjects. Parenthetically, similarities may occur in the activity pattern observed in the startle response elicited by infants and the distribution of muscle activation in humans. However, this possibility is purely hypothetical at this time.

Hypothesis III: Abnormal Synergies Result From Cortical Reorganization

A third and potentially most intriguing mechanism underlying altered spatial patterns of muscle organization in stroke is that these abnormal muscle synergies emerge as an obligatory and adverse consequence of the functional reorganization of cortex that follows stroke-induced cerebral lesions. For example, lesions of the middle cerebral artery territory produce the most severe adverse effects in the distal portions of the upper extremity. This is especially manifested in hand movements and wrist extension.

By following lesions either to peripheral sensory structures or to the cerebral cortex, relatively swift and sometimes extensive functional reorganization of cortex occurs (Cohen et al. 1993). This may result in adjacent regions of cortex assuming the control of spinal motoneuron pools and other segmental circuits, and of their appropriate muscles.

For lesions of the middle cerebral artery territory, the adjacent portions of cortex that could be responsible would include those responsible for controlling the activity in the shoulder and, to some extent, neck and axial muscles. A subject desiring to activate elbow flexors may be unable to do so directly but could achieve the desired forces by commanding shoulder elevation and abduction or some similar movements. Although the evidence here is quite oblique, it does appear that subjects do use this type of strategy when attempting to activate elbow flexors on command.

This hypothesis has no direct confirmation as yet. Potentially, a number of methods are available to explore this possibility further. For example, the use of functional magnetic resonance imaging should allow researchers to observe whether activation occurs in the damaged areas coupled with significant activation of shoulder areas, inducing significant elbow flexion torque. It may also be possible to map the peripheral muscular distribution of cortical projections systematically using localized magnetic stimulation applied to the scalp surface.

Conclusion

This chapter has identified three possible mechanisms that could contribute to the altered spatial and temporal patterns of muscle activation associated with the severe motor impairments arising in hemiparetic stroke. Although difficult to validate, these hypotheses are potentially distinguishable using experimental interventions, some of which have been outlined in this chapter. Furthermore, it is very likely that no single mechanism is responsible but that several combine to induce the adverse effects. Indeed, all three disturbances may coexist simultaneously so that altered reflex pathways, reliance on primitive ventromedial pathways, and cortical reorganization might each contribute to the total clinical picture.

Several factors make separating the various options difficult. In addition to the aforementioned cortical plasticity, adaptive and even pathological changes may occur in the spinal cord, muscles, and peripheral connective tissues. Muscle may atrophy, even in supraspinal lesions. Associated changes often occur in the muscle fiber type and in the amount and distribution of muscle connective tissues. Profound alterations also occur in the excitability state of segmental reflex pathways, including a significant increase in flexor reflex responsiveness. Reorganization of segmental circuitry resulting from the denervation of spinal motoneurons may also occur.

References

Bourbonnais, D., Vanden Noven, S., Carey, K.M., Rymer, W.Z. (1989) Abnormal spatial patterns of elbow muscle activation in hemiparetic human subjects. *Brain*, 112: 85–102.

Brunnstrom, S. (1970) *Movement Therapy in Hemiplegia*. New York: Harper & Row.

Buchanan, T.S., Almdale, D.P.J., Lewis, J.L., Rymer, W.Z. (1986) Characteristics of synergic relations during isometric contractions of human elbow muscles. *Journal of Neurophysiology,* 56: 1225–1241.

Cohen, L.G., Brasil-Neto, J.P., Pascual-Leone, A., Hallett, M. (1993) Plasticity of cortical motor output organization following deafferentation, cerebral lesions, and skill acquisition. *Advances in Neurology,* 63: 187–200.

Dewald, J.P.A., Pope, P.S., Given, J.D., Buchanan, T.S., Rymer, W.Z. (1995) Abnormal muscle coactivation patterns during isometric torque generation at the elbow and shoulder in hemiparetic subjects. *Brain,* 118: 495–510.

Eccles, R.M., Lundberg, A. (1959) Supraspinal control of interneurons mediating spinal reflexes. *Journal of Physiology (London)*, 147: 565–584.

Flanders, M., Soechting, J.F. (1990) Arm muscle activation for static forces in three-dimensional space. *Journal of Neurophysiology,* 64: 1818–1837.

Fisher, M.A., Shahani, B.T., Young, R.R.(1979) Electrophysiologic analysis of the motor system after stroke: The flexor reflex. *Archives of Physical Medicine and Rehabilitation,* 60: 7–11.

Holmqvist, B., Lundberg, A. (1959) On the organization of the supraspinal inhibitory control of interneurons of various spinal reflex arcs. *Arch. ital. Biol.,* 97: 340–356.

Holmqvist, B., Lundberg, A. (1961) Differential supraspinal control of synaptic actions evoked by volleys in the flexion reflex afferents in alpha motoneurons. *Acta Physiol. Scand.,* 54: 1–51.

Rothwell, J.C., Vidailhet, M., Thompson, P.D., Lees, A.J., Marsden, C.D. (1994) The auditory startle response in progressive supranuclear palsy. *Journal of Neural Transmission,* 42: 43–50.

Sherrington, C.S. (1910) Flexion-reflex of the limb, crossed extension reflex and reflex stepping and standing. *Journal of Physiology (London),* 40: 28–121.

10

CHAPTER

From Bernstein's Physiology of Activity to Coordination Dynamics

J.A. Scott Kelso

Program in Complex Systems and Brain Sciences,
Center for Complex Systems, Florida Atlantic University,
Boca Raton, FL, U.S.A.

It is an honor to be invited to contribute to a volume recognizing *Bernstein's Traditions in Motor Control*. The word "tradition" stems from the Latin *traditio*, a handing down, as in the handing down of opinions, doctrines, rites and so forth, as from parents to children, or from one age to another. It is impossible to do justice to the remarkable range of insights passed on by Nikolai Aleksandrowitsch Bernstein (1896–1966) to students of motor function. The transmission of Bernstein's ideas, and the fact that they are now so widely recognized, is due, in this country, to the seminal efforts of people like Peter Greene, Curtis Boylls, Michael Arbib and Michael Turvey, as well as the members of the famous Moscow school, notably Israel Gelfand, Viktor Gurfinkel, Mark Shik, Misha Berkinblit, Anatol Feldman, and others. Additional credit is due to John Whiting, then in Amsterdam, for reprinting Bernstein's best known book *The Coordination and Regulation of Movements* and arranging for cogent commentary (Whiting 1984). This surely helped perpetuate Bernstein's name and galvanized the field.

In this paper I shall try to do the following: First, I will remind the reader briefly of some of the major themes that Bernstein has passed on to us and that have influenced mine and others' research since the 1970s. It is, of course, impossible to be inclusive in this regard; many of the gaps will no doubt be filled by other contributors in the present volume. Second, I will put these themes in a modern context, here a self-consistent theory of coordination dynamics that has emerged over the last decade or so. I will try to show that this rather primitive dynamical scheme embraces and formalizes several of Bernstein's concepts (in a way that I hope would have pleased him), as well as yields new insights of its own that have been subject to extensive experimental test. My main goal, however, is to show how the strategy of coordination dynamics can be elaborated in a number of new directions, both with respect to the brain and aspects of perceptual-motor behavior not much considered by Bernstein. In this way, I hope to extend and, in particular, challenge existing theoretical models of coordination.

Bernstein's Coordination of Motor Function: "The Physiology of Activity"

Here I remind the reader briefly of some of Bernstein's great themes (Bernstein 1967). First, Bernstein's physiology of activity dealt with goal-directed acts, not reflexes like paw withdrawal or knee jerks (Bernstein 1967, page144). In modern times, one might think this statement trivial. But in days when psychology was dominated by associationist thinking and Pavlov's physiology of conditioned and unconditioned reflexes ruled the roost, Bernstein's stress on the active, not reactive action of the organism on its environment, can scarcely be overstated. Second, for the production of any purposeful act, a "most important biomechanical premise" was the mastery of the enormous number of degrees of freedom possessed by "the motor organ," and their conversion into a controllable system. This means, of course, that whatever form such "mastery" takes, it must be contingent on the goals of the organism. Third, Bernstein recognized that a solution to the so-called degrees of freedom problem lay in the presence of widespread synergies. Whenever he looked closely at a complex motor act like chiseling or locomotion, Bernstein observed precise intercorrelations among a multiplicity of joints in space and time. Fourth, all movements were seen to possess a clearly defined *structure* or topology, as well as a *metric*. Bernstein viewed these aspects of coordination as quite independent of each other, as when a person's handwriting remains the same (the topology) regardless of whether it is large or small (mere changes in the metrical aspect). Fifth, and relatedly, Bernstein believed that contained in the nervous system was ". . . Some form of brain trace corresponding to the whole process of the movement in its entire course in time" (Bernstein 1967, page 37). Such hypothesized formulae or engrams were conceived to be abstract, high level, and not related to muscle structure per se. Handwriting again was offered as a salient example of so-called motor equivalence (see also Wiesendanger in this volume). Finally, for Bernstein, coordination was based on a "determinate *organization* (his emphasis) among common elements." It was very much this *"organization of the control* of the motor apparatus" (Bernstein's emphasis) that defined coordination.

From Bernstein to Coordination Dynamics

Bernstein was fascinated about the relation between the sciences of cybernetics and physiology, viewing movement coordination as a rich source of material for building cybernetic analogs and models. Yet in his later writings, he raised the intriguing possibility that "the honeymoon" between cybernetics and physiology may be coming to an end. New developments in mathematics and physics (e.g., of stochastic, anti-entropic processes, tendencies toward structuration in open systems, etc.) might not bring about a divorce in his lifetime, but the seeds, Bernstein speculated, were there. In this respect, Bernstein was captivated by, for want of a better word, boundaries—between the natural and the artificial, the animate and inanimate—a theme that permeates contemporary science (and science fiction). A chief puzzle for Bernstein was where and how the organism demarcates the boundary between active and reactive processes (Bernstein 1967, page 180), an issue that I shall return to later on, because recent evidence suggests organisms live right on this boundary.

Coordination dynamics has its ancestry in the Bernsteinian notion of coordinative structure or functional synergy, highly evolved task-specific ensembles of neuromuscular and skeletal components constrained to act as a single unit (Boylls 1975; Edelman 1987; Turvey et al. 1978). Evidence for functional synergies has accumulated steadily. A conspicuous signature of a functional synergy or coordinative structure is relative timing, seen not only in actions such as locomotion and posture, but in voluntary discrete movements of the upper limbs (Kelso et al. 1979). But how are functional synergies formed and how do they change? Inspired, as was Bernstein, by developments in contemporary theories of pattern formation in physics (Haken 1977; Kugler et al. 1980 for review) basic insights into these questions were obtained from experimental observations of spontaneous pattern formation and change in rhythmic interlimb movements (Kelso 1981, 1984; Kelso and Scholz 1985). Coordination dynamics emerged, in its most elementary form, from a mapping of these observations onto the concepts of synergetics, using the mathematical tools of nonlinear dynamical systems (Haken et al. 1985). These first steps were later elaborated into a fully stochastic treatment (Schöner et al. 1986), and there have been many empirical and theoretical developments of coordination dynamics since, only a few of which I can mention here (for recent reviews see Haken 1996; Kelso 1995; Swinnen et al. 1994; Turvey 1994). The basic concepts of coordination dynamics are the following:

1. Coordination is hypothesized to be a self-organized, synergetic process. As such, it is a generic feature of living things, exhibiting the phase space reduction of coupled, nonlinear dynamical systems. In the prototypical case of bimanual coordination, for example, the phase space is reduced from four (positions and velocities

for each hand) to one, the relative phase between the hands. The resulting lower-dimensional dynamics may, nevertheless, exhibit complicated behavior, such as the coexistence of several coordination states (multistability), bifurcations, hysteresis, intermittency, etc. depending on where the system lives in the space of its parameters.

2. Essential collective or coordination variables that characterize states of coordination are task- or function-specific. Such coordination variables, though physically realized and measurable, are informational in nature (Kelso 1994b). That is, they represent the coupling relation between (usually) material things, such as neurons, other kinds of cells, body parts, organisms, and so forth, during coordinated functions. Such (meaningful) informational coupling constitutes an essential difference between self-organization in the living and the nonliving. Couplings in the latter are mediated by physical forces with the usual dimensions of mass, length, and time, whereas one of the key variables identified in studies of coordination, relative phase, is strictly a mathematical (and, as I say, meaningful informational) quantity. Take studies of the moving room as a case in point (Lee and Lishman 1975). The moving room can be defined in terms of physical quantities such as frequency and amplitude. Likewise, the postural state of an organism can be defined in terms of physical quantities such as center of mass. Such descriptors are quite adequate for the components alone, but not when the components are functionally coupled. Experiments show that under certain conditions the room and the organism are coordinated together, and this coordination is captured best by informational quantities such as the phase relation and its dynamics (Dijkstra et al. 1994). This suggests a connection between optic flow and phase variables (Kelso 1986).

3. Control parameter(s) lead the system through different coordination states. Such parameters may be unspecific to the system's dynamical behavior, as in standard bifurcation theory, or specify a particular coordination state, as in studies of intentional change and learning. The distinction between control parameters and coordination variables is not rigid, but depends on level of description. For example, in multilimb or multi-effector coordination, frequency may be a control parameter at the coordinative level, but a collective variable at the level of the interacting components.

4. As other theories of motor control such as the generalized motor program (GMP) have recognized, relative timing is a conspicuous feature of coordinated behavior. But whereas invariant (or almost so, see Schmidt, this volume) relative timing has been used to bolster the GMP hypothesis and related schema notions, in coordination dynamics, relative timing has come to reflect the central theoretical concept of temporal stability, namely, the ability to maintain a coordinated state despite fluctuations or external perturbations. The reasons temporal stability is such a key concept in coordination dynamics are: first, that there are now many experiments in different kinds of systems showing that relative timing patterns are differentially stable (e.g., in-phase versus anti-phase coordination; see the following); second, that systematic shifts in relative timing occur depending on the components being coordinated and their biomechanical (e.g., eigenfrequency) properties; third,

that loss of stability gives rise to new (or different) coordination states and/or switching between them when control parameters (independent variables in the language of experimental design) are continuously varied over a sufficiently wide range. Fluctuations play a key conceptual and practical role in coordination dynamics, probing the stability of coordinative states and enabling the system to discover new coordinative states according to internal and external demands; and fourth, that additional biomechanical degrees of freedom may even be recruited and others suppressed while preserving temporal stability among interacting components (see below).

5. Finally, in coordination dynamics, the same equations of motion (Haken et al. 1985; Kelso et al. 1990; Schöner et al. 1986) have been shown to govern the coordination among different kinds of components and even across different coupling media. Thus, coordination dynamics is a law-based mathematical structure describing and predicting the coordination activity of a system, whether it be the central nervous system, an organism interacting with its environment, or even shared actions between organisms (Kelso 1994a, b; Kelso and Haken 1995; Turvey 1994).

Some Recent Extensions of Coordination Dynamics

How Do Biological Systems Stabilize Unstable States?

Everyone knows that complex systems, from marriages to the economy, can be unstable. Walking itself has been characterized as an inherently unstable process. One way to control an unstable system is to anticipate its behavior so that corrective adjustments can be made before a catastrophic event occurs. How do we understand this kind of intelligent control? Inspired by Dr. Seuss's famous *Cat in the Hat* character, I had the idea to study balancing, as in balancing a broomstick on the end of your finger. In four dimensions (x, y, z, and t), this is already quite complicated. Graphic representations of the end of the finger look like a tangled lump of spaghetti. A key strategic element of coordination dynamics is to create experimental windows that simplify complicated situations without destroying the very essence of the problem. With this in mind, Paul Treffner and I designed an apparatus that allowed us to study balancing in one spatial coordinate. Subjects gripped a base between their fingers that was fixed to a ball bearing constrained to slide along a track. The task was to balance a rod or pole that was connected to the base by a very narrow fixture (practically a knife edge). Without active control, of course, the pole, whose properties, (e.g., mass, length, and so forth) can be varied, falls down.

This experimental paradigm taps the essence of what it means to control an inherently unstable system. Notice that the subject is perturbing the very element he or she is trying to control. For the present discussion, we consider only successful balancing behavior, not the equally interesting issue of how balancing is learned. For ease of description, let us call the motion of the base the "action" component, and the motion of the pole that the subject monitors the "perceptual" component.

Optotrack measures of the base show that motion never settles into a simple, smooth periodic trajectory, as, for example, in controlling chaos (Kelso and Ding 1993). No limit cycles are observed, only wiggles within wiggles.Detailed analysis reveals a kind of constrained randomness—the trajectories never repeat but exhibit a fantastic geometry. Holding phases during which the pole is finely balanced over the base are interspersed with escape phases during which the hand must make large amplitude, high velocity motions to prevent the rod from falling. The motion is never stable in any asymptotic sense. But it is, nevertheless, stabilized in terms of its function.

From an examination of the action component alone, one only obtains a limited view. However, when viewed as a hand-pole system, one sees an intimate connection or coupling between the perceptual component (the pole) and the action component (hand motion). Not only do we observe a "structure" between the coupled components, given by a phase-like quantity, but also the magnitudes or metrical aspects are intimately coupled as well. Moreover, at one point the start of the action (the zero crossing of the hand velocity in figure 10.1) is slightly delayed with respect to the motion of the rod; in the next moment, the action is in advance of the pole.

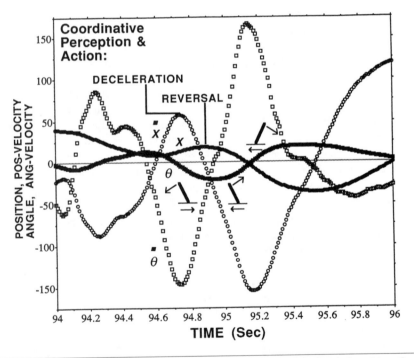

Figure 10.1   The position and velocity *(x* and its derivative) of the hand plotted together with the angle and angular velocity (theta and its derivative) of the pole during a successful two-second sample of the one-dimensional, pole-balancing task. Notice how the two sets of variables are precisely coordinated, spatially and temporally.

Here we touch on Bernstein's recognition of the decisive importance of advance or anticipatory information. Intelligent control turns out to be a subtle blend of reaction and anticipation: To perform best, one can't rely on only one mode or the other. Better to live in the boundary between reactive and anticipatory modes (see also Engström et al. 1996).

What information is used to perform this task? Our preliminary work suggests that subjects, regardless of the physical dimensions of the pole, are sensitive to a tau-like quantity, analogous to the time-to-contact variable in many studies of optical flow (Treffner and Kelso 1995). Here, tau refers to the time to return the pole to the (vertical) unstable balance point, and is the ratio of pole angle over pole velocity. Whereas engineers, neural network modelers, and others often talk about the need for "internal, predictive models" to explain this kind of behavior, we are pursuing the hypothesis that tau-like quantities (e.g., time to balance and its derivative) may provide the perceptual basis for ongoing, active control of inherently unstable systems. Although much more research is necessary, already we see that broadening the level of description to include both action and perceptual components (analogous to the moving room example) shifts the focus of analysis from within (internal models and the like) to between (in terms of action-perception as a coupled, pattern-forming dynamical system).

Coupling Action and Perception: Experiment and Theory

Let's try to pin this idea of coupled, pattern-forming dynamics down a little bit. To do that, I want to take the even simpler situation of syncopating or synchronizing with an external stimulus. This allows us to achieve at least three goals. First, to explore the region between reactive and anticipatory control more deeply (Engström et al. 1996). Second, to actually identify a minimal form of the coordination dynamics. Third, it is a paradigm through which we may probe the brain (i.e., study the selective engagement of different brain areas of time in relation to task or environmental changes).

Briefly, let me remind you of the paradigm. The subject's task is to synchronize peak flexion of the index finger with an auditory (or visual) metronome, in two modes of coordination: on the beat (synchronize) or off the beat (syncopate). Starting in each mode of coordination, the metronome frequency is systematically increased (or in another condition, decreased). Several different behaviors are observed. By far the most dominant one is that subjects spontaneously switch from syncopation to synchronization at a critical metronome frequency. At higher frequencies, synchronization is often lost: the relative phase between stimulus and hand movement wraps continuously. This means that the subject can no longer keep up with the metronome. A very interesting phenomenon occurs near where synchronization is lost: the relative phase slips but holds for brief periods of time. There is a tendency to maintain phase attraction, even though the components (hand and stimulus) are no longer one-to-one frequency locked.

What is the simplest theoretical model that can accommodate all the observed effects and predict others? Equation 10.1 (Kelso et al. 1990) should be familiar to the reader,

$$\dot{\Phi} = \delta\omega - a\sin\Phi - 2b\sin 2\Phi + \sqrt{Q}\Psi t \qquad (10.1)$$

because it not only captures the present case of coupled perception-action coordination but successfully accounts for coupled movements of hand-held pendula and even coupled movements of two people (Kelso 1994a; Turvey 1994). Quite rightfully, equation 10.1 has taken its place as an elementary law of coordination, a sort of minimal coordinative structure that possesses a beauty in terms of the basic laws governing coordination and control sought by Bernstein. Notice the following features captured by this theoretical model:

1. The collective variable or order parameter, Φ, characterizes the coordination states of the coupled system. Coordination is thus a self-organized synergetic process in which (at least for this case, but see below) the interaction among component motions is reduced to a one-dimensional equation for the relative phase alone.

2. The control parameters, b/a and $\delta\omega$ lead the system through coordination states. In the scenario shown here, $\delta\omega$, which represents the uncoupled frequency difference between the stimulus and the finger, is fixed. This breaks the symmetry in two senses: stimuli and finger are obviously different on many levels, not least of which are the neuroanatomical areas and neural mechanisms involved in auditory processing and movement generation. But also the data show that at some point the finger cannot keep up with the stimulus and lags behind. The opposite is not true. So not only is neuroanatomical symmetry broken, but the symmetry of the dynamics is broken too.

3. Scanning figure 10.2 from left to right we see how the layout of the dynamics is modified by parametric changes. On the left is an illustration of a key concept in

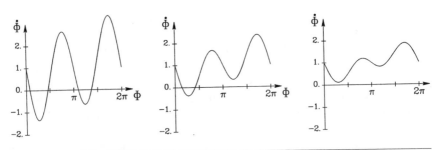

Figure 10.2 The elementary coordination law, equation 10.1, for different parameter regimes. Stable (unstable) fixed points of the coordination dynamics correspond to negative (positive) slopes of the function when the derivative of phi passes through the x-axis. Notice how the layout of stable and unstable fixed points changes from left to right. On the extreme right, no fixed points exist. However, attraction to the ghost or remnant of the previously stable fixed point still remains.

coordination dynamics, mentioned earlier, namely temporal stability. States of coordination near 0 and π are maintained despite fluctuations or external perturbations.

4. Notice that equation 10.1 constitutes a multi-stable coordinative structure. The system is able to stabilize several different relationships among stimulus and movement components under the same task conditions. Here at least two solutions, corresponding to stable coordination states, are possible for the same parameter values. Bernstein might agree that such multistability or multifunctionality provides a theoretical mechanism for the flexibility and adaptability of the coordination process.

5. Still another source of flexibility is the ability to switch at critical values of a parameter. In the middle part of figure 10.2, the state near anti-phase has disappeared and only the state near in-phase remains (the syncopate to synchronize transition). Notice how the slope of the function flattens as coordination states disappear, indicative of loss of stability. (Here, we observe that the disappearance of the internal, predictive model for syncopation is a dynamical effect due to loss of stability!)

6. Finally, on the right part of figure 10.2, we see that there are no longer any stable coordination states available, but there is still attraction to previously stable states. Elsewhere, this regime has been shown to correspond to a "relatively" but not absolutely coordinated structure (see Kelso 1995, chapter 4 for review). The reader will remember that von Holst (1939/1973, page 29) saw this relative coordination as, "Rendering visible the operative forces of the central nervous system." Elsewhere, I have called this the principle of attraction without attractors, as defined in the usual asymptotically stable sense of mathematics (Kelso 1995).

It is worth reiterating that equation 10.1 governs the coordinative behavior of different kinds of physical elements and is indifferent to the medium through which they are coupled. In coordination dynamics, the coupling medium is not, or rather does not have to be mechanical, as in the coupling between two clocks on a wall or in terms of reactive forces between articulatory segments. Nor is the coupling necessarily neurophysiological, as in some kind of neural cross-talk between the hemispheres controlling the hands that is thought to underlie mirror movements and bimanual key pressing errors. Crucial experiments, as when a subject must syncopate or synchronize with an auditory or visual signal (Kelso et al. 1990; Wallenstein et al. 1995) or when two people coordinate with each other (Schmidt et al. 1990), show all the predicted effects of the coordination dynamics, attesting to the fundamentally abstract and informational nature of the nonlinear coupling.

Additional evidence for abstract spatial coupling comes from studies of rhythmic motions between the hand and foot (Baldissera et al. 1991; Carson et al. 1995), between the knee and elbow joints (Jeka and Kelso 1995; Kelso and Jeka 1992), and even between the joints of a single limb (Kelso et al. 1991b). Hand and foot coordination is more stable, for example, when movements are produced in the same direction, necessarily involving the coordinated activity of non-homologous muscles. Likewise, in the "waiter" experiments of Kelso et al. (1991b), the most stable coordination state is when wrist and elbow rotations are in the same direction, regardless of initial postural orientation (whether the forearm is pronated or supinated). Once again, all the main predicted features of coordination dynamics (loss of stability, switching, and so forth) have been observed in all these situations.

To reiterate, the three main take-home messages of this section are: First, coordination, the collective intercorrelated activity between task components, is not only a mere product of biomechanical coupling. Rather, the coupling, which can be precisely calculated, is abstract, spatial, and relational. Second, the quantities that characterize coordination, the so-called collective variables or order parameters, are best thought of as informational in nature. Bernstein might have referred these facts to a high-level, muscle-independent "engram." The third, more theoretical point, is that the "exact fomulae of (coordinated) movement" (Bernstein's *Bewegungsformein*) (Bernstein 1967), take the form of a coupled, nonlinear dynamical system, in which one and the same nonlinear coupling produces all the observed effects.

Probing the Brain's Coordination Dynamics

Near the end of his 1967 book, Bernstein speculated in his usual inspired fashion that the "active process of perception and action in the brain is not determined by the nature of some hypothetical content of the cells, synapses, interstitial tissues, etc. but by the very dynamic form of its organization and connections" (page 167). Armed with new technology, experimental paradigms that study the coevolution of brain and behavioral dynamics in space and time, and the theoretical concepts of pattern forming dynamical systems, we can now examine Bernstein's hypothesis in more detail. In particular, I refer to recent research on dynamic brain imaging using large-scale EEG and SQuID arrays in conjunction with the bimanual and sensorimotor coordination paradigms employed by my colleagues and I. In each case, subjects produce either in-phase or anti-phase coordination in response to a stimulus whose frequency is systematically increased. The following behaviors are observed: first, coherent power spectra in brain activity corresponding to both anti-phase and in-phase conditions are present in numerous brain regions; second, relative phase (or timing) between stimuli and brain signals is "almost invariant" across a range of (low) stimulus rates; and third, a dynamic reorganization of brain activity occurs at a critical rate as seen in abrupt changes in phase and frequency relations of recorded brain signals. Moreover, very recent analyses show that anti-phase (syncopated) behavior engages midline, centrally located cortical structures such as motor cortex and SMA, whereas in-phase (synchronized) behavior engages prefrontal and frontal brain areas (Meaux et al. 1996).

What causes this dynamic reorganization or switching? Evidence obtained from both large-scale EEG (Wallenstein et al.1995) and SQuID arrays (Fuchs et al.1992; Kelso et al. 1991, 1992, 1994) indicates that relative timing measures of brain electrical and magnetic signals exhibit both critical fluctuations and critical slowing down prior to switching. The former is indicated by systematic growth in the variance of the anti-phase pattern before the onset of the transient switching process. The latter is revealed by the observation that after a perturbation, the brain takes longer and longer to return to the pre-perturbation timing pattern as the transition draws near.

Several points are in order. Importantly, these and other data provide direct evidence of stable, relative timing in ensembles of cortical neurons in the human brain during movement coordination tasks, at least for certain values of task parameters. Given that relative timing is one of the key variables hypothesized by both coordination dynamics and the generalized motor program theory of motor control (Schmidt 1988), this is surely a positive result. Notice also that the present work shows that coherent brain and behavioral states are captured by the same kinds of collective variables. So again, it's not so much the co-relation between "internal" variables and "external" variables that holds the key to the brain-behavior relation. Rather, it is finding the relevant coupling quantities that bridge the two domains. As a final comment on this section, observations of nonequilibrium phase transitions in global patterns of brain activity suggest a new mechanism for the coordination of cortical neurons in the brain. Notice that loss of stability in cortical activity is a predicted feature of coordination dynamics, not the GMP. Stability and instability are dynamical concepts that are foreign to motor programming notions. Moreover, experimental results showing the onset of spatially coherent patterns, instability, transitions, etc. strongly suggest that active dynamical processes in the human brain are the result of self-organization, and are not (as in the computer metaphor) preprogrammed. Bernstein's notion of perception and action as an active process determined by a very dynamic form of brain organization is consistent with these new results and their theoretical interpretation in terms of coordination dynamics.

The Recruitment and Suppression of Biomechanical Degrees of Freedom

Dynamic paradigms that study transitions between reactive, syncopated, and synchronized modes of coordination may allow us to understand how information is routed selectively among brain areas. However, selective, task-dependent recruitment and suppression processes occur on several levels of description and in many different kinds of behavior. Although a dominant feature of biological coordination, the flexible recruitment-annihilation of biomechanical degrees of freedom is seldom studied systematically. Whereas, bifurcations from one ordered coordination state to another within an already active set of components are well-known (e.g., bimanual transitions), much less understood are the dynamical mechanisms underlying the recruitment of previously quiescent degrees of freedom. Earlier, we had shown, in a simple bimanual paradigm that allows motion in both the horizontal (x) and vertical (y) dimensions, a sequence of transitions within and across planes of motion as a control parameter (frequency) is varied (Kelso and Scholz 1985). More detailed empirical study and theoretical analysis reveals that these spatial transitions can be understood as a result of two consecutive Hopf bifurcations (Kelso et al. 1993; Buchanan et al. 1997b), as when a fixed point (e.g., no motion on the y-plane) turns into a limit cycle (oscillatory motion in the y-plane) and vice versa.

Such generic dynamical mechanisms appear to be used by the CNS to maximize stability (i.e., reduce fluctuations) while still fulfilling task requirements.

We (John Buchanan and I) wondered if a similar hypothesis might pertain to the coupled pendulum paradigm studied by Turvey and colleagues (Kugler and Turvey 1987; Turvey 1994). Although equation 10.1 (the elementary coordination law) was formulated specifically for a perception-action system (Kelso et al. 1990), it has received its most rigorous testing from studies of coupled pendulum motion in which a seated person oscillates hand-held pendula whose lengths and masses can be systematically manipulated. The task is to swing the pendula back and forth on the sagittal plane by adducting (backward motion) and abducting (forward motion) around the axes of motion of the wrist(s). Other joints are assumed to be fixed. With respect to equation 10.1, the parameter $\delta\omega$ may be manipulated by altering the length-mass characteristics of each pendulum, thereby yielding frequency differences $(\delta\omega = \omega_L - \omega_R)$ between the coordinating limbs. Frequency or rate of coupled motion (b/a) can be varied using a metronome that the subject must follow in a 1:1 fashion. Many of the predictions of the elementary coordination dynamics (Kelso 1994a), including mean phase shifts proportional to $\delta\omega$, enhancement of phase shifts with movement rate at $\delta\omega > 0$, differential amounts of phase shifts in anti-phase versus in-phase modes, enhancement of phase variability as a function of increasing $\delta\omega$, greater variability in the anti-phase mode as both $\delta\omega$ and b/a are increased, and so forth, have been tested and observed (Turvey 1994).

One of the puzzling findings from the coupled pendula paradigm (some 30-odd published studies), is the absence of observed transitions between anti-phase and in-phase modes of coordination. This is a major prediction of the law; indeed, multistability and transitions are at the very basis of its existence. One frequently cited reason is that the control parameter has not been varied over a sufficiently large range as to induce transitions. However, recent work by Schmidt et al. (in press) scaled movement rate to high frequencies yet was still unable to induce transitions. The possibility explored by Buchanan and me was to consider the role of effects that have not been experimentally measured, in particular, the possible recruitment and supression of biomechanical degrees of freedom.

When swinging a single pendulum or coupled pendula, great care is taken to instruct subjects to use only wrist ulnar adduction-abduction and to keep the pendulum on the plane parallel to the body's sagittal plane. Yet the degree to which such task constraints are satisfied has not been investigated. Through detailed measurement and analyses of pendulum trajectory and various musculoskeletal components, we explored the hypothesis that recruitment of degrees of freedom occurs as movement rate is increased, thereby precluding the necessity to switch between coordinative modes. Specifically, recruitment at the component level (e.g., wrist flexion-extension) helps stabilize the task-defined coordination state, thereby obviating the need to switch.

Our experiments (Buchanan and Kelso, submitted) required subjects to swing hand-held pendulums either individually or in pairs. Movement frequency was parametrically scaled in eight steps of 0.15 Hz (two below the calculated eigenfrequency and five above it), with 10 cycles per frequency plateau. The experiments were conducted with the forearms strapped (by two Velcro bands to the armrest) and unstrapped, on two separate days at least one week apart. Essentially three major results are pertinent here:

1. As frequency was increased, we observed a bifurcation from planar to spherical motion in the pendulum. Such recruitment was produced by flexion-extension of the wrists.

2. A sudden or gradual increase in elbow flexion-extension was also frequently observed—phase and frequency entrained with the pendulum motion. This onset of oscillations in elbow motion compensated for the decrease (due to increasing frequency) in the amplitude of wrist motion, thereby helping stabilize the anti-phase coordination state at higher frequencies.

3. As in all the coupled pendula studies conducted so far, no switching between coordinative modes was observed, even though subjects were instructed to let the pattern of coordination change if it felt comfortable to do so. Moreover, although variability increased modestly with increases in movement rate, it did so for both in-phase and anti-phase coordination modes. These results strongly suggest that the one-dimensional phase equation (equation 10.1)—although it has predicted many observed effects—is nevertheless an incomplete account of all the phenomena observed in the coupled pendulum paradigm, particularly the presence of recruitment-suppression mechanisms. Apparently, in so-called redundant multidegree of freedom systems, recruitment at the component level enables the CNS to stabilize coordination states that would otherwise be unstable or unavailable under current conditions. Why switch modes when you can recruit other components and thereby fulfill task requirements?

Trajectory Formation

Finally, let's consider a topic of much recent interest, trajectory formation. Current studies of trajectory formation in, for example, multijoint arm movements, have identified various constraints and (putative) invariants as clues to what the CNS plans and controls. None of these schemes, however, incorporates the naturally occuring, sudden, and flexible recruitment of new degrees of freedom, nor (crucially!) the perceptual basis of such recruitment.

Using their own multijointed arm and moving in the vertical plane, we (Buchanan et al. 1997) required subjects to rhythmically trace an arc whose shape (as specified by a curvature control parameter) was systematically varied from high (nearly semicircular) to low (almost straight) and vice versa. The main results were as follows: First, over a wide range of arc shape, only a few coordinative states were observed (i.e., wrist-elbow coordination was always in-phase at high curvatures and anti-phase at low curvatures). Second, when 1:1 frequency entrained, shoulder and elbow were in-phase at all curvature values. Third, transitions between the two stable coordination states were observed to occur at a critical curvature value. Fourth, transitions were always accompanied by a significant reduction in wrist amplitude, irrespective of the direction of change in the curvature parameter.

Such behavior points to an amplitude-mediated dynamic mechanism for switching (Haken et al. 1985). Notice again that this behavior of phase-locking among the joints and the more radical course of spontaneously switching off the wrist degree of freedom, serves to reduce the effective space of joint angles and angular velocities and is the dynamical equivalent of reducing biomechanical redundancies in multidegree of freedom movements. A model of three coupled, nonlinear oscillators (one for each moving segment) parameterized by curvature captures all observed constraints on multijoint trajectory formation (DeGuzman et al. 1997). A key feature of our model is the sudden shift (observed experimentally) in the wrist-elbow relative phase as the amplitude of the wrist decreases. Such transitions are at the heart of flexibility.

In summary, in trajectory formation (and we suppose for coordination in general) this recent evidence suggests five facts: First, only a few coupled coordination states among interacting components are stabilized by the CNS. Second, transitions and switching processes are exploited by the CNS to flexibly meet changing task demands. Third, recruitment (e.g., shoulder) and suppression (e.g., wrist) of degrees of freedom at critical control parameter values provide additional sources of flexibility for the CNS to exploit. Fourth, a theoretical mechanism for transition and recruitment processes is reduction in component amplitudes. Fifth, via perception, the CNS controls component amplitudes thereby "selecting" a coupled, coordinated state to produce a stable trajectory in space. Obviously, this work on trajectory formation again involves (as does much other research, e.g., on multicomponent and multifrequency coordination, learning, perception, etc.) an extension of coordination dynamics, certainly beyond the elementary law of coordination described earlier.

Conclusion

In closing, let me return to the great man Nikolai Bernstein himself and address his aspirations for the field of movement science. In particular, I'd like to comment on Bernstein's stated hope that "the science of human movement and theories of motor coordination" might emerge out of obscurity to play a central role, not merely in human and comparative physiology, but in the sciences of life and mind itself. The position here is that they have not. One reason is that biology has moved over the last 60 years from the study of the big to the very small. Great successes have followed from the atomizing of biology that have changed the discipline, perhaps irreversibly. A second reason is that movement science—at least through the 1980s— like the rest of the cognitive and neurosciences, has been dominated by the computer metaphor. Now, in large part because of the developments in coordination dynamics and ecological psychology, the hypothesis is on the table that cognition is a subset of dynamical systems, not a subset of computation (Kelso 1995; Port and van Gelder 1995; Thelen and Smith 1994). It may be time to recognize that all computers are dynamical systems, but not all dynamical systems are computers. A third reason is that to many, human movement is governed by the laws of motion itself. That is to say, Newtonian mechanics. So, to many, the science of human movement is, for the most part, the application of classical mechanics, physics at the terrestrial scale.

Now I am not saying that studies motivated by the computer metaphor (motor programs and the like) have not been useful, and in many cases enlightening. Nor do I question the very helpful, and often brilliant applications of Newtonian mechanics to the many facets of human movement. What I am saying is that motion is not the same as coordination, and that new concepts, tools, and methods are needed to understand this fundamental feature of living things. Whereas, the science of movement has not played and will not likely play a central role in physiology as Bernstein avidly hoped (physiology itself has virtually ceased to exist as an autonomous discipline), the science of coordination may yet play a pivotal role in an emerging science of pattern and structure formation that deals with how living things are (self) organized, including, fundamentally, the (circularly) causal relationship between mind, brain, and behavior.

Research in coordination dynamics shows that, time and time again, generic dynamical mechanisms are exploited by living things to provide both flexibility and stability of motor function. Why, one asks, are these generic mechanisms so ubiquitous? One might propose they appear so frequently because organisms and environments have co-evolved in such a way that those regularities and styles of change that work are selected. To the extent that dynamical laws of coordination such as equation 10.1 characterize the behavior of different kinds of systems and are instantiated through various means, this, I would argue, is due to task or functional equivalence. That is, whether it be the coordination of neural ensembles, fingers or people, all share the common functional property of generating order in time. Such a view opens up new opportunities to uncover the dynamical mechanisms underlying so-called motor equivalence, arguably the single most universal phenomenon in movement science (or, indeed, biology and psychology). May Bernstein's ghost smile on such endeavors.

Acknowledgments

Much of the research described in this article was supported by NIMH Grants MH 42900 and KO5 MH01386, U.S. Office of Naval Research Grant N00014-88-J-119 and NSF Grant SBR9511360.

References

Baldissera, F., Cavallari, P., Marini, G., Tassone, G. (1991) Differential control of in-phase and anti-phase coupling of rhythmic movements of ipsilateral hand and foot. *Experimental Brain Research,* 83: 375–380.

Bernstein, N.A. (1967) *The Coordination and Regulation of Movements.* Oxford: Pergamon.

Boylls, C.C. (1975) A theory of cerebellar function with applications to locomotor behavior in the cat. *COINS Tech. Rep.,* (U. Mass., Dept. of Computer and Information Science), Technical Report: 76–1.

Buchanan, J.J., Kelso, J.A.S. (submitted) To switch or not to switch: recruitment of degrees of freedom as a stabilization process in rhythmic pendular movements.

Buchanan, J.J., Kelso, J.A.S., DeGuzman, G.C. (1997a) Self-organization of trajectory formation: I. Experimental evidence. *Biological Cybernetics,* 76: 257–273.

Buchanan, J.J., Kelso, J.A.S., DeGuzman, G.C., Ding, M. (1997b) The spontaneous recruitment and annihilation of degrees of freedom in rhythmic hand movements. *Human Movement Science* 16: 1–32.

Carson, R.G., Goodman, D., Kelso, J.A.S., Elliot, D. (1995) Phase transitions and critical fluctuations in rhythmic coordination of ipsilateral hand and foot. *Journal of Motor Behavior,* 27: 211–224.

DeGuzman, G.C., Kelso, J.A.S., Buchanan, J.J. (1997) The self-organization of trajectory formation: Theoretical model. *Biological Cybernetics,* 76: 275–284.

Dijkstra, T.M.H., Schöner, G., Giese, M.A., Gielen, C.C.A.M. (1994) Frequency-dependence of the action-perception cycle for postural control in a moving visual room: Relative phase dynamics. *Biological Cybernetics,* 71: 489–501.

Edelman, G.M. (1987) *Neural Darwinism: The Theory of Neuronal Group Selection.* New York: Basic Books.

Engström, D.A., Kelso, J.A.S., Holroyd, T. (1996) Reaction-anticipation transitions in human perception-action patterns. *Human Movement Science,* 15: 809–832.

Fuchs, A., Kelso, J.A.S., Haken, H. (1992) Phase transitions in the human brain: Spatial mode dynamics. *International Journal of Bifurcation and Chaos,* 2: 917–939.

Haken, H. (1977) *Synergetics: An Introduction.* Berlin: Springer-Verlag.

Haken, H. (1996) *Principles of Brain Functioning.* Berlin: Springer-Verlag.

Haken, H., Kelso, J.A.S., Bunz, H. (1985) A theoretical model of phase transitions in human hand movements. *Biological Cybernetics,* 51: 347–356.

Jeka, J.J., Kelso, J.A.S. (1995) Manipulating symmetry in the coordination dynamics of human movement. *Journal of Experimental Psychology: Human Perception and Performance,* 21: 360–374.

Kelso, J.A.S. (1981) On the oscillatory basis of movement. *Bulletin of the Psychonomic Society,* 18: 63.

Kelso, J.A.S. (1984) Phase transitions and critical behavior in human bimanual coordination. *American Journal of Physiology: Regulatory, Integrative and Comparative Physiology,* 15: R1000–R1004.

Kelso, J.A.S. (1986) Pattern formation in multidegree of freedom speech and limb movements. *Experimental Brain Research Supplement,* 15: 105–128.

Kelso, J.A.S. (1994a) Elementary coordination dynamics. In *Interlimb Coordination: Neural, Dynamical, and Cognitive Constraints.* S.P. Swinnen, H. Heuer, J. Massion, P. Casaer (Eds.), New York: Academic Press, 301–318.

Kelso, J.A.S. (1994b) The informational character of self-organized coordination dynamics. *Human Movement Science,* 13: 393–413.

Kelso, J.A.S. (1995) *Dynamic Patterns: The Self-Organization of Brain and Behavior.* Cambridge, MA: MIT Press (Paperback edition, 1997).

Kelso, J.A.S., Bressler, S.L., Buchanan, S., DeGuzman, G.C., Ding, M., Fuchs, A., Holroyd, T. (1991) Cooperative and critical phenomena in the human brain revealed by multiple SQuIDS. In *Measuring Chaos in the Human Brain.* D. Duke, W. Pritchard (Eds.), New Jersey: World Scientific, 97–112.

Kelso, J.A.S., Bressler, S.L., Buchanan, S., DeGuzman, G.C., Ding, M., Fuchs, A., Holroyd, T. (1992) A phase transition in human brain and behavior. *Physics Letters A,* 169: 134–144.

Kelso, J.A.S., Buchanan, J.J., DeGuzman, G.C., Ding, M. (1993) Spontaneous recruitment and annihilation of degrees of freedom in biological coordination. *Physics Letters A,* 179: 364–371.

Kelso, J.A.S., Buchanan, J.J., Wallace, S.A. (1991b) Order parameters for the neural organization of single, multijoint limb movement patterns. *Experimental Brain Research,* 85: 432–444.

Kelso, J.A.S., DelColle, J.D., Schöner, G. (1990) Action-perception as a pattern formation process. In *Attention and Performance XIII.* M. Jeannerod (Ed.), Hillsdale, NJ: Erlbaum, 139–169.

Kelso, J.A.S., Ding, M. (1993) Fluctuations, intermittency and controllable chaos in biological coordination. In *Variability and Motor Control*. K.M. Newell, D.M. Corcos (Eds.), Champaign, IL: Human Kinetics, 291-316.

Kelso, J.A.S., Fuchs, A., Holroyd, T., Cheyne D., Weinberg, H. (1994) Bifurcations in human brain and behavior. *Society for Neuroscience*, 20: 444.

Kelso, J.A.S., Haken, H. (1995) New laws to be expected in the organism: Synergetics of brain and behavior. In *What is Life? The Next 50 Years*. M. Murphy, L. O'Neill (Eds.), Cambridge, England: Cambridge University Press, 137–160.

Kelso, J.A.S., Jeka, J.J. (1992) Symmetry breaking dynamics of human multilimb coordination. *Journal of Experimental Psychology: Human Perception and Performance*, 18: 645–668.

Kelso, J.A.S., Scholz, J.P. (1985) Cooperative phenomena in biological motion. In *Complex Systems: Operational Approaches in Neurobiology, Physical Systems and Computers*. H. Haken (Ed.), Berlin: Springer, 124–149.

Kelso, J.A.S., Southard, D.L., Goodman, D. (1979) On the nature of human interlimb coordination. *Science*, 203: 1029–1031.

Kugler, P.N., Kelso, J.A.S., Turvey, M.T. (1980) On the concept of coordinative structures as dissipative structures: I. Theoretical lines of convergence. In *Tutorials in Motor Behavior*. G.E. Stelmach, J.E. Requin (Eds.), Amsterdam: North-Holland, 1–47.

Kugler, P.N., Turvey, M.T. (1987) *Information, Natural Law and the Self-Assembly of Rhythmic Movement*. Hillsdale, NJ: Erlbaum.

Lee, D.N., Lishman, J.R. (1975) Visual proprioceptive control of stance. *Journal of Human Movement Studies*, 1: 87–95.

Meaux, J.R., Wallenstein, G.V., Bressler, S.L., Fuchs, A., Kelso, J.A.S. (1996) Cortical dynamics of the human EEG associated with behavioral phase transitions in an auditory- motor task. *Society for Neuroscience*, 22: 890.

Port, R.F., van Gelder, T. (Eds.) (1995) *Mind as Motion: Explorations in the Dynamics of Cognition*. Cambridge, MA: MIT Press.

Schmidt, R.A. (1988) *Motor Control and Learning*. Champaign, IL: Human Kinetics.

Schmidt, R.C., Bienvenu, M., Fitzpatrick, P.A., Amazeen, P.G. (in press) Effects of frequency scaling and frequency detuning on the steady state properties and breakdown of coordinated rhythmic movements. *Journal of Experimental Psychology: Human Perception and Performance*.

Schmidt, R.C., Carello, C., Turvey, M.T. (1990) Phase transitions and critical fluctuations in the visual coordination of rhythmic movements between people. *Journal of Experimental Psychology: Human Perception and Performance*, 16: 227–247.

Schöner, G., Haken, H., Kelso, J.A.S. (1986) A stochastic theory of phase transitions in human hand movement. *Biological Cybernetics*, 53: 442–452.

Swinnen, S., Heuer, H., Massion, J., Casaer, P. (Eds.) (1994) *Interlimb Coordination: Neural, Dynamical and Cognitive Constraints*. New York: Academic Press.

Thelen, E., Smith, L.B. (1994) *A Dynamic Systems Approach to the Development of Cognition*. Cambridge, MA: MIT Press.

Treffner, P.J., Kelso, J.A.S. (1995) Functional stabilization of unstable fixed-points. In *Studies in Perception and Action, III.*.B.G. Bardy, R.J. Bootsma, Y. Guiard (Eds.), Hillsdale, NJ: Erlbaum.

Turvey, M.T. (1994) From Borelli (1680) and Bell (1826) to the dynamics of action and perception. *Journal of Sport and Exercise Psychology*, 16: S128–S157.

Turvey, M.T., Shaw, R.A., Mace, W. (1978) Issues in the theory of action: Degrees of freedom, coordinative structures and coalitions. In *Attention and Performance VII*. J.E. Requin (Ed.), Hillsdale, NJ: Erlbaum, 557–595.

von Holst, E. (1939/1973) Relative coordination as a phenomenon and as a method of analysis of central nervous function. In *The Collected Papers of Erich von Holst*. R. Martin (Ed.), Coral Gables, FL: University of Miami, 33–135.

Wallenstein, G.V., Kelso, J.A.S., Bressler, S.L. (1995) Phase transitions in spatiotemporal patterns of brain activity and behavior. *Physica D*, 84: 626–634.

Whiting, H.T.A. (Ed.) (1984) *Human Motor Actions: Bernstein Reassessed*. Amsterdam, North Holland.

11

CHAPTER

Optical Flow Fields and Bernstein's "Modeling of the Future"

Nam-Gyoon Kim and M.T. Turvey

Center for the Ecological Study of Perception and Action,
University of Connecticut, Storrs, CT, U.S.A.

"Each significant act is a solution (or an attempt at one) of a specific problem of action. But the problem of action, in other words the effect which the organism is striving to achieve, is a something which is not yet, but which is due to be brought about. *The problem of action, thus, is the reflection or model of future requirements* (somehow coded in the brain); and a vitally useful or significant action cannot be either programmed or accomplished if the brain has not created a prerequisite directive in the form of the future requirements that we have just mentioned." (Bernstein 1967, page 171, italics added)

Introduction

Bernstein's physiology is a physiology of activity, not a physiology of reaction (Bernstein 1967, 1996). The relevant issues in Bernstein's view are those implicated in the solving, by activity, of the multiple, ever changing problems posed by the environment. As he remarked, an organism and its environment are in a never-ending game, "a game where the rules are not defined and the moves planned by the opponent are not known" (Bernstein 1967, page 173). Contrary to the sentiment of Pavlov's school, which Bernstein often criticized (Feigenberg and Latash 1996), organisms rarely react to the environment. The surroundings with respect to which organisms behave are not to be understood as collections of so many triggering stimuli and the behaving organisms are not to be construed as collections of so many conditioned and unconditioned reflexes. Bernstein saw that, rather than simply reacting to triggering stimuli, organisms cleverly exploited the information about their surroundings and their movements to control their actions both retrospectively ("after the fact") and prospectively ("before the fact").

In retrospective control, adjustments are made in respect to what has occurred; in prospective control, adjustments are made in respect to what will occur (if current conditions persist) (Turvey 1992). There are many situations encountered in daily activity for which "retrospective control becomes practically impossible" (Bernstein 1967, page 141). The class of so-called ballistic movements are only controllable prospectively: throwing at a target, jumping a brink or a barrier, hammering a nail, and so on. Similarly requiring of prospective control are "movements that forestall others" (Bernstein 1967, page 141) such as the predator moving to an extrapolated point in the prey's trajectory and the perceptually analogous achievements of catching a moving object with a hand, passing a ball to a running teammate in basketball or soccer, intercepting the ball with a striking implement in tennis or baseball or cricket, and so on.

Modeling the Future and the Problem of Induction

Whereas retrospective control for Bernstein could be readily addressed through the ring principle that was to become the hallmark property of cybernetical devices (involving set points, feedback, and corrective signals), prospective control required that Bernstein speculate on the forms of extrapolation by which an animal might anticipate or model the future given present and past conditions. There are serious limitations, however, to extrapolation as the basis for prospective control.

Any form of extrapolation to future states is an induction, that is, it is the determination of some general proposition based on instances. In the prospective control of action, the kinds of general propositions to be reached by induction are of the form "If I keep doing what I am doing, then so-and-so will occur" and "If that thing keeps doing what it is doing, then so-and-so will occur." In formal terms, induction seeks to establish $\forall x P(x)$ (in words, "for all x, or for any x, or for every x, x has the property P; $P(x)$ is true") on the basis of particular instances of the set X for which x_i having the property P is known (Rosen 1991). If the subset of X that is experienced is S, then given the knowledge $(\forall s)P(s)$ do we then have the knowledge that $(\forall x)P(x)$? It has long been respected that the answer in the general case is "no." With respect to any arbitrary chosen property P, no sample entails anything about an unsampled instance, defining, thereby, what is referred to as the problem of induction (e.g., Goodman 1965).

The problem of induction can be resolved if the sampled property P manifests a property called *contagion*, that is, if the truth of $P(x_i)$ implies the truth of $P(x_j)$ for some other x_js in X. If the sampled property is embodied with a contagious property, induction can be carried out with the value of P with the first element of the set X, $P(1)$, and a rule which embodies an entailment relationship between the value $P(n)$ and its successor $P(n + 1)$. A series whose value at n, $P(n)$, entails the next value $P(n + 1)$ is called recursive. If a series is recursive, then extrapolation of the entire series from a sample is just a matter of a simple mechanical operation. Hence, through recursion the future can be extrapolated given the knowledge of the present. In mathematics, a device for doing precisely this has been known for some time— Taylor's theorem, a magical device which converts synchrony (what happens at an instant) into diachrony (what happens over a series of instants) (Rosen 1991, page 78). Specifically, the conversion of the present to the future is accomplished in Taylor's theorem through derivatives. Derivatives contain information about the future as well as the past (Rosen 1991).

Despite the favorable property of contagion, it is nonetheless the case that recursion *tout court* is an unsatisfactory solution to the general problem of modeling the future. Although recursion ensures the entailment relationship between two values at two successive instants of a given series, the initial value out of which a series is generated is completely independent of the recursive rules of the series. Depending on which initial value is chosen, a completely different outcome of a series is obtained. Moreover, the choice of initial value is completely arbitrary. For a mechanical particle whose trajectory is solely determined by an external force acting upon it, an entailment between two successive states assures a complete description of its behavior. However, as the amount of force varies, the path traveled by the particle similarly varies. At every instant, then, there would be infinitely many different trajectories a particle can traverse given that the range of force values can be infinite. A living organism, on the other hand, is capable of aiming itself toward a target by authoring its own initial conditions (Shaw and Kinsella-Shaw 1988). In fact, the ability to select a path in anticipation of future goal states is the hallmark of living things, which is why this particular class of systems is called goal-directed.

Indeed, the recognition that animal movements are goal-directed rather than simple reactions to environmental stimuli was the rationale behind Bernstein's call for a physiology of activity rather than a physiology of reactions. Specifically, goal-directed movements are movements which are directed to "what must be done" instead of "what is" (Bernstein 1967, page 147). To address this problem, Bernstein (1967, page 161) had to resort to probabilistic extrapolation as the basis for modeling the future. It is highly questionable, however, whether probability estimates based on past experiences can provide sufficient constraint for the prospective control of movements. Patently, experience with a given situation and its contingent events can lead to an expectation of what is likely to occur when the situation reoccurs and such expectations must shape, to an important degree, the selection and conduct of actions (see Bernstein 1996, pages 223–225). But an *expectation of what might occur* is not the same as *specification of what will occur* and it is the latter that is necessary for successful prospective control of everyday, commonplace actions. Given a stop sign at the junction of two roads, one expects on the basis of very particular experiences that the likely action of approaching automobiles will be a slowing, and eventual arresting, of forward motion. Executing the activity of bringing the vehicle to a halt at the stop sign, however, requires specification of the appropriateness of the current decelerative forces brought about by the drivers of their respective automobiles. We suspect that such specification is much more deep-seated and universal—that is, grounded in aspects more fundamental than individual experiences—than is the basis of probabilistic extrapolation.

Arguably, Bernstein was forced to resort to probabilistic extrapolation as a model for the future because he adhered to the classical view that perception can provide awareness of "what is" but not an awareness of "what must be done." On this view, perception is an inadequate basis for resolving "the problem of action . . . the reflection or model of future requirements" (Bernstein 1967, page 171). This classical view of perception is based in the notion of sensations (Gibson 1966, 1986; Turvey et al. 1981). A major contrasting view of perception takes as its departure point energy distributions ambient to an animal that are (a) uniquely determined by environmental layout and (b) uniquely transformed by changes in the layout and by the animal's movements. This spatio-temporal structuring imposed on the ambient energy distributions is *information* for Gibson (1966, 1986) and the corresponding theory is an information-based theory of perception. Because information in Gibson's sense is lawfully determined by the environmental layout and by one's own movements, information is said to be specific to the layout and the movements that structured it. Hence, an animal with the requisite machinery for registering the lawfully determined structure of ambient energy distributions can, in principle, perceive the properties that they specify.

In the present chapter, we address how the transforming light structure at a moving point of observation (called optical flow) can be informative of future conditions of action providing, thereby, a physiology of activity or prospective control.

Seeing Where One Is Heading in Curvilinear Locomotion

Arguably, the most fundamental of all activities is locomotion (by wing, fin, limb, or undulation). For most creatures with eyes, patterned light distributions are the primary basis for guiding or steering locomotion in cluttered surroundings. A central question, therefore, is what is the form of the information contained in optical distributions that subserves prospective control of locomotion?

The concept of *point of observation* refers to a place in a transparent medium densely filled with light that has been multiply reflected from substantial opaque surfaces. The light ambient to an observation point consists of different intensities (or differences in intensities) in different directions and is referred to as an optic array. One useful construal of the optic array is as an indefinite nesting of optical solid angles forming a sphere with origin at the point of observation (Koenderink 1986). The base of each solid angle is a face or facet of the environment and the apex of each angle is at the observation point (Gibson 1986). At different observation points, the optic arrays differ. For a moving point of observation there will be a transformation of the optic array specific to the movement and to the environmental layout. The changing optical structure has been referred to as *optical flow* (Gibson 1950, 1966, 1986). Creatures with eyes can occupy observation points and, in principle, register (as they locomote) the adjacent and successive structure of optical flow.

When an animal moves forward rectilinearly, the flow lines of optical flow form a "melon-shaped family of curves" (Gibson 1950, 1966, 1986). Given this melon-shaped family of curves as optical flow, the question of immediate interest is: How does it constrain the steering of rectilinear locomotion? When the observer looks in the direction of heading during rectilinear forward locomotion, the image pattern on the retina, called retinal flow (Cutting et al. 1992; Warren 1995), flows radially from a point referred to as the focus of expansion (Gibson 1950; see figure 11.1). The observer can steer his or her locomotion in the direction of the focus of expansion. In everyday locomotion, however, negotiating curved paths is more commonplace, more natural, than negotiating strictly rectilinear paths. During locomotion along a curved path, however, there is no singularity in the retinal flow analogous to the focus of expansion during linear translation (see figure 11.2). How then is an animal able to find its way through a cluttered environment, faced as it is by the challenge of negotiating various types of curved paths? A treatment of optical flow as a form of fluid dynamics might provide an answer.

As a point of observation moves along a circular path parallel to the ground plane, the resulting velocity flow field can be understood as each surface element tracing a circular trajectory with an instantaneous velocity tangential to the trajectory or, equivalently, as the ground plane rotating beneath a stationary observer (figure 11.2b). That is, for each point on the ground with coordinates x, y at the instant of time t, the velocity of the element passing through the point can be represented as a vector $v(x, y, t)$.[1] Suppose that the point of observation is moving at a constant speed, meaning that the velocity vector at any point does not change with time. Then the flow field would be independent of time. In hydrodynamics, time-independent flow is called *steady flow*. In the case of the steady motion of a fluid, for example, the trajectories of particles can be determined by the fact that the velocity of a particle is everywhere tangent to the trajectory. Consequently, trajectories can be obtained by plotting curves such that the curve direction at each point agrees with the direction of the velocity vector v at that point. These curves are called *streamlines*. There exists exactly one streamline passing through each point in the plane. Each streamline can be represented by a function $\Psi(x, y) = c$. Assigning different

---

[1] The ground surface is represented in a coordinate system anchored at the nodal point of the eye with the x-axis corresponding to the observer's transverse plane and the y-axis corresponding to the observer's sagittal plane (depth direction). Since the observer's movement is assumed to be parallel to the ground plane, the vertical distance of the ground plane is fixed at one eye height unit below the observer. This dimension which could be denoted as the z-axis is not shown in the vector notation to avoid a possible confusion with a complex variable denoted by z used later in this section. However, in the subsequent sections in which the environment is represented in an eye coordinate system, depth dimension is denoted by the z-axis, whereas the vertical dimension is denoted by the y-axis.

a

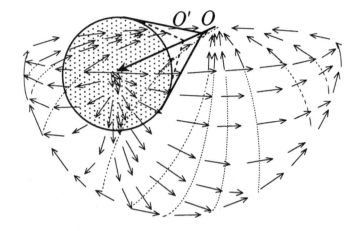

b

Figure 11.1 *(a)* The observer's locomotion along a rectilinear path gives rise to a "melon-shaped family of curves" as optical flow (Gibson 1986). The solid angle depicts the field of view, the portion of optical flow entering the eye. *(b)* The cross section of the field of view consists of velocity vectors all radiating from the focus of expansion, which corresponds to the observer's heading.
Reprinted from Kim and Turvey 1996.

a

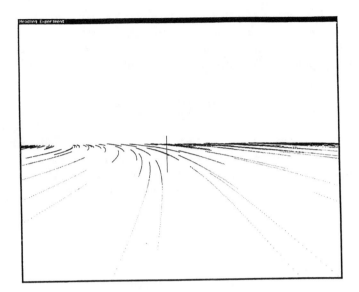

b

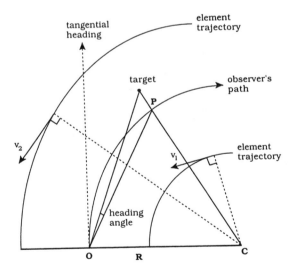

Figure 11.2   *(a)* A field of view of an optical flow field due to an observer's translation along a circular path with $R = 160$ m. The field of view corresponds to the observer's tangential heading. The vertical line identifies the observer's circular heading.
(b) Schematic diagram for circular locomotion. (O) the observation point, (C) the center of rotation of the observer's path, (R) the radius of curvature of the observer's path, and (P) the point on the observer's path line used to define heading angle.

values to c in the function gives a family of streamlines. The totality of these curves is called a one-parameter family of curves with c the parameter of the family and Ψ the *stream function*. If the stream function is known, then the velocity of the flow at a given point can be expressed in terms of the derivatives of the stream function

$$u = \frac{\partial \psi}{\partial y} \text{ and } v = -\frac{\partial \psi}{\partial x} \tag{11.1}$$

When there is no internal friction between the particles, the fluid particles do not rotate, that is, the fluid motion is everywhere free from rotation. The resulting fluid motion is called *irrotational flow*.[2] When the flow is irrotational, it can be shown that there always exists another function $\Phi(x, y)$, called the *velocity potential*, such that the components of the velocity are expressed as

$$u = \frac{\partial \Phi}{\partial x} \text{ and } v = \frac{\partial \Phi}{\partial y} \tag{11.2}$$

Further if it is assumed that variations in the density of moving particles is negligible, then the fluid flow is said to be *incompressible*.[3] Inspection of the velocity components of $\Phi(x, y)$ and $\Psi(x, y)$ of an irrotational and incompressible flow reveals that $\Phi(x, y)$ and $\Psi(x, y)$ satisfy the Cauchy-Riemann equations

$$\frac{\partial \Phi}{\partial x} = \frac{\partial \psi}{\partial y} \text{ and } \frac{\partial \Phi}{\partial y} = -\frac{\partial \psi}{\partial x} \tag{11.3}$$

When Φ and Ψ satisfy equation 11.3, it follows that Φ and Ψ constitute the real and imaginary parts of a single analytic function, that is, a differentiable function of a complex variable z such that

$$F(z) = F(x + iy) = \Phi(x, y) + i\psi(x, y) \tag{11.4}$$

This function is called the *complex potential* of the flow (Keldys 1963; Kreyszig 1962). The curves $\Phi(x, y) = c$ are the *equipotential lines*, and the curves $\Psi(x, y) = c$ are the streamlines. The curves of these two families are mutually orthogonal, and their trajectories are the orthogonal trajectories of the given curves. The velocity of the flow can be obtained by differentiating equation 11.4 and using the Cauchy-Riemann equations

---

[2] In vector calculus, an irrotational flow is expressed as curl $\mathbf{v} = 0$ in which curl $\mathbf{v}$ at a point is a measure of the angular velocity of the fluid in the neighborhood of the point. Since the photons making up the optical flow are viscosity-free, unlike fluid particles making up the fluid flow, it can be safely assumed that there is no internal friction between optical particles, hence, the resulting flow is irrotational.

[3] An incompressible flow means that every region in the flow is free of sources or sinks. In vector calculus, an incompressible flow is expressed as div $\mathbf{v} = 0$ in which div $\mathbf{v}$ is a measure of the amount of the substance that flows out of a given volume. Since photons are mass-less, optical motions cannot create or destroy optical elements. It follows that the resulting flow field must be incompressible.

$$F'(z) = \frac{\partial \Phi}{\partial x} + i \frac{\partial \psi}{\partial x} = \frac{\partial \Phi}{\partial x} - i \frac{\partial \Phi}{\partial y} = u - iv$$

Taking the complex conjugate gives the velocity vector of the flow

$$\mathbf{v} = u + iv = \overline{F'(z)} \tag{11.5}$$

In this way, the notion of analytic function is a potentially powerful tool for investigating various flow patterns because the real and imaginary parts of any analytic function of the complex variable z satisfy the Cauchy-Riemann conditions and may be considered as the velocity potential and stream function, respectively, of an irrotational steady flow of an incompressible fluid. Hence, by taking any analytic function and splitting it into real and imaginary parts, an infinite variety of patterns of streamlines can be obtained (Sutton 1957).

With respect to locomotion and optical flow, the streamlines of the flow field will correspond to the observer's path of locomotion, regardless of whether the path is linear or circular. By equation 11.1, the streamlines of any flow field are determined by the stream function Ψ. Considering Ψ as a higher-order property of optical flow, it can be hypothesized that perception of heading direction is based on the detection of Ψ (Kim and Turvey 1994). Importantly, there is experimental data to suggest that when Φ (the velocity potential) and Ψ are decoupled, observers can perceive translational and circular heading on the basis of the streamlines of the flow field determined by Ψ. Specifically, Warren et al. (1991a) randomized the magnitude of each vector component of the flow fields (translational and circular), while preserving the directional component. The researchers found reliable perception of heading based solely on the direction components of the velocity vector field, a result that was interpreted as indicating the ability of the visual system to tolerate a great deal of noise in optical flow. By the present analysis, the result of Warren et al. implicates the significance of Ψ to perceiving heading direction. Rather than suggesting an ability of the visual system to tolerate high noise levels, their result might be taken as suggesting an ability of the visual system to attend selectively to Ψ. Detection of Ψ would give the observer awareness of the entire flow field with a concomitant tolerance to local perturbations induced by various forms of noise, such as those manipulated in the experiment of Warren et al.

We have extended the preceding investigations to elliptical paths. In our experiment, we simulated, by means of pixel motions on a computer screen, movements parallel to the ground plane. We simulated three circular trajectories and four elliptical trajectories. The curvilinear path was controlled as a ratio between two semi-axes in which a is along the x-axis (along the observer's frontal plane) and b is along the y-axis (along the tangential heading direction). The aspect ratios varied randomly among $\pm 120/120$ m, $160/160$ m, $320/320$ m, $120/160$ m, $160/320$ m, $320/160$ m, and $160/120$ m (positive values correspond to a right-hand turn and negative values correspond to a left-hand turn). The first three ratios corresponded to circular paths, and the last four corresponded to elliptical paths.

Heading accuracy was assessed in terms of heading angle, the same measure used in Warren, Mestre et al. (1991b). Heading angle is the visual angle defined by a target on the ground surface, the point of observation and the point at which the observer's path would pass the target (see figure 11.2b). Heading angle varied randomly among ±0.5°, 1°, 2°, and 4°. Due to the curvilinear patterns of the flow fields, flow velocity was controlled as angular velocity, which was held constant at 3.3 deg/s. Consequently, tangential velocity varied depending on the radius of the curvature of the flow line on which a given particle traveled; for the three circular movements, the corresponding tangential velocities were 6.9 m/s (25 mph) for the 120 m radius condition, 9.2 m/s (33 mph) for the 160 m radius condition, and 18.4 m/s (66 mph) for the 320 m radius condition. In order to effect the separation of Φ (the velocity potential) and Ψ (the stream function), the magnitude of each flow vector was randomized at values between 0 and 5.94 deg/s in steps of 0.66 deg/s.[4] Because the display was updated every 15 ms (66 frames/s), the preceding meant that each dot rotated by an amount ranging from 0° to 0.09° in steps of 0.01° along its streamline.

The subject was told that he or she would be watching displays depicting the appearance of the ground when running or driving on a curvilinear (turning or winding) path or road. The subject initiated the displays through a key press. The subject watched the display until it stopped and a blue vertical bar appeared on the simulated ground surface. At that time, the subject pressed one key if the path of travel seemed to be to the left of the vertical bar and a different key if the path of travel seemed to be to the right of the vertical bar. (No feedback was provided during the experiment.) The results revealed that a separation of 2.1° between the locomotory path and the vertical marker was sufficient, on the average, for subjects to make a correct judgment 75% of the time, and that this threshold value did not differ for circular and elliptical paths. This level of accuracy, achieved with the local velocity vectors randomly varying in time, was comparable to the level achieved when the velocity potential function was consistent with the stream function (that is, when the local velocity vectors varied systematically in time). The results of our experiment suggest a perceptual sensitivity to a property of optical flow defined globally, consistent with the hypothesis that perceived direction of heading on different curvilinear paths may be based on the selective detection of Ψ.

Actively Changing the Field of View

The preceding suggests that optical flow fields characterized through equations 11.1–11.4 and 11.6 can provide information satisfying an important "modeling of

---

[4]In the case of circular locomotion, the resulting flow field can be described in terms of a complex potential, $F(z) = (K/2\pi i) \ln z$, where K is a real constant. From $z = re^{i\theta}$, the velocity potential and the stream function can be computed as $\Phi = (K/2\pi)\,\theta$ and $\Psi = -(K/2\pi) \ln r$, respectively, with the velocity of the flow, $\mathbf{v} = -(K/2\pi i)(1/\bar{z})$. Hence, the streamlines are the circles $r = c$, whereas the velocity is constant on every streamline.

the future," namely, information specific to the places in the visible environment that one will arrive at if one continues with the current locomotory dynamics. That is to say, there is some support for the hypothesis that the streamlines of the flow field determined by Ψ comprise information for directing locomotion in cluttered surroundings. Animals such as humans, however, are equipped with frontal eyes enabling them to sample only a portion of the ambient optical flow. This portion referred to as the *field of view* spans about 180° horizontally and 150° vertically in the case of humans. Accordingly, a person must constantly look around if potentially harmful collisions with other objects are to be avoided or if objects of interest that pass from the momentary field of view need further scrutiny. Bernstein (1967, page 117) was very appreciative of the significance of scanning and searching to the achievement of "maximally full and objective perception." He commented that this searching "is an entirely active process, and the effector side of the organism is here employed in a manner completely analogous to . . . afferentation in the control of movements." In like fashion, Gibson (1966) has emphasized how the exploratory activity of perceptual systems grounds the performatory activity of action-perception systems (Gibson 1966; Reed 1982).

Effects of Looking Around While Locomoting

As the observer looks around sampling different portions of the ambient optical flow, the structure of the cross-section of the fields of view, or retinal flow, changes. In other words, there is a non-trivial relation between optical flow and retinal flow. The availability of information in optical flow does not necessarily guarantee that the observer can extract the relevant information as desired. Instead, the optical flow field must be explored constantly in search of the information required to regulate movement. How can the observer find the needed information? In the case of locomotion on a linear path, retinal flow contains the focus of expansion only if the observer looks in the direction of heading, that is, if gaze direction coincides with heading direction. As soon as the gaze deviates from the direction of heading, the focus of expansion vanishes. For example, when the observer maintains fixation on an object of interest while moving rectilinearly, the eyes rotate. Since the fixated object is projected onto the fovea, the fixation point becomes another singularity in retinal flow replacing the focus of expansion and the image motion becomes curved (see figure 11.3a). The further the fixated object is located away from the direction of locomotion, the more the eyes rotate, and the more curvilinear are the image trajectories on the retina. If the observer is using the singularity in retinal flow, the strategy presents an ambiguous situation. Only when the observer looks in the direction of linear translation, can the singularity in retinal flow, that is, the focus of expansion, be informative of her heading. Despite this ambiguity, however, the observer can achieve nominal awareness of heading direction, or heading with respect to gaze, by means of motion parallax (Cutting et al. 1992; Kim et al. 1996a, 1996b).

In the case of pursuit fixation, the rotation is about a principal axis of the fixated object. That is, the image of the fixated object is projected on the fovea, whereas the images of the surrounding objects rotate about the axis of the image of the fixated object. Another type of rotation can be induced when the observer scans across the ambient optical flow. As the observer moves forward and the eyes sweep across the equator plane, the entire retinal image displaces in the opposite direction of rotation

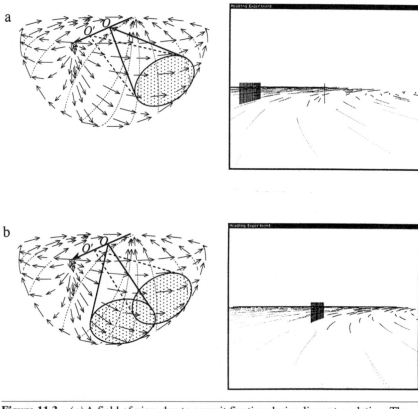

Figure 11.3   *(a)* A field of view due to pursuit fixation during linear translation. The base of the solid angle is fixed, while the apex is displaced along the path. The corresponding retinal flow, change of the image in the cross section of the field of view, is generated by fixation at the top of an object 1.6 m in height lying 15° to the right of the path. The cluster of vertical lines depicts the displacement of an object's image. The object is initially 25 m away from an observer along the observer's path. The focus of expansion is replaced by another singularity at the fixation point. Nevertheless, perception of heading can still be achieved through motion parallax that provides "nominal information (left or right) about the direction of movement with respect to gaze" (Cutting et al. 1992, page 59). *(b)* A field of view due to an execution of continuous eye rotation during linear translation. The base of the solid angle sweeps across the ambient flow field. Retinal flow is generated by rotating the eye to the right at the rate of 0.1°/frame.

Reprinted from N.G. Kim and M.T. Turvey 1996.

and there is no singularity in retinal flow (see figure 11.3b). Identification of heading direction using visual displays alone under these transformations turns out to be extremely difficult if not impossible (Royden et al. 1992; 1994).

Locomotion along a circular path in densely cluttered surroundings, such as a wooded area or the inside of a lecture theater, gives rise to a torus-like optical flow field, as shown in figure 11.4. In figure 11.5, curvilinear locomotion is depicted for the more simple case of traversing a ground plane essentially devoid of obstacles. Figure 11.5a reminds us that when the observer is looking in the direction of tangential heading, retinal flow is inherently curvilinear and there is no singularity in retinal flow (compare with figure 11.1b). Nonetheless, perception of curvilinear heading is sufficiently accurate for purposes of locomotory control, as reported in the "Seeing Where One Is Heading in Curvilinear Locomotion" section in this chapter. We make note here that the conditions adopted in the experiments reported in the above noted section were such that the subjects were allowed to rotate their eyes freely during the experiments. The setup was necessary, however, because in translation along a linear path the curvature of the path remains fixed so that the deviation between the path and the line of sight can be controlled as a measure of heading accuracy. In translation along a curved path, on the other hand, the curvature changes constantly. Accordingly, the subjects were allowed to scan freely the display area to identify the path whose direction of turn changed as a part of an experimental condition (i.e., turn either left or right). There is some evidence, however, that the field of view corresponding to tangential heading during translation along a curved path may not be sufficient to permit accurate heading. A radius effect was reported in Warren, Mestre et al. (1991b), which was further replicated in our study reported in the "Seeing Where One Is Heading in Curvilinear Locomotion" section. That is, the subjects in both studies were poor in heading judgment in the small radius (or large curvature) condition. Note that the smaller the radius of a curved path the bigger is the amount of deviation between the path and tangential heading.

We examined whether different fields of view resulted in different degrees of heading accuracy during translation along a circular path by simulating circular locomotion parallel to the ground with the ground composed of random dots. We further introduced a fixed amount of eye rotation to the simulations of circular locomotion—a method used by Warren and his colleagues in their heading studies during linear translation (Warren and Hannon 1988; Warren et al. 1988) and referred to by Cutting et al. (1992) as a fixed-camera-angle technique. The amount of eye rotation varied among values of $\pm 0°$, $6°$, and $12°$. Positive values indicate that gaze direction is right and negative values indicate that gaze direction is left with respect to the observer's tangential heading. The observer's circular path was defined in terms of the radius of a circle which varied randomly among values of ± 120 m, 160 m, and 320 m. Positive values correspond to a right-hand turn and negative values correspond to a left-hand turn. A constant tangential velocity of 13.2 m/s (or 29.7 mph) was used for each trial. Finally, heading angle varied randomly among $\pm 1°$, $2°$, and $4°$.

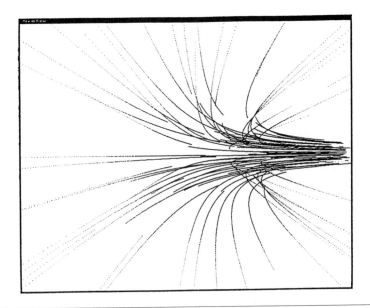

Figure 11.4   A torus-like circular flow field with $R = 320$ m seen through its cross section defined by a 10×10 m region.

The subject watched the display until it stopped and a blue vertical bar appeared on the ground surface. At that time, the subject pressed a key if the path of travel seemed to be to the left of the vertical bar or a different key if the path seemed to be to the right of the vertical bar. Displays were blocked by eye rotation and presented within-subject in two sessions, with the zero eye rotation condition (see figure 11.5a) in the first session and the two eye rotation (6° and 12°) conditions (see figure 11.5b, c) in the second session. There was no feedback throughout the experiment.

Results showed that performance was quite accurate in the 0° eye rotation condition with a mean threshold of 1.4°, consistent with the results reported in the previous studies (Warren et al. 1991a, b). (Cutting (1986) estimates that a heading accuracy of 4.2° is required for braking appropriately with respect to upcoming obstacles when driving at 13.2 m/s.) The accuracy of perceived heading degraded, however, with increased eye rotation. That is, the more the eyes were deflected from the path of locomotion, the less precise was the perception of heading, with the precision least for the circular path of greatest curvature.[5]

The results also revealed that not all eye rotations gave rise to the same degree of accurate judgment of heading direction. When direction of turn matched direction of gaze (left-left or right-right), overall performance was 76% in the left turn and 77% in the right turn (figure 11.5c). When direction of turn did not match

---

[5]The line of sight rotates as the observer moves along a circular path. From a coordinate system anchored at the eye, however, gaze direction is always fixed, simply being deflected from the sagittal plane by a fixed amount. Nonetheless, we refer to this manipulation as eye rotation for lack of a better expression.

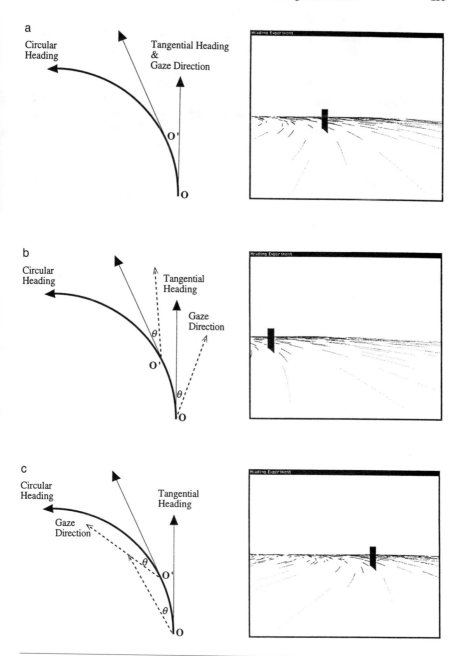

Figure 11.5   Fields of view generated by translation along a circular path with $R = 120$ m and direction of turn to the left. *(a)* Gaze direction coincides with tangential heading. *(b)* Gaze direction is deflected $12°$ from tangential heading in the opposite direction from circular heading. *(c)* Gaze direction is deflected $12°$ from tangential heading in the same direction as circular heading.

direction of gaze (left-right or right-left), performance was 69% and 64%, respectively (figure 11.5b). Clearly, this interaction demonstrates that some directions of looking are better than others when negotiating curved roads.

Optical flow resulting from rectilinear translation is depicted as a "melon-shaped family of curves" (Gibson 1950, 1986). Hence, the flow field is symmetrical with respect to focus of expansion. Whether one looks to the left or to the right, the field of view (the sampled region of optical flow by the eye) would be similar. Recall that optical flow resulting from circular translation was described in terms of a one-parameter family of concentric circles with the radius r as the parameter, that is, the streamfunction Ψ due to circular translation was defined as $-(K/2\pi) \ln r$. Hence, the flow field is asymmetrical with respect to tangential heading. For the same degree of rotation, the field of view is not the same gazing to the right and gazing to the left of heading direction. Put differently, the regions of optical flow sampled by the eye contain different kinds of information. In the case of circular heading perception, perception is enhanced when direction of gaze coincides with the direction in which the curved path turns. In the opposite case, in which gaze direction mismatches circular heading direction, perception is impaired. This result is consistent with the observation that when following a roadway with visible edge lines, drivers fixate the point on the inside of the curve that is tangent to the line of sight (Land and Lee 1994).

Effects of Looking at an Object While Locomoting

What happens when an object is fixated during curvilinear locomotion? We can examine this question concerning the effect of eye rotation through simulations graphically represented in figure 11.6. Note that the motion trajectories of surface elements projected on the retina due to circular locomotion are hyperbolic (Lee and Lishman 1977; Warren et al. 1991b). A clear contrast is revealed between the conditions of circular translation with fixation on an object lying away from the circular path (figure 11.6a, b) and fixation on an object lying on the path (figure 11.6c). Specifically, in the former, all velocity vectors in the image plane are curved with some degree of motion parallax around the fixated object. In the latter, all the image trajectories are linearized. Moreover, each velocity vector appears to lie in the direction tangent to the corresponding flow line. A tangent to a curve corresponds to the instantaneous direction of a velocity vector. Hence, by fixating on an object lying on her circular path, the observer can induce a tangent vector field on the retina. In a tangent vector field, curves formed by connecting the velocity vectors with the same direction define streamlines of a flow field (figure 11.6c).

The potentially deep significance of figure 11.6c is that it demonstrates that by actively rotating the eye, the observer can, in effect, *recover the ambient optical flow field from the motion in the image plane*, that is, from retinal flow. Of possible theoretical significance is the fact that this behavior of the eyes in sampling the flow field renders a nonlinear flow linear and, thereby, renders an apparently complicated problem of information pickup into an apparently simpler problem. In circular locomotion, the observer's path is specified by the optic flow line which passes from

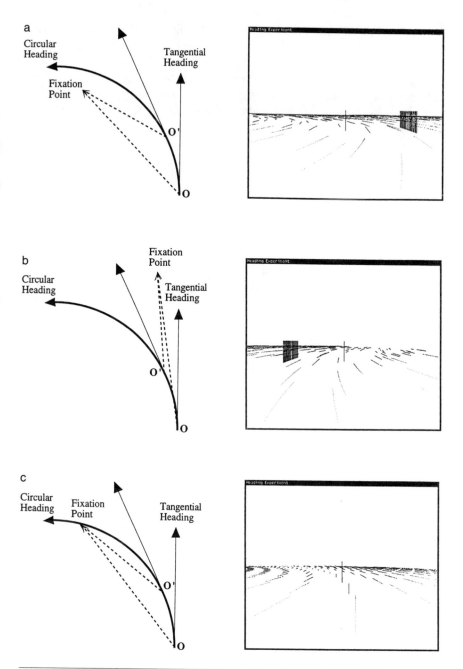

Figure 11.6   Pursuit fixation during translation along a circular path. (a) The fixated object lies inside of the circular path. (b) The fixated object lies outside of the circular path. (c) The fixated object lies on the circular path. The perpendicular velocity vectors identify the optic flow line corresponding to the observer's path.

view directly beneath her (Lee and Lishman 1977). This optic flow line can be easily determined by connecting all the tangent vectors in figure 11.6c whose directions lie perpendicular in the image plane—the projection of the observer's tangential heading in the image plane produces perpendicular vectors in the image plane. This method can also be used to determine the observer's path during linear translation. A line is a special case of a circle and rectilinear locomotion is a special case of circular locomotion.

The Essential Role of Eye Movements

It was noted in the "Seeing Where One Is Heading in Curvilinear Locomotion" section that an optical flow field due to the observer's locomotion along a linear path as well as a curved path can be effectively described by methods adopted from fluid dynamics. Specifically, it was shown that an optical flow field, regardless of path of locomotion, can be described in terms of a stream function and a velocity potential. The stream function of the flow field was proposed as the information specific to direction of heading. In the present section, we have concentrated on the fact that animals such as humans are equipped with frontal eyes, limiting sampling to only a portion of the ambient flow field. This ocular feature is particularly significant in relation to the notion that animals might use singularities contained in the retinal flow, specifically in the region of the flow field sampled by the eye, such as the focus of expansion. As noted, this singularity disappears by virtue of the observer's eye rotation during linear translation or during translation along a curved path, or it is replaced by another singularity as shown in the cases of pursuit fixation. For example, as is depicted in figure 11.3b, the combined effect of eye rotation and linear translation on the retinal flow pattern is the curved velocity vectors in the image plane, under which perception of heading turned out to be extremely difficult (Royden et al. 1992, 1994).

It is evident from the preceding that focusing the eyes on certain regions of the ambient optical flow field enhances detection of the nested structure included in the sample and lessens the detection of the nested structure outside of the sample. As the eyes turn away from the direction of locomotion—to examine more closely an environmental aspect in the field of view—perception of the exact place to which one is traveling becomes less precise. The further away the eyes are turned from the direction of heading, the more coarse becomes the perception of the place being approached. Pursuit fixation off the path of locomotion enhances *exteroperception* (perceiving distal objects) at the expense of *exproprioperception* (perceiving one's orientation to the surrounding surface layout) (Lee 1974). When steering vehicles, people do in fact pay attention to those regions in the optical flow field that are most relevant to their needs. For example, during driving along the open, straight road, drivers appear to maintain fixation for approximately 90% of the time within $\pm4°$ from the focus of expansion (Shinar 1978). Even in the more demanding case of negotiating a curve, drivers tend to fixate the point on the inside of the curve that is tangent to the line of sight (Land and Lee 1994; see also Shinar et al. 1977, for a similar finding). Clearly, an important research strategy (one that is in concert with

Bernstein's understanding of the significance of exploratory activity supporting performatory activity) is to study the different fields of view and their corresponding retinal flows with respect to a certain perception-action tasks as opposed to the more abstract concern for mechanisms that can separate optical flow components from retinal flow regardless of eye rotation (see summary in Warren 1995).

Seeing and Controlling the Severity of Encounters With Upcoming Surfaces

Another ubiquitous motor problem that animals face every day is controlling the severity of collisions with upcoming surfaces. Consider a mother walking toward her child. If she intends to stop and embrace the child, then she must decelerate sufficiently such that at the moment of embrace no forward momentum remains. How does she achieve the control of her locomotion which ensures nothing less than gentle contact with the child?

Consider the means by which she regulates her locomotion. The step cycle of her leg consists of two basic phases: The swing phase in which the foot is off the ground and moving forward, and the stance phase in which the foot is on the ground and the leg is moving backward. Furthermore, she can change her speed of locomotion by varying the propulsive force developed during the stance phase, while the duration of the swing phase remains relatively constant throughout the walk. It has been found that human subjects regulate step length during walking and running by reducing the horizontal braking impulse (Patla et al. 1989). In principle, then, the observer can bring about an intended contact with the approaching surface by regulating propulsive force during the stance phase of the step cycle. How does she determine how much braking force she needs to apply for an intended contact with the surface? Too much braking may result in stopping short of reaching the surface. Too little would lead (in the case of mother and child) to a disastrous consequence.

The Dynamics of a Transpiring and Eventual Collision Event

As the observer approaches a surface at a speed of $V(t)$, this would give rise to a kinetic energy of $1/2mV(t)^2$. To ensure a gentle contact with the surface, the forces applied to brake her forward motion must, over the successive steps, dissipate the kinetic energy she possesses due to her approach velocity. That is, she has to perform work (Force $\times$ Distance) over the remaining distance, $Z(t)$, so as to cancel the current kinetic energy such that there would be no kinetic energy left when (or before) she reaches the surface. Since Force = Mass $\times$ Acceleration, the observer can achieve the intended contact by regulating the amount of propulsive force she applies during the stance phase of her walking. That is, given access to the dynamics, the observer would know what kind of impact of collision would arise from the following relations:

if $\frac{1}{2} m \times V(t)^2 \leq m \times D(t) \times Z(t)$ then a soft collision or no collision results;

if $\frac{1}{2} m \times V(t)^2 > m \times D(t) \times Z(t)$ then a hard collision results.

If the observer bases her control of collision on the above relations, she needs three types of information, namely, the remaining distance from the surface, $Z(t)$, her current speed, $V(t)$, and her current deceleration, $D(t)$. There is the possibility, however, of a much simpler solution.

To be specific, as an observer moves toward an obstacle, the optical solid angle subtended by the obstacle expands. The derivative of the inverse of the relative expansion of the optical solid angle with respect to time, or $\tau(t)$, is specific to the time that will elapse before the observer contacts the surface (Lee 1976, 1980). $\tau(t)$ is given as follows

$$\dot{\tau}(t) = -\frac{Z(t)}{V(t)} \tag{11.6}$$

Differentiating equation 11.6 with respect to time yields

$$\dot{\tau}(t) = -1 + Z(t) \times \frac{D(t)}{V(t)^2} \tag{11.7}$$

For a soft collision,

$$\frac{1}{2} m \times V(t)^2 \leq m \times D(t) \times Z(t) \tag{11.8}$$

Rewriting equation 11.8 gives

$$Z(t) \times \frac{D(t)}{V(t)^2} \geq 0.5 \tag{11.9}$$

From equations 11.7 and 11.9,

$$\dot{\tau}(t) \geq -0.5$$

In summary, the observer can, in principle, simply detect $\dot{\tau}$ (the rate of change of the inverse of the relative rate of expansion of the optical solid angle subtended by the approaching surface) in order to regulate, with appropriate precision, the accelerative or decelerative forces of her locomotion (figure 11.7). That is, when $\dot{\tau} \geq -0.5$, the corresponding optical state specifies that the impending collision will be soft (i.e., the current level of deceleration is adequate to stop in front of the surface). When $\dot{\tau} < -0.5$, the corresponding optical state specifies that the impending collision will

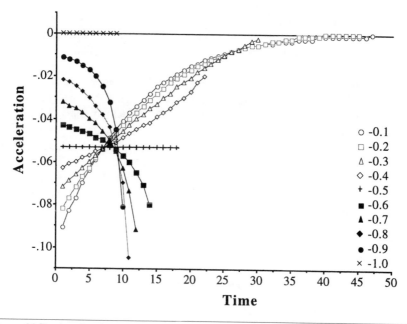

Figure 11.7 Acceleration shown as a function of time for values of $\dot{\tau}$ ranging between −1.0 and −0.1 using simulated data. Note that the units of acceleration and time are arbitrary. Each simulation was conducted with a fixed value of $\dot{\tau}$ maintained over the time course of the collision. All simulations were generated by the same initial conditions: an initial distance of 10 units and an initial velocity of 1 unit of distance/unit of time. Simulations were terminated either when contact occurred or after 50 iterations, which-ever came first. Soft collisions $(\dot{\tau} \geq -0.5)$ are characterized by decreasing deceleration, whereas hard collisions $(\dot{\tau} < -0.5)$ are characterized by increasing deceleration. The critical value of −0.5 stands out as a case of constant deceleration. The value of −1.0 also stands out as a case of constant velocity (zero deceleration). Soft collisions are also characterized by longer duration.

be hard (that is, unsafe or a crash). Because the preservation of $\dot{\tau}$ at any value greater than −1 will mathematically result in a safe stop, it has been argued that the preced-ing distinction between hard and soft collisions is invalid (Kaiser and Phatak 1993). It is the case, however, that the infinite decelerations required by the mathematical argument are physically impossible. When the fact of finite decelerative changes is noted, a value of −0.5 defines a pragmatic boundary between safe and crash states (Kim et al. 1993; Yilmaz and Warren 1995; Zaal and Bootsma 1995). Returning to the example of a mother approaching her child, it can be hypothesized that on the basis of $\dot{\tau}$ and its critical value of −0.5, the mother may have a reliable basis for regulating her current stepping so as to achieve a future state of a soft or safe contact with her child.

It has been demonstrated that animals and humans alike control their braking by regulating $\dot{\tau}$ around the critical value during safe approaches to obstacles. Human participants in a study by Yilmaz and Warren (1995) viewed closed-loop displays of approach to an object and manipulated a "brake" consisting of an optical mouse that moved in a slide attached to a spring-loaded handle, with dynamics modeled on an automobile brake. The mean value of $\dot{\tau}$ during braking was –0.51. In another study reported by Wann et al. (1993) in which human subjects approached a surface in a variety of running and touching tasks, the manner of deceleration was consistent with a constant $\dot{\tau}$ of –0.45 to –0.5. In a passive perceptual judgment task, human observers were shown to be able to partition various optical states corresponding to collision events and simulated as computer displays into two distinct categories, soft and hard contacts, at $\dot{\tau} = -0.5$ as predicted (Kim et al. 1993). Data also indicate that animals utilize $\dot{\tau}$ values between –0.5 and –1.0 to achieve different manners of contact with a surface such as controlled collision (Lee et al. 1993). For example, in an investigation of hummingbirds, Lee et al. (1991) found that docking at a feeding tube occurred with a mean $\dot{\tau}$ value of –0.71, that is appropriate for entering the feeding tube rather than stopping at it (see also Wann et al. 1993).

Other Action Categories Based on Critical Values of $\dot{\tau}$

$\dot{\tau}$ is a dimensionless number. In physics, dimensionless variables have proven to be extremely useful in classifying certain physical states into distinct energetic states at critical values (e.g., Reynolds and Rayleigh numbers). As shown above, $\dot{\tau}$ parses optical energy fluxes into two distinct informational states of hard and soft collisions at the value of –0.5. In common with the principal dimensionless variables investigated in physics, $\dot{\tau}$ has other critical values with different categorical effects.

First, as noted above at $\dot{\tau} = -0.5$, deceleration becomes constant, thereby partitioning hard from soft collisions. When $\dot{\tau} > -0.5$, deceleration decreases, resulting in a soft collision at impact, whereas when $\dot{\tau} < -0.5$, deceleration increases, resulting in a hard collision.

Second, when $\dot{\tau} = -1.0$, the approach velocity of collision becomes constant. That is, when $\dot{\tau} > -1.0$, the collision is manifested by some degree of deceleration, whereas when $\dot{\tau} < -1.0$, the collision is manifested by some degree of acceleration (figure 11.8). Accordingly, if the observer intends to accelerate as a target is approached (think of tackling an opposing player in American or rugby football), then this intent can be satisfied by holding $\dot{\tau} < -1.0$. If, on the other hand, the observer intends to slow down, this intent can be satisfied by holding $\dot{\tau} > -1.0$. Although it may appear to be a trivial task, locomoting at a constant velocity is indeed regulated by visual information such as $\dot{\tau}$. Bernstein in fact recognized that the major muscle impulse that throws the leg forward ensuring, thereby, strides of uniform length, is guided by anticipatory "sensory correction," an ability absent in the very young (1996, page 224).

Third, when $\dot{\tau} = 0$, at every instant the value of τ remains the same, thereby separating actual contact with the target from no contact (contact with or without collision). If $\dot{\tau}$ is maintained at 0 and the approach is instantiated with the condition that the second step is exactly half the size of the first step, then each successive step becomes exactly half of the previous step taken, thereby creating Zeno's paradox

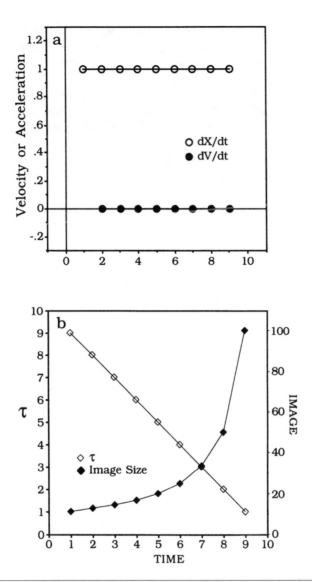

Figure 11.8   A simulation of an approach toward an obstacle with $\dot{\tau} = -1.0$. The simulation was conducted with an initial distance of 10 distance units and an initial step size of 1 distance unit. It takes 1 time unit to complete a step cycle, resulting in an initial velocity of 1 distance unit/time unit. The object is 10 units by 10 units, and the image plane is located 1 unit (distance) away from the nodal point, which gives an initial image of 1 distance unit. τ is computed in terms of $[\text{Image}/(\Delta I/\Delta t)]^{-1}$, that is, the inverse of the relative rate of expansion of the object image. Simulation terminated after the tenth iteration, when "contact" is made between the point of observation and the object. At that point, the image has expanded infinitely. *(a)* An approach with $\dot{\tau} = -1$ is noted for its constant velocity (zero deceleration). *(b)* The time to contact the obstacle is accurately specified by τ.

(figure 11.9). When $\dot{\tau} = 0$, the target becomes an asymptote (a point which cannot be reached). The absolute distance between the point of observation and the contact point shrinks, but, paradoxically, the relative distance in terms of "time-to-contact" remains the same as when the first step was taken toward the goal. When $\dot{\tau} > 0$, the goal cannot be reached; time-to-contact increases, even though the point of observation nears the goal. On the other hand, when $\dot{\tau} < 0$, contact will occur between the point of observation and the goal, and time-to-contact decreases as the point of observation approaches the goal. Hence, $\dot{\tau} = 0$ is another singularity which partitions collision events into contact and no-contact categories. Practically, this value can be used to maintain a safe following distance when driving a car behind another moving vehicle (Lee 1976).

Paradoxically, as $\dot{\tau}$ increases beyond 0 $(\dot{\tau} > 0)$, the time to contact the intended target increases even though the distance between the point of observation and the target shrinks. That is, despite the shrinking distance between the observer and the target, the relative distance in terms of "time-to-contact" increases, indicating that the observer is getting farther away from the target. As $\dot{\tau}$ reaches the value of 1, another interesting property is revealed. When $\dot{\tau} = 1$, the rate of optical expansion becomes constant (figure 11.10). Hence, not every approach results in a looming optical pattern, as has been commonly assumed. Approaches with $\dot{\tau} > 1$, at least theoretically, will result in a decreasing rate of optical expansion.

Seeing the Size of the Step to be Taken

Another motor problem faced by terrestrial animals, one that mandates Bernstein's physiology of activity, is negotiating uneven terrain. Ordinarily, the ground surface is uneven; it is also cluttered with obstacles of various sizes and substances. Adjusting gait to match the terrain is a crucial problem. Even a very simple and monotonous movement such as maintaining uniform footsteps by a trained athlete, according to Bernstein, cannot be attained by a motor formula or a motor cliche, i.e., the brain sending identical motor impulses to the muscles (Bernstein 1996, pages 180–181). To the contrary, such movements are achieved through the vigilant tracking of movements with sensory corrections (Bernstein 1996, page 196).

How do animals control gait or step length during locomoting on an uneven ground surface while at the same time balancing their postures? Consider long jumping, a task that demands precise control of gait to a take-off board while maintaining maximum controllable velocity (Berg et al. 1994; Lee et al. 1982). The final few strides are crucial to achieving accurate foot placement at take-off. Specifically, jumpers' gaits have been shown to follow a rather stereotypical pattern in which stride length progressively increases with acceleration. However, the jumpers' final few strides to the board are qualitatively different and seem to be adjusted on the basis of visual information about the target. According to Lee et al. (1982), long jumpers adjust their final strides (stride-lengths) by regulating the vertical impulse of each stride. To explore how the regulation of vertical impulse can bring

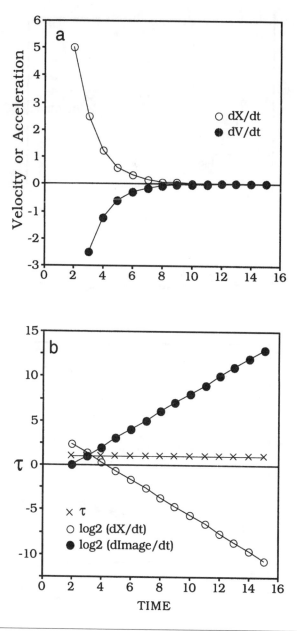

Figure 11.9 A simulation of an approach with $\dot{\tau} = 0$, initial distance of 10 distance units, and initial step of 5 distance units, or initial velocity of 5 distance units/time unit. Simulation was terminated after the fifteenth iteration, when the approach speed became too "slow." *(a)* Changes in velocity and deceleration. *(b)* The predicted value of τ becomes constant at every step. The rate of log base 2 of the approach velocity and the rate of log base 2 of the rate of image expansion become constant as well.

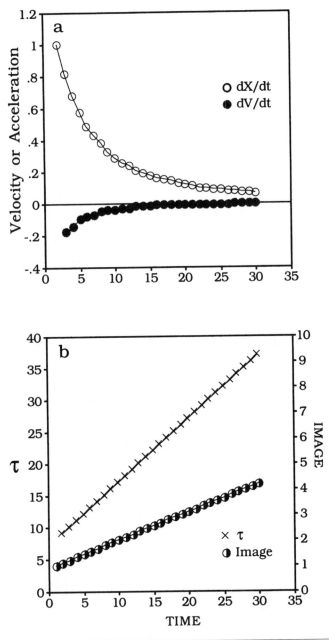

Figure 11.10 A simulation of an approach with $\dot{\tau} = 1$ with initial distance of 10 distance units, and initial step of 1 distance unit (the initial velocity of 1 distance unit/time unit). *(a)* Changes in velocity and deceleration. *(b)* When $\dot{\tau} = 1$, the image expands at a constant rate.

about adjustments in step length, Warren et al. (1986) conducted a study in which subjects were asked to run on an uneven surface while stepping on a series of irregularly spaced targets. Their analysis showed that runners' adjustments in step length were made by modifying step time. Warren et al. proposed that step time regulation can be achieved by determining the time interval between two targets on the irregular ground surface, where time-to-contact each surface target was specified by τ. This time interval is referred to as *tau gap*. In other words, step time for each stride is specified by tau gap which, in turn, is used to regulate vertical impulse of each stride by the control function

$$I = mg\Delta\tau \tag{11.10}$$

where I is the vertical impulse, m the runner's mass, g the gravitational constant, and $\Delta\tau$ tau gap.

Tau gap may indeed specify the correct step time needed for each step as one runs over an uneven surface; however, extending this type of analysis to the long jumping case may be inappropriate. Recall that long jumpers' regulation of their final strides to the take-off board was distributed over a minimum of three steps. Hence, long jumpers need access to at least three separate tau gaps, one corresponding to each of the final three steps. Lee and his colleagues speculate that the final approach phase might be regulated as a single functional unit based on an average value of τ, distributing their total adjustment in a consistent way over the three strides (Lee et al. 1982; Lee and Thomson 1982). However, their data do not support this conclusion. First, the three strides that constituted the zeroing-in phase were not equal; instead, they followed a general pattern of short step, long step, short step, jump. Second, even if we assume that this pattern is also stereotyped or preprogrammed, the amount of variance in each stride length during the final phase was substantially larger than the variance in the steps making up the approach phase. Taken together, there is little indication that suggests that step regulation during the final zeroing-in phase is an act of preprogramming. Furthermore, the final phase of the long jump was not always accomplished in three steps. Instead, the number of steps in the final phase was determined by the consistency of the approach phase (Lee et al. 1982). In sum, adjustment of step length during the final phase of the long jump can be better construed as a result of continuous coupling between perception and action rather than as a single ballistic adjustment. Instead of relying on a single measure of information in terms of τ, detected prior to the initiation of the final phase in the long jump, long jumpers may rely on a variety of measures of target-related information as the basis for fine-tuning their final approach. Accordingly, the online correction process may vary with each jump, resulting in high variance of step length during the final phase, an example of a continuous coupling of perception and action (cf. Bootsma and van Wieringen 1990). How do long jumpers determine exact vertical impulse for each stride in the zeroing-in phase such that they can strike the take-off board with maximum precision and minimum loss of velocity?

τ *as a Recursive Function*

One possible solution is provided by recursive functions. Continuing the discussion of such functions in the "Introduction" section, the formal device used to construct a recursive formula is Taylor's theorem, which expresses what happens near a point in terms of what happens at the point. Consider the following expression:

$$f(t_0 + h) = f(t_0) + hf'(t_0) + \frac{h^2}{2!} f''(t_0) + \; . \; . \tag{11.11}$$

where $f'(t_0), f''(t_0), \ldots$ are the successive derivatives of $f(t)$, evaluated at t_0, provided that the necessary successive derivatives exist. A function f can be evaluated "near" any point, entirely in terms of its behavior at the point.

Similarly, a recursive formula of τ can be formulated as

$$\tau(t_0 + h) = \tau(t_0) + h\tau'(t_0) + \frac{h^2}{2!} \tau''(t_0) + \; \ldots \tag{11.12}$$

where $\tau'(t_0), \tau''(t_0), \ldots$ are the successive derivatives of $\tau(t)$ evaluated at t_0, provided that the necessary successive derivatives exist.

Recall that earlier we rejected recursion as a way to model the future or goal-directed purposeful behavior exhibited by animals. It was rejected because, despite successful establishment of entailment between successive values in a series, the series is completely stripped of any significance by the arbitrariness of the initial condition. The appropriateness of recursion in the present section is due to the fact that intention fixes the derivative—in order to fulfill the intent of a particular type of contact with an upcoming surface, the informational variable $\dot{\tau}$ is maintained constant (or practically so) at a particular value throughout the approach (cf. Lee et al. 1991; Wann et al. 1993; Yilmaz and Warren 1995). When $\dot{\tau}$ is held constant, its higher order derivatives become all zeros, which simplifies equation 11.12 as follows:

$$\tau(t_0 + h) = \tau(t_0) + h\dot{\tau}(t_0) \tag{11.13}$$

By varying h, equation 11.13 provides all potential times-to-contact of an approach event (with h, in the case of bipedal or quadrupedal locomotion, defined in units of locomotor cycles, e.g., von Hofsten and Lee 1994). An approach in which $\dot{\tau}$ is held constant has an added benefit. Solving equation 11.12 requires the availability of higher order derivatives of $\tau(t)$ ad infinitum and, therefore, infinite perceptual resolution. It is standard practice in physical analyses to truncate a Taylor expansion after the first few terms and usually after the first two terms. Any arbitrary truncation, however, replaces a series by an approximation with inherent errors. The intent to bring about a particular type of contact with an approaching surface is doubly significant: it eliminates the necessity for an infinite degree of perceptual resolution and it provides a principled truncation. Equation 11.12 can be replaced by equation 11.13.

Returning to the long jump example, the large amount of variance—as well as the varying number of steps—in the final phase suggests that long jumpers may be regulating their strides by using a series of "tau gaps" or $\Delta\tau$. Utilizing equation 11.13, the series of tau gaps can be obtained from the available series of τs (e.g., $\Delta\tau_3 = \tau_3 - \tau_2$, $\Delta\tau_2 = \tau_2 - \tau_1$, and $\Delta\tau_1 = \tau_1 - \tau_0$, where τ_3, τ_2, and τ_1 correspond to the values of the times-to-contact with the take-off board from each anticipated footstep position). Since $\Delta\tau$ specifies the step time of each stride and step time is proportional to vertical impulse (Warren et al. 1986), long jumpers can achieve accurate step length by adjusting vertical impulse for the next series of steps on the basis of the values of $\Delta\tau$.

Similar patterns of movement adjustment have been reported in other goal-directed activities. For example, Wann et al. (1993) observed that, in a variety of running and touching tasks, participants changed their approach pattern prior to making the required kind of hand or head contact with a door. Participants employed two different approach patterns, a transport phase and a homing phase. The homing phase was initiated about two or three arm reach lengths from the target, a pattern similar to the final stride adjustment observed in the study of long jumpers. One can speculate that the adjustment of the homing phase in these tasks also is regulated by detecting a series of τs and tau gaps.

A Simulation of the Long Jump Approach

In order to demonstrate how the recursive function of $\tau(t)$ specifies prospective information about time-to-contact, we simulated an approach to a surface at a constant deceleration; an approach which results in a safe collision. The simulation was conducted with an initial distance of 10 distance units and an initial step size of 1 distance unit. It takes 1 time unit to complete a step cycle, resulting in an initial velocity of 1 distance unit/time unit (figure 11.11a). The object is 10 units by 10 units, and the image plane is located 1 unit (distance) away from the nodal point, which gives an initial image of 1 distance unit (figure 11.11b). τ is computed in terms of the inverse of the relative rate of expansion of the object image. The simulation terminated after the ninth iteration, when "contact" is made between the point of observation and the object. At that point the image has expanded infinitely.

When an observer moves toward a surface at a constant deceleration, $\dot{\tau}$ remains constant at -0.5. The value of τ at the next step cycle can be obtained by adding the values of τ and $\dot{\tau}$ at the current step, in conformity with equation 11.13. In figure 11.11b, the slope of the curve depicting τ coincides with the value of $\dot{\tau}$. Note that, after the first step, time-to-contact as specified by τ is 4; i.e., τ predicts that contact would occur 4 time units later. However, actual contact occurs 8 time units later. Thus, time-to-contact is underestimated by τ alone. In contrast, τ as a recursive function as in equation 11.13, accurately predicts when contact would occur. We know that, if a contact is to occur, the predicted value of τ will be 0. In the example shown in figure 11.11, after the first step is taken the current values are $\tau = 4$ and $\dot{\tau} = -0.5$. By substituting these values in equation 11.13 and solving, we obtain a

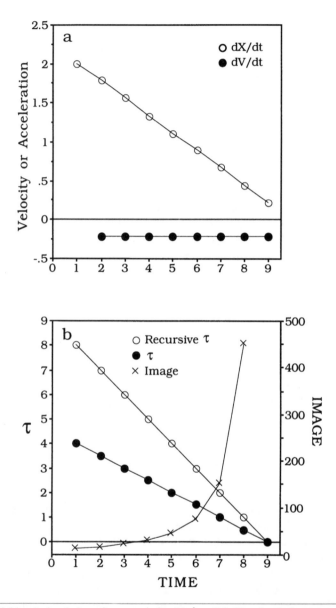

Figure 11.11 A simulation of an approach with $\dot{\tau} = -0.5$. The simulation was conducted with initial distance of 10 distance units and initial step size of 1 distance unit (the initial velocity of 1 distance unit/time unit). Simulation was terminated after ninth iteration, when a contact is made between the point of observation and the object. *(a)* The collision event induced by $\dot{\tau} = -0.5$ is noted for the constant deceleration. *(b)* τ specific to distance/velocity underestimates the actual time-to-contact under the non-constant velocity approach (e.g., a constant deceleration approach). The recursive function of τ, on the other hand, accurately predicts the time-to-contact of the obstacle at every instant.

value for h of -8. That is, contact will occur 8 steps later, in agreement with the actual time of contact shown in figure 11.11. With an increase in step cycle, h, the "prospectiveness" of information (i.e., the ability to see into the future) increases.

In the present section we have characterized the prospective nature of perceptual information through a recursive formulation of $\tau(t)$ generated by applying Taylor's theorem and have shown in a simulation how the future states of τ can be specified given current values of τ and $\dot{\tau}$. We also explored how the observed variance in the final strides taken by long jumpers might be explained as the continuous coupling of perception and action supported by advance availability of perceptual information through τ formulated as a recursive function. The recursive formulation of τ allows long jumpers to access future values of τs, as in equation 11.13, so that they can adjust their strides in a future-looking manner, thereby achieving precise regulation—an example of prospective control. After each step is taken, further adjustments can be made if needed by repeating the same process—an example of the continuous coupling of perception and action. It remains to be seen whether recursive formulations of τ are indeed the basis for prospective control, as conjectured. It would be also interesting to see whether recursive formulations of other kinds of perceptual information, given the steadying and selective influences of intention, offer equally appealing possibilities in terms of prospective control.

Seeing Where and When to Catch a Ball

When a fly ball is hit toward him, a skilled baseball player reacts immediately, moving smoothly and continuously to the landing site. For a proper catch, however, the problem is complicated by the facts that not only does the fielder have to be at the projected landing site but, in addition, he has to be there at the right time. Ordinary observations of skilled players and experimental data on less skilled performers confirm that people are quite competent at catching baseballs (McLeod and Dienes 1993, 1996; Michaels and Oudejans 1992). With ball motion limited to just the vertical dimension, experiments have revealed that fielders perceive the future landing point. They never break in the wrong direction, that is, they move forward to catch a ball that is going to drop in front or backward to catch a ball that is going to fall behind their present positions. These strategic adjustments were made on the order of 500 ms or less (McLeod and Dienes 1996; Michaels and Oudejans 1992). The experimental results suggest that in the first few hundred milliseconds of the ball's flight there is optical information to guide the outfielder's locomotion.

Figure 11.12 depicts the trajectory of a fly ball projected with initial velocity of 22 m/s at a 45° angle toward a fielder. In an ideal environment without wind resistance, the ball flies about 49 m horizontally for a duration of 3.2 s. For an outfielder standing inside of the arc of the trajectory (he has to retreat to catch the ball), the angle of elevation of gaze from the fielder to the ball, or α, increases eventually reaching 90° when the ball passes right over the fielder, then decreases until it lands on the ground. For the outfielder standing outside of the arc (he has to advance to

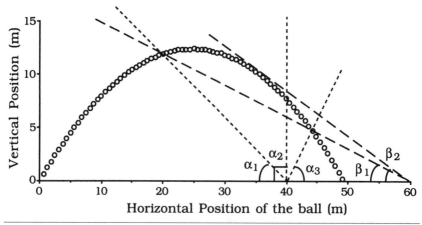

Figure 11.12   The trajectory of a fly ball projected with initial velocity of 22 m/s at a 45° angle toward a fielder in an ideal environment without wind resistance.

catch the ball), α increases reaching the maximum value when the ball is at the point where the trajectory is tangent to the line of sight, after which α decreases. Figure 11.13a depicts the change in α for outfielders standing at 40 m, 45 m, 49 m, 55 m, and 60 m, respectively, away from the launching site. When α decreases, the solution is simple regardless of the position of the fielder—simply move in the direction of the ball. The real challenge, however, lies in identifying the direction of movement during the first few hundred milliseconds of the ball's flight when it has just started to rise upward.

The Information for Catching

One possible source of information the fielder might use is the rate of change in the tangent of the elevation angle, i.e., $d(\tan \alpha)/dt$. (Michaels and Oudejans (1992) use the vertical optical velocity of the ball, i.e., dy/dt, instead of $d(\tan \alpha)/dt$. These latter two terms are mathematically equivalent—one is defined in terms of image plane variables whereas the other is defined in terms of optical angle variables.) Specifically, Chapman (1968) noted that $\tan \alpha$ increases at a constant rate for a fielder standing at the landing site (that is, $d(\tan \alpha)/dt$ = constant or $d^2(\tan \alpha)/dt^2 = 0$). He also noted that for the fielder standing inside of the arc of the trajectory $\tan \alpha$ increases at an increasing rate (i.e., $d^2(\tan \alpha)/dt^2 > 0$), whereas for the fielder standing outside of the arc, $\tan \alpha$ increases at a decreasing rate (i.e., $d^2(\tan \alpha)/dt^2 < 0$). Figure 11.13$b$, depicting the values of $d^2(\tan \alpha)/dt^2$ for the first second after launch, clearly demonstrates that this is indeed the case. Given these observations, Chapman (1968) suggested that the fielder can catch the fly ball if he moves in such a way as to maintain a constant rate of increase of $\tan \alpha$. Behavioral data further suggested that, indeed, skilled ball players kept $d^2(\tan \alpha)/dt^2 = 0$ as they ran toward the ball for an interception (McLeod and Dienes 1993, 1996; Michaels and Oudejans 1992).

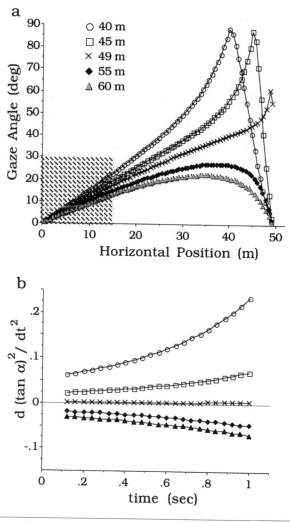

Figure 11.13   *(a)* The angle of elevation of gaze from outfielders standing at 40 m, 45 m, 49 m, 55 m, and 60 m, respectively, away from the launching site as the fielders track a fly ball projected with initial velocity of 22 m/s at a 45° angle toward them. *(b)* Change in $d^2(\tan \alpha)/dt^2$ for the first 1 s after launch.

A Simulation of Catching

In order to illustrate the preceding claim, simulations were conducted to examine whether an interception can be made based on the strategy of canceling out vertical optical acceleration of the fly ball. For example, if $d^2(\tan \alpha)/dt^2 > 0$, then the point of observation O moved in the forward direction. If, on the other hand, $d^2(\tan \alpha)/dt^2 < 0$, then O moved in the backward direction. The simulations were updated every 40 ms.

For each update, if the same condition continued, speed was increased by 2.5 m/s (0.1 m per each update). If the condition was reversed, speed was decreased by 2.5 m/s. Despite changes in the position of O, if two successive values of $d^2(\tan \alpha)/dt^2$ remained the same sign, speed was increased by an additional 1.25 m/s (0.05 m per each update). In all of these simulations, O started to move 0.5 s after launch.

Two types of ball trajectories were used. One trajectory was generated with initial velocity of 22 m/s projected at 45° and the other with initial velocity of 24 m/s projected at 55°. In the 22 m/s, 45° case, the ball traveled about 49 m for a duration of 3.2 s, whereas, in the 24 m/s, 55° case, the ball traveled about 55 m for a duration of 4 s. In the 22 m/s, 45° case, when O started 55 m from the launch site, following the above rule, O arrived at the landing site with a velocity of 4 m/s (figure 11.14a). When O started 43 m from the launch site, O arrived at the landing site with a velocity of 4.8 m/s (figure 11.14b). In the 24 m/s, 55° case, where O started 60 m from the launch site, O reached the landing site with a velocity of 3.9 m/s (figure 11.15a). When starting 45 m from the launch site, O reached the landing site with a velocity of 5.8 m/s (figure 11.15b). In all of these simulations, the final position of O was within ±0.2 m away from the landing position of the ball, which is well within the reach of an average human adult. Hence, they all can be considered as successful catches. Moreover, the final speed at the moment of interception, which ranged from about 4 m/s to 5.8 m/s, appears to be compatible with the running speed reached by actual fielders after running 8–10 m (McLeod and Dienes 1996).

Seeing When the Paths of Others Will Intersect: The Third Person Perspective

The environment consists of animate as well as inanimate objects. For many activities, it is crucial for an animal to be able to perceive not only the properties of the environment but also the properties of other animals taken with reference to itself. Animals, however, are constantly in motion. Accordingly, it is equally important for one animal to be able to perceive the trajectory of other animals' motions so as to control and coordinate its activity prospectively with respect to other animals. That is, prospective control is not limited to the environment of inanimate objects but extends to the other members of the society sharing the same environment with the individual. For example, for safe driving, a driver should be able to control and coordinate his driving with the road ahead of him. At the same time he should be able to anticipate the driving of other drivers who share the same road with him to avoid a collision. That means that the driver must not only be aware of which direction he is heading, but also he must be aware of which direction the driver in the next lane is heading.

At each observation point in the environment, there is an ambient optic array whose structure is lawfully determined by the surrounding environment. However, only one observer can occupy a point of observation at a given moment. Hence, information contained in a point of observation is private to the observer

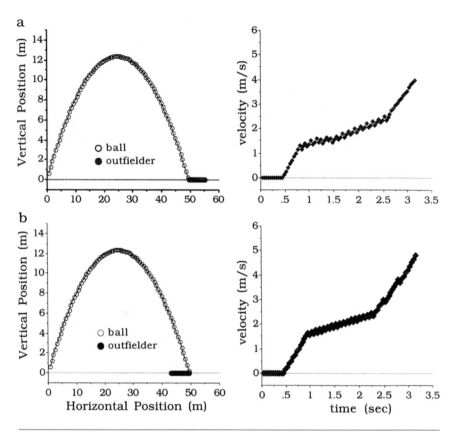

Figure 11.14   Simulations of interceptions by an outfielder based on the strategy of canceling out vertical optical acceleration of the fly ball. The ball was launched with initial velocity of 22 m/s at a 45° angle. *(a)* An outfielder starting 55 m from the launch site advanced for a catch. *(b)* An outfielder starting 43 m from the launch site retreated for a catch.

who occupies that point of observation. But observers are constantly in motion. Therefore, whoever occupies a given point of observation can sample the same information available at that point of observation. That is, the observer can have the same perceptions as another observer could have. To the extent that other observers can occupy the same point of observation, information can be said to be public. As information is public, so is perception. By sharing the same information, accordingly, each observer can be aware of the shared environment (Gibson 1982a, 1986).

The examples presented in the preceding sections can be considered as private information, that is, information taken from the point of view of an observer participating in the event. As the observer moves in the environment, there is information in optical flow specifying the observer's direction of heading. As the observer approaches

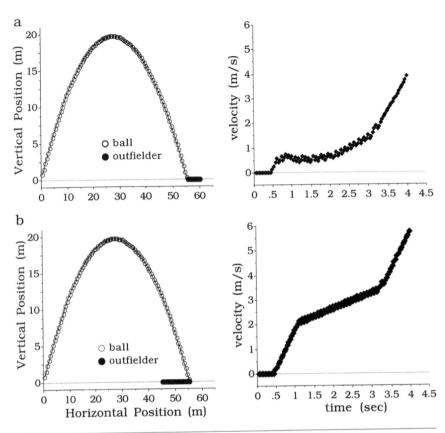

Figure 11.15 Simulations of interceptions by an outfielder. *(a)* An outfielder starting 60 m from the launch site advanced for a catch. *(b)* An outfielder starting 45 m from the launch site retreated for a catch.

an obstacle, there is information specifying the impact of collision. As a projectile rises up in the air, there is information specifying the landing site of the projectile. Can the observation of an event by a spectator be as accurate as that of a participant in the event? If so, what is the optical basis for the spectator's judgment?

Gibson (1982a, page 412) noted that: "If you see a head-on view of a bounding tiger and I see a side view, you are in greater danger than I am; but we both see the same tiger. We also see the same event: You see him approaching you and I see him approaching you." To the observer watching the bounding tiger head-on, the impending collision is specified by the dilating pattern of the optical solid angle projectively linked to the tiger. To the observer watching the tiger from a side view, on the other hand, the imminent collision is specified by the contracting gap between two optical solid angles, those corresponding to the tiger and the other person. Despite different optical patterns, the observers can perceive an impending collision and imminent danger.

An observer acting as participant can control her own movement prospectively by detecting information specifying the future state of affairs. An observer acting as spectator can intervene in the event if needed based on the same perception as the participant has. For example, a mother seeing the imminent danger as she watches a bicycle hurtling toward her child can aim herself at the bike, intercepting it and protecting her child. That is, the ability to see from different perspectives is as important as the ability to see from one's own perspective. The sophistication of this ability would seem to be the hallmark quality of those athletes who are said to be "great readers of the game." Indeed, even with respect to running track, the dexterity of a master runner lies not only in his ability to anticipate his own movements in the next few seconds with respect to the uneven sections of the track, of the turns, and other possible changes imposed by the environment, but also in his ability to anticipate the opponent's actions (Bernstein 1996, page 225).

The Stationary Spectator

We have investigated whether a spectator can, indeed, perceive the same event as accurately as the event's participant. The events were depicted as a (moving) object colliding into another (stationary) object. If one were to view the event from the perspective of the colliding object (as in Kim et al. 1993), the optical transformations were various patterns of optical dilation in accordance with the chosen value of t. From the perspective of a spectator, however, the optical transformation would be various patterns of constriction of an optical gap. Note that the spectator's vantage point can vary with respect to the collision event. That is, the path of collision trajectory may lie parallel to the observer's transverse plane or in an oblique angle to the observer. For the experiment, six collision trajectories were adopted as shown in figure 11.16a. At each end of each line was positioned an object which was roughly the size of a compact car. At the initiation of the subject, one object moved toward the other object which remained stationary throughout the trial. The object started to move with initial velocity of either 9.9 m/s or 13.2 m/s. Initial velocity was controlled as a between-subject variable. After each frame, the moving object decelerated or accelerated according to a value of t chosen for the trial. The values of t adopted were -0.3, -0.5, -0.7, -0.9, and -1.1. Hence, except for -1.1, all the other values of t resulted in varying degrees of deceleration. Among these values, however, -0.3 and -0.5 resulted in soft collisions, whereas -0.7, -0.9, and -1.1 resulted in hard collisions. The subject watched either the first 85% or 90% of a collision event (the final 15% or 10% of the event was blocked out depending on the experimental condition) and was asked to indicate what the impact of collision might have been, judging from the way the moving object approached the stationary one. The results were quite consistent with those reported in Kim et al. (1993) in which subjects watched collision events from a participant's perspective. Trials in which $t \geq -0.5$ were recognized more frequently as soft collisions (86%) and trials in which $t < -0.5$ were recognized more frequently as hard collisions (89%).

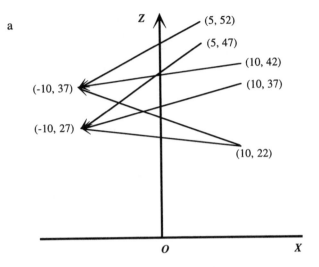

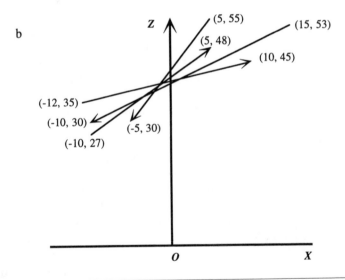

Figure 11.16 *(a)* Six collision trajectories in the *x-z* plane relative to the observer used in the first experiment reported in the section entitled "Seeing When the Paths of Others Will Intersect: The Third Person Perspective." All units are in meters. Another six trajectories were obtained by reflecting along the *z*-axis. The object located at the end of the line collided into another object located at the tip of the arrow according to the value of *ṫ* chosen for each trial. The observer was stationary throughout the collision. *(b)* Four collision trajectories used in the second experiment reported in that same section. Another four trajectories were obtained by reflecting along the *z*-axis. The observer moved during the collision at a speed of 1.85 m/s.

The Moving Spectator

Animals rarely remain stationary. As such, spectators may be in motion when they observe an event taking place. From an optics point of view, however, the fact of a moving observer further complicates the description. Suppose that the observer is approaching a collision event in which one object is colliding into a stationary object as is depicted above. As the collision of the moving object with the stationary object unfolds, the event is specified by the constricting optical gap subtended by the leading edges of each object. The observer's locomotion toward the collision scene, on the other hand, brings about optical expansion. Hence, the optical gap which is constricted by the collision is negated by the observer's forward locomotion.

As a further complication of the description, the footfalls and leg extensions of a moving observer generate not only forward movement but also vertical and horizontal oscillatory movement of the head and eyes, termed bounce and sway, respectively. Only when the observer is transported passively (e.g., when a passenger in a vehicle), are these oscillatory bounce and sway effects eliminated. Despite these complex patterns induced on the optics by the translatory movement combined with bouncing and swaying movements, the observer's perception of the surface layout and his orientation to it continues to be reliable for the insignificant effect of bounce and sway with respect to heading judgment (see Cutting et al. 1992 and Kim et al. 1996a). What is the informational support underlying this achievement?

In another experiment, we examined whether an event can be perceived accurately from a moving spectator's perspective. The collision trajectories adopted in the experiment are shown in figure 11.16b. Two objects, each the size of a compact car, were positioned at the starting points of the two trajectories. Once the collision event was initiated (by the subject pressing a key), one object moved toward the other object according to a designated value of τ. The designated values of τ were -0.35, -0.5, -0.65, -0.8, and -0.95. Trials in which $\tau \geq -0.5$ resulted in soft collisions, whereas trials in which $\tau < -0.5$ resulted in hard collisions. Two initial velocities of 9.9 m/s and 13.2 m/s were used for the experiment, which were controlled as a between-subject variable.

Finally, the observer's movement along the z-axis was added to the preceding display. Simulated forward speed was 1.85 m/s. The amount of bounce and sway was 5.6% of simulated eye height for bouncing movements and 4.8% for swaying movements. Displays with bounce and sway conformed to real walking and displays without these motions conformed to a smooth dollying movement. (For a detailed description of these manipulations, see Cutting et al. 1992 or Kim et al. 1996a.) In an environment made of two objects, both objects (moving or stationary) are displaced because of bouncing and swaying movements. In order to determine whether the presence of a limited background would facilitate the observer's ability to distinguish the local transformation (the collision between two objects) from the global transformation (the observer movement), half of the stimuli were generated with a background composed of the horizon, indicated by a horizontal line, and three stick-figure trees. (The horizon remained at a fixed position 100 m from the point of observation. A tree was depicted by a vertical trunk 2 m in length and two branches, the left 0.5 m in length and the right 0.95 m in length.) The results were still comparable to those reported above as well as those reported in Kim et al. (1993). Trials in which $\dot{\tau} \geq -0.5$ were recognized 85% of the time as soft collisions, whereas trials in which $\dot{\tau} < -0.5$ were recognized 87% of the time as hard collisions. Furthermore, no other variables affected performance except $\dot{\tau}$.

Common Information for Spectator and Participant

The results of these experiments, as well as those reported by Kim et al. (1993), suggest that the information specifying the severity of collision may not lie in the specific form of optical structure in which the event unfolds. Patterns of dilation (as in Kim et al. 1993) and patterns of constriction are both identified in accordance with the $\dot{\tau}$ hypothesis. Moreover, the additional global transformation resulting from the observer's movement and the accompanying bouncing and swaying movements had no effect on the observers' perceptions of the impact of collision when they watched the collision from a spectator's perspective. Despite these varying optical structures, the consistency of the results of these studies suggests that the information specific to the harshness of collision may be a quantity that remains invariant over variations in the optical structure. A clue about this quantity may be found by examining how $\dot{\tau}$ was instantiated in these experiments. For the first person perspective (Kim et al. 1993), $\dot{\tau}$ took the form of the rate of change of the inverse of the relative rate of dilation of an optical solid angle. For the third person perspective, $\dot{\tau}$ took the form of the rate of change of the inverse of the relative rate of constriction of a gap (either horizontal or vertical) between two objects. What is common to both of the preceding cases is *the rate of change of (the inverse of) a relativized rate of change*. In other words, what remains invariant over the varying roles (participant or spectator) of the perceiver, the varying vantage points of the perceiver, and the motion (stationary or locomoting) of the perceiver, is the rate of change of the relative rate of structural changes—regardless of the form in which those changes are instantiated. There is an invariant pattern of change over time.

To reiterate, in both global (in the participant case) and local (in the spectator case) transformations, the impact of collision is specified by the rate of the relative rate of the optical transformation in the optic array—an invariant pattern in the optic array disturbance. This is reminiscent of what Gibson (1982b) called *formless invariants*. As a point of observation changes, there is a change in the perspective appearance of an object, that is, the corresponding optical solid angle. Despite the varying optical form of the object as the vantage point occupies a different spatial location, the observer sees the same object, a phenomenon known as shape constancy (Gibson 1950). For Gibson, "an object is specified by invariants under transformation. Far from being forms, these invariants are quite 'formless'; they are invariants of structure" (Gibson 1986, page 178).

It is also true that, as we move about during the course of our daily activities, we encounter different events, described by Gibson (1986) as changes of surface layout, color or texture, or existence. Despite our viewing these changes from varying perspectives, we recognize the underlying events. And this is true despite the structural characteristics of the object that is undergoing the change (Kim et al. 1995). This phenomenon has been called change constancy (Mark et al. 1981). Because the invariant pattern responsible for the perception of change constancy lies in the optical disturbance produced over time, this invariant is not a formless (or structural) invariant, but a timeless (or transformational) invariant (Kim et al. 1995; Pittenger and Shaw 1975; Shaw and Pittenger 1977). The formless (structural) invariant preserves the adjacent order of the properties of the object; in contrast, the timeless (transformational) invariant preserves the successive order in the changing optical transformation. In short, the formless invariant refers to structural properties preserved over temporal variation, whereas the timeless invariant refers to temporal patterns preserved over structural variation.

Taken together, the results of the experiments reported support Gibson's (1966, 1986) hypothesis that the basis of perception is not the perspective structure, which changes point by point and moment to moment, but the invariant structure under the changing perspective structure. Gibson called the invariant structure under the changing energy flux *information*. It was Gibson's contention that one sees the world by detecting information. Gibson further suggested that by detecting information, different observers can see the same world, despite different perspective appearances of the world. The present experiments demonstrated that information such as τ allows different observers to see the same event from different perspectives.

Conclusion

The underlying theme of the present chapter has been the conception that at every point of observation in a transparent medium (air, water) there is an ambient energy distribution that is (a) uniquely determined by the environmental layout and (b) uniquely transformed by changes in the layout and by displacements of the observation point. This transforming energy distribution at a moving point of observation, or optical flow, in turn, is said to be specific to the layout and the movements that structured it.

We began with the notion that optical flow generated by an observer's movement in the environment contains information specific to the very movement that generated it. It was shown that, by detecting information about heading direction, the observer can be aware of the consequences of her locomotory activity. Furthermore, we presented arguments that concepts such as stream function and velocity potential developed in fluid dynamics apply to the transforming energy distributions at the moving point of observation.

Although optical flow provides the basis of visual perception in the service of activity, detection of the information contained in optical flow is further constrained by the anatomical structure of the ocular device of the animal. Animals with frontal eyes can sample only a portion of the ambient optical flow, requiring them to scan continuously across the flow field to be fully aware of their surroundings. Scanning, on the other hand, introduces additional transformations on the retinal image, or retinal flow. Behavioral data indicate that observers generally look in the direction of their heading. Graphic simulations of pursuit fixation during translation along a circular path revealed that fixating on an object lying on one's circular path linearizes all velocity vectors in the image plane or induces a tangent vector field on the retina. In a tangent vector field, curves formed by connecting the velocity vectors with the same direction define streamlines of a flow field. In short, the active rotation of the eye is not an additional source of ambiguity. To the contrary, it is, in effect, a solution by the animal in extracting the information contained in the ambient optical flow field from the motion in the image plane, that is, from retinal flow.

Animals faced with the problem of controlling the severity of collisions with upcoming surfaces may need information about the remaining distance from the surface, the approach velocity, and the current deceleration. A much simpler solution can be found in the rate of change of the inverse of the relative rate of expansion of the optical solid angle subtended by the approaching surface or $\dot{\tau}$ in regulating the accelerative or decelerative forces of locomotion. That is, when $\dot{\tau} \geq -0.5$, the corresponding optical state specifies that the impending collision will be soft (or safe). When $\dot{\tau} < -0.5$, the corresponding optical state specifies that the impending collision will be hard (that is, unsafe or a crash). Indeed, animals and humans alike appear to control their braking by regulating $\dot{\tau}$ around the critical value during safe approaches to obstacles. In short, the future consequence of one's approach toward an obstacle in terms of impact of contact is specified in the optical quantity $\dot{\tau}$, which in turn can be utilized in controlling one's approach toward an obstacle.

In a task that demands regulating several steps ahead of time rather than the next single step, $\dot{\tau}$ plays an additional role in conjunction with τ. In such a task, one needs not just a single τ but several τs for an anticipation of future regulation. Under the intent of a particular type of contact with an upcoming surface, τ and $\dot{\tau}$ comprise a minimal recursive function. The recursive function of τ can then provide the temporal demand of a future step to be taken such as the amount of impulse force to be exerted during that step. Through the continuous coupling of perception and action, a reformulation of the recursive function of τ can be obtained after each step for further adjustments if needed.

A proper catch of a fly ball demands the catcher to be not only at the right place but also at the right time. The requisite information appears to be in the form of the rate of change in the tangent of the elevation angle or the vertical optical velocity of a projectile. Not only behavioral data but also simulations demonstrate that canceling out vertical optical acceleration of a fly ball does indeed bring about a successful catch of the fly ball.

Successful interactions with other animals (think of people driving cars, walking in crowded malls, or playing team sports such as soccer) demand not only the ability of being able to see from one's own perspective but also from the perspectives of others. For example, the optical basis of collision differs for a spectator and for a participant. In the former, it is varying patterns of constriction of an optical gap, whereas in the latter it is varying patterns of optical dilation. Despite these contrasting optical structures, people appear to be quite accurate in perceiving the impact of collision in conformity to τ. There is, at a more abstract level, the same invariant structure under the differently changing energy flux—Gibson's (1986) information. By detecting the same information, it appears that different observers can perceive the same events, despite their obvious differences in perspective.

Collectively, the preceding examples show how information in Gibson's specificational sense (e.g., Kugler and Turvey 1987) provides the basis for prospective control or "modeling of the future," an ability that is foundational to Bernstein's physiology of activity. Gibson's sentiments with regard to activity and its control echo Bernstein's. As Gibson (1986, page 225) remarked: "Locomotion and manipulation are neither triggered nor commanded but *controlled*. They are constrained, guided, or steered, and only in this sense are they ruled or governed. And they are controlled not by the brain but by information, that is, by seeing oneself in the world." A thoroughgoing understanding of activity requires a thoroughgoing understanding of prospective information. By advancing our understanding of prospective information we will further our understanding of the dexterity that so enchanted Bernstein: *a motor ability to quickly find a correct solution for a problem in any situation* (Bernstein 1996, page 210).

References

Berg, W., Wade, M.G., Greer, N.L. (1994) Visual regulation of gait in bipedal locomotion: Revisiting Lee, Lishman, and Thomson (1982). *Journal of Experimental Psychology: Human Perception and Performance*, 20: 854–863.

Bernstein, N.A. (1967) *The Coordination and Regulation of Movements*. London: Pergamon.

Bernstein, N.A. (1996) On dexterity and its development. In *Dexterity and its Development*. M.L. Latash, M.T. Turvey (Eds.), Mahwah, NJ: Erlbaum, 3–244.

Bootsma, R.J., Wieringen, P.C.W. van (1990) Timing an attacking forehand drive in table tennis. *Journal of Experimental Psychology: Human Perception and Performance*, 16: 21–29.

Chapman, S. (1968) Catching a baseball. *American Journal of Physics*, 36: 868–870.

Cutting, J.E. (1986) *Perception with an Eye for Motion*. Cambridge, MA: MIT Press.

Cutting, J.E., Springer, K., Braren, P.A., Johnson, S.H. (1992) Wayfinding on foot from information in retinal, not optical, flow. *Journal of Experimental Psychology: General*, 121: 41–72.

Feigenberg, I.M., Latash, L.P. (1996) N.A. Bernstein: The reformer of neuroscience. In *Dexterity and its Development*. M. L. Latash, M.T. Turvey (Eds.), Mahwah, NJ: Erlbaum, 247–275.

Gibson, J.J. (1950) *The Perception of the Visual World*. Boston, MA: Houghton Mifflin.

Gibson, J.J. (1966) *The Senses Considered as Perceptual Systems*. Boston, MA: Houghton Mifflin.

Gibson, J.J. (1982a) Note on perceiving in a populated environment. In *Reasons for Realism: Selected Essays of James J. Gibson*. E. Reed, R. Jones (Eds.), Hillsdale, NJ: Erlbaum, 411–412.

Gibson, J.J. (1982b) On the concept of "formless invariants" in visual perception. In *Reasons for Realism: Selected Essays of James J. Gibson*. E. Reed, R. Jones (Eds.), Hillsdale, NJ: Erlbaum, 284–288 (Original work published 1973).

Gibson, J.J. (1986) *The Ecological Approach to Visual Perception.* Hillsdale, NJ: Erlbaum (Original work published 1979).

Goodman, N. (1965) *Fact, Fiction and Forecast.* Indianapolis: Bobbs-Merrill.

Kaiser, M.K., Phatak, A.N. (1993) Things that go bump in the light: On the optical specification of contact severity. *Journal of Experimental Psychology: Human Perception and Performance,* 19: 194–202.

Keldys, M.V. (1963) Functions of a complex variable. In *Mathematics: Its Contents, Methods and Meaning, Part 3.* A.D. Aleksandrov, A.N. Kolmogorov, M.A. Lavrent'ev (Eds.), Providence, RI: American Mathematical Society, 143–200.

Kim, N.-G., Effken, J.A., Shaw, R.E. (1995) Perceiving persistence under change and over structure. *Ecological Psychology,* 7: 217–256.

Kim, N.-G., Growney, R., Turvey, M.T. (1996a) Optical flow not retinal flow is the basis of wayfinding by foot. *Journal of Experimental Psychology: Human Perception and Performance,* 22: 1279-1288.

Kim, N.-G., Turvey, M.T. (1994) Optical foundations of perceived ego motion. *Behavioral and Brain Sciences,* 17: 322–323.

Kim, N.-G., Turvey, M.T, Carello, C. (1993) Optical information about the severity of upcoming contacts. *Journal of Experimental Psychology: Human Perception and Performance,* 19: 179–193.

Kim, N.-G., Turvey, M.T., Growney, R. (1996b) Wayfinding and the sampling of optical flow by eye movements. *Journal of Experimental Psychology: Human Perception and Performance,* 22:1314-1319.

Koenderink, J.J. (1986) Optic flow. *Vision Research,* 26: 161–180.

Kreyszig, E. (1962) *Advanced Engineering Mathematics.* New York: Wiley.

Kugler, P.N., Turvey, M.T. (1987) *Information, Natural Law, and the Self-Assembly of Rhythmic Movement.* Hillsdale, NJ: Erlbaum.

Land, M.F., Lee, D.N. (1994) Where we look when we steer. *Nature,* 369: 742–744.

Lee, D.N. (1974) Visual information during locomotion. In *Perception: Essays in Honor of J.J. Gibson.* R.B. MacLeod, H. Pick (Eds.), Ithaca, NY: Cornell University Press, 250–267.

Lee, D.N. (1976) A theory of visual control of braking based on information about time-to-collision. *Perception,* 5: 437–459.

Lee, D.N. (1980) Visuo-motor coordination in space-time. In *Tutorials in Motor Behavior.* G. Stelmach, J. Requin (Eds.), Amsterdam: North Holland, 281–295.

Lee, D.N., Davies, M.N.O., Green, P.R., van der Weel, F. R. (1993) Visual control of velocity of approach by pigeons when landing. *Journal of Experimental Biology,* 180: 85–104.

Lee, D.N., Lishman, J.R. (1977) Visual control of locomotion. *Scandinavian Journal of Psychology,* 18: 224–230.

Lee, D.N., Lishman, J.R., Thomson, J.A. (1982) Regulation of gait in long jumping. *Journal of Experimental Psychology: Human Perception and Performance,* 8: 448–459.

Lee, D.N., Reddish, P.E., Rand, D.T. (1991) Aerial docking by hummingbirds. *Naturwissenschaften,* 78: 526–527.

Lee, D.N., Thomson, J.A. (1982) Vision in action: The control of locomotion. In *Analysis of Visual Behavior.* D. Ingle, M.A. Goodale, R.J.W. Mansfield (Eds.), Cambridge, MA: MIT Press, 411–433.

Mark, L.S., Todd, J.T., Shaw, R.E. (1981) The perception of growth: How different styles of change are distinguished. *Journal of Experimental Psychology: Human Perception and Performance,* 7: 355–368.

McLeod, P., Dienes, Z. (1993) Running to catch the ball. *Nature,* 362: 23.

McLeod, P., Dienes, Z. (1996) Do fielders know where to go to catch the ball or only how to get there? *Journal of Experimental Psychology: Human Perception and Performance,* 22: 531–543.

Michaels, C.F., Oudejans, R.R.D. (1992) The optics and actions of catching fly balls: Zeroing out optical acceleration. *Ecological Psychology,* 4: 199–222.

Patla, A.E., Robinson, C., Samways, M., Armstrong, C.J. (1989) Visual control of step length during overground locomotion: Task-specific modulation of the locomotor synergy. *Journal of Experimental Psychology: Human Perception and Performance,* 15: 603–617.

Pittenger, J.B., Shaw, R.E. (1975) Aging faces as viscal-elastic events: Implications for a theory of non-rigid shape perception. *Journal of Experimental Psychology: Human Perception and Performance,* 1: 374–382.

Reed, E.S. (1982) An outline of a theory of action systems. *Journal of Motor Behavior,* 14: 98–134.

Rosen, R. (1991) *Life itself.* New York: Columbia University Press.

Royden, C.S., Banks, M.S., Crowell, J.A. (1992) The perception of heading during eye movements. *Nature,* 360: 583–585.

Royden, C.S., Crowell, J.A., Banks, M.S. (1994) Estimating heading during eye movements. *Vision Research,* 34: 3197–3214.

Shaw, R.E., Kinsella-Shaw, J. (1988) Ecological mechanics: A physical geometry for intentional constraints. *Human Movement Science,* 7: 155–200.

Shaw, R.E., Pittenger, J.B. (1977) Perceiving the face of change in changing faces: Implications for a theory of object perception. In *Perceiving, Acting, and Knowing.* R. Shaw, J. Bransford (Eds.), Hillsdale, NJ: Erlbaum, 103–132.

Shinar, D. (1978) *Psychology on the Road: The Human Factor in Traffic Safety.* New York: Wiley.

Shinar, D., McDowell, E.D., Rockwell, T. (1977) Eye movements in curve negotiation. *Human Factors,* 19: 63–71.

Sutton, O.G. (1957) *Mathematics in Action.* 2d ed. New York: Dover.

Turvey, M.T. (1992) Affordances and prospective control: An outline of the ontology. *Ecological Psychology,* 4: 173–187.

Turvey, M.T., Shaw, R.E., Reed, E.S., Mace, W.M. (1981) Ecological laws of perceiving and acting: In reply to Fodor and Pylyshyn (1981) *Cognition,* 9: 237–304.

von Hofsten, C., Lee, D.N. (1994) Measuring with the optic sphere. In *Perceiving Events and Objects.* G. Jansson, S.S. Bergstrom, W. Epstein (Eds.), Mahwah, NJ: Erlbaum, 455–467.

Wann, J.P., Edgar, P., Blair, D. (1993) Time-to-contact judgment in the locomotion of adults and pre-school children. *Journal of Experimental Psychology: Human Perception and Performance,* 19: 1053–1065.

Warren, W.H. (1995) Self motion: Visual perception and visual control. In *Handbook of Perception and Cognition.* Vol. 5. W. Epstein, S. Rogers (Eds.), San Diego: Academic Press, 263–325.

Warren, W.H., Blackwell, A.W., Kurtz, K.J., Hatsopoulos, N.G., Kalish, M.L. (1991a) On the sufficiency of the velocity field for perception of heading. *Biological Cybernetics,* 65: 311–320.

Warren, W.H., Hannon, D. (1988) Direction of self-motion is perceived from optical flow. *Nature,* 336: 162–163.

Warren, W.H., Mestre, D.R., Blackwell, A.W., Morris, M.W. (1991b) Perception of circular heading from optical flow. *Journal of Experimental Psychology: Human Perception and Performance,* 17: 28–43.

Warren, W.H., Morris, M.W., Kalish, M. (1988) Perception of translational heading from optical flow. *Journal of Experimental Psychology: Human Perception and Performance,* 14: 644–660.

Warren, W.H., Young, D.S., Lee, D.N. (1986) Visual control of step length during running over irregular terrain. *Journal of Experimental Psychology: Human Perception and Performance,* 12: 371–383.

Yilmaz, E.H., Warren, W.H. (1995) Visual control of braking: A test of the τ hypothesis. *Journal of Experimental Psychology: Human Perception and Performance,* 21: 996–1014.

Zaal, F.T.J.M., Bootsma, R.J. (1995) The topology of limb deceleration in prehension tasks. *Journal of Motor Behavior,* 27: 193–207.

12
CHAPTER

Bernstein's Legacy for Motor Development: How Infants Learn to Reach

Esther Thelen
Department of Psychology, Indiana University,
Bloomington, IN, U.S.A.

One discovers in the writings of Nikolai Bernstein many fundamental insights for understanding motor development. A careful reading of Bernstein should be required of any student of human movement. Those studying development should be required to reread Bernstein at least once a year thereafter, preferably on the same day. This day could then be thought of as a disciplinary "Day of Atonement". We could then recall all the times during the past year when we thought we had an original idea only to be reminded of the encompassing legacy of Bernstein's genius.[1]

This chapter focuses on one of Bernstein's foundational principles, a principle that has changed the field of motor development over the last decade or so. Adapting a Bernsteinian perspective (with a strong dose of dynamic systems) allows researchers to ask different and more interesting questions about how people come to control their bodies, particularly about how infants learn to reach. Adopting a developmental perspective, in turn, provides an important clue about how the adult motor system works. A developmental perspective shows how motor behavior is naturally acquired—what changes and what stays the same.

---

[1]Mark Latash has suggested that the Bernstein "Day of Atonement" be on September 9, which is "World Motor Control Day," to provide a moment of sober self-reflection amid the festivities.

Motor Development From the Outside In

In the middle of his chapter entitled "Biodynamics of Locomotion" in the seminal *The Coordination and Regulation of Movements* (Bernstein 1967; available in Whiting 1984), Bernstein described in detail the subtle changes of dynamic forces during walking and running during the first few years of life. Hidden in the middle of a paragraph is one of my favorite quotes:

> The reorganization of the movement begins with its biomechanics, that is to say, with the *peripheral parts* of the process . . . this biomechanical reorganization sets new problems for the central nervous system, to which it gradually adapts, mirroring that adaptation in subsequent changes in the longitudinal dynamic curves. (Whiting 1984, page 197, emphasis in original)

The startling importance of that buried statement can be appreciated only when contrasted with mainstream, Western views of motor development at that time. Because the pioneering motor development researchers such as Myrtle McGraw (1943) and Arnold Gesell (1945) drew their lineage from neuroembryology (rather than biomechanics or bioengineering), they saw the issue strictly inside out—waiting for the brain to ripen and allow better body control. Their deficiencies were not so much in their assumption that motor behavior provided a good readout of the status of the central nervous system. Researchers still use that assumption today. Rather, those pioneers failed to appreciate how much biomechanical challenges facing infants and their solutions to those challenges sculpt the brain. Bernstein, more than half a century before the recent discovery of brain plasticity, fully understood the bidirectionality of change. Researchers now must ask not only how structural changes in the central nervous system allow and support body control but also how moving limbs and torso in a world of information and forces determine the connections in the brain.

What Is Involved in Motor Development?

Learning to move is not a trivial problem if you are a baby. Consider the problem of learning to reach and grasp objects, a skill fundamental to all aspects of motor and cognitive development. Infants are faced with biomechanical and neural problems of enormous complexity. The task is to get the end point, the hand, to a visually specified target in three-dimensional space. This requires the apparent transformation of visual space to a body-centered coordinate system. Infants then have to generate an appropriate, smooth, straight pathway from start to target, with a rising and then falling velocity. The trajectory must be produced by muscle commands that both support the arm against gravity and generate it forward and out. At the same time, the system must take into account the passive and elastic forces generated by the movement itself, which, especially in fast movements, contribute significantly to the motor outcome. Researchers are hotly debating the issues about which of these neuromotor organizational levels is actively controlled by the central nervous system, what is computed, and what comes along for free. Is the central nervous system working on a bell-shaped velocity profile (Morasso 1981), biphasic muscle bursts (Gottlieb et al. 1990), a final or moving equilibrium point, (Bizzi 1992; Feldman and Levin 1995; Latash and Gottlieb 1991), a smooth path of acceleration or higher derivatives (Hogan 1984), or an appropriately decreasing hand-to-target optical variable (Zaal and Bootsma 1995)?

To make the developmental problem even more intractable, infants come into this world seemingly ill designed for adaptive movement, especially as bipedal creatures on the surface of the earth. They have large, heavy heads, narrow shoulders, short legs, and weak muscles. They have spent the last nine months in an aquatic environment, floating around oblivious to the demands of gravity on muscles and bones. True, they have been moving for much of that time underwater, but their movements are not calibrated to their new environment. Equally vexing is that the masses, segment lengths, and moments of inertia that constitute babies' Newtonian universe change very, very rapidly in the first months of life. This guarantees that whatever mapping of force to outcome is acquired, it will not last long (Guo et al. 1991). Finally, infants are born with a wide range of individual movement styles— some active, vigorous, and jerky; others languid, smooth, and slow (Thelen et al. 1993). The nervous system simply cannot have anticipated all these problems beforehand by hard-wiring solutions genetically. The system must be designed to learn by interacting with the world.

Learning to Reach

A developmental analysis allows some understanding of these issues. It provides a long window when movements are just emerging, poorly coordinated, and controlled but changing over time in the direction of higher skill. My colleagues and I have followed infants from their earliest weeks of life to a year old, when reaching is swift and accurate but not fully adult-like (Thelen et al. 1996). Our results support and expand upon Bernstein's insight that the reorganization of movement begins with its biomechanics. In what follows, I first give an outline of the process of learning to reach, and then fill in the picture with what we have learned about the mechanisms of control and change.

The study involved four normal babies, Nathan, Gabriel, Hannah, and Justin, who came into the laboratory each week from weeks 3 until 30 and then every other week thereafter. Infants were seated in a specially designed infant seat in an almost vertical position with their torsos supported while we presented toys to them at midline and shoulder height. We tracked, by means of a four-camera WATSMART system and surface electromyography, the movement of both arms in three-dimensional space and associated muscle patterns of biceps, triceps, anterior deltoid, and trapezius muscle groups. Off-line, we obtained hand and joint kinematics and calculated torques using inverse dynamics. The details of the data collection and analysis are available in Thelen et al. (1993), Corbetta and Thelen (1995), and Thelen et al. (1996).

How Reaching Improves

The developmental profile of the changes in reach trajectory control variables constitute the changes to be explained. When infants first reach out to touch and grab objects, their movements are tortuous and indirect, tracing a sinuous path to the target and consisting of several velocity bumps (peaks) and valleys (Halverson 1931; Hofsten 1991). Figure 12.1 shows examples of two infants' early reaches and their subsequent improvement some months later. The path of the hand became straighter. The number of velocity bumps (called movement units (Hofsten 1991)) decreased such that a single large movement unit dominated. Figure 12.2 depicts two conventional measures of the goodness of infant reaching over the entire first year for the four infants: how straight the pathway was from start to target and how smooth the movement was, measured by the number of movement units. These variables indicate an increasing level of arm control.

As expected, the infants' reaches were not very good at the beginning. They were poorly controlled and quite variable. In this early stage, babies sometimes directly swiped and sometimes took long meandering paths with many starts and stops. In contrast, around 30–34 weeks of age, roughly about seven months, all four infants seemed to have discovered a fairly stable and reliable reach configuration. These were not as smooth and direct as adult reaches, but they were pretty good. The infants did not noticeably improve trajectory control over this long period. What was surprising, however, was that during the period before this stable era, infants' skill development was not increasing in a linear fashion. Instead, it showed interesting plateaus and even regressions, times when previous pattern stability appeared lost. Note, for instance, Hannah's dramatic regression in control during weeks 29–36 and her equally dramatic discovery of a more stable trajectory thereafter.

As it turned out, these nonlinearities provided at least some clues to the underlying control processes and especially to the force-environment issues that, as Bernstein had hinted, form the basis of skill development. When I designed this longitudinal study, I was motivated by an important concept from dynamic systems principles. In particular, changes in variables that index control of reaching movements, either linear or discontinuous, must arise from modifications of the multiple systems components that are continuous in time. That is, the new patterns or forms do not arise de novo but from changes in the self-organization of the component elements (Kelso 1995; Thelen and Smith 1994). The experimental task, then, was to track these ongoing changes and to link them to observable changes in the reach trajectory control. Thus, we looked not only at the usual trajectory variables describing the reach itself but also at preferred movement patterns when infants were not reaching. These included spontaneous movements before the emergence of target-directed reaching at three or four months and when reaching developed, the

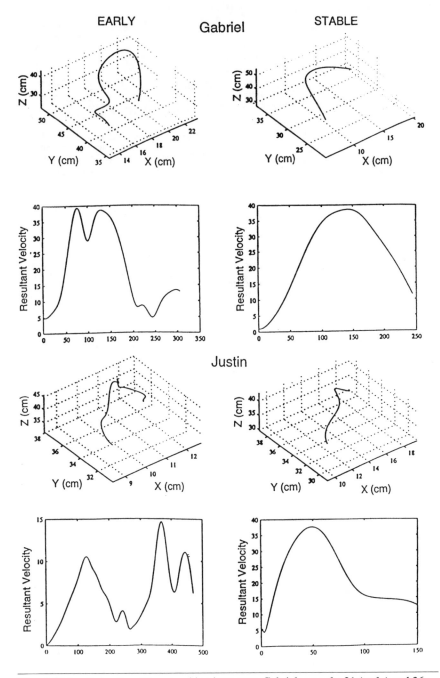

Figure 12.1   Examples of how reaching improves. Gabriel at weeks 21 (early) and 36 (stable) and Justin at weeks 27 (early) and 52 (stable). The three-dimensional plot shows the hand trajectory. Below is the corresponding hand speed.

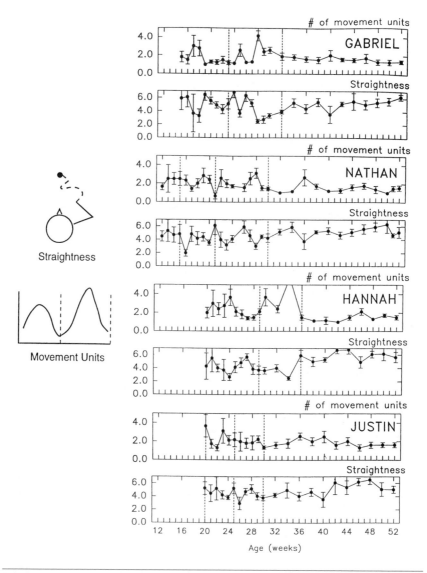

Figure 12.2 Longitudinal measures of reaching improvement over the first year in four infants. Means and standard errors of the number of movement units and straightness index as a function of age. Number of movement units, a measure of hand path smoothness, was determined by an algorithm that identified above-threshold increasing and decreasing hand speeds. A movement unit was defined as a speed maximum between two minima, where the difference between the maximum speed and both minima exceeded 1 cm/s. The straightness index was the ratio between the virtual path, a straight line from the three-dimensional coordinates of reach initiation to toy contact, and the actual hand path length. The obtained 0 to 1 interval ratio was then standardized using the following Z-transform equation: $z(x) = 2\ln((1 + x)/(1 - x))$. Increasing values indicate straighter paths. Adapted from Thelen, Corbetta, Kamm, Spencer, Schneider, and Zernicke 1993.

nonreaching movements infants made before and after the reach. Would there be some relationship between movements that were successful as reaches and all the other movements that constituted the infants' motor repertoire?

The Transition to Reaching

The first transition of interest was from not reaching at all to the first times when the infants reliably got their hands to the toy offered (Thelen et al. 1993). We discovered that infants' first reaching movements bore the signatures of their earlier movement dynamics, particularly in the level of characteristic movement speed or the individual force parameters. Figure 12.3 compares the average speed of the infants' prereaching movements to those of their reaching movements at the first, second, and third weeks of reaching. Gabriel, for instance, was a particularly strong and active infant. He generated fast, vigorous movements before learning to reach. His reaches were also fast, although he learned to slow his movements somewhat during an actual reach. Contrast his movement speeds with those of Hannah, who had slow, smooth, and deliberate movements both before and after reach onset. The other two babies also had characteristic speed profiles that spanned the important transition to a new form of control.

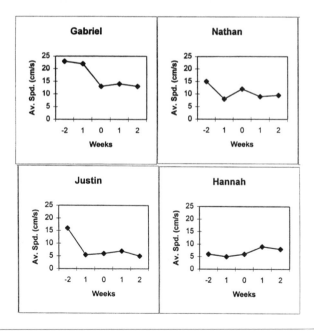

Figure 12.3   Average speed of infants' movements across the transition of learning to reach. Week 0 is the onset of reaching, which differed among infants (see text). Negative numbers are the two weeks previous to onset (pre-reaching); positive numbers are the two weeks following onset (reaching).

Adapted from Thelen, Corbetta, Kamm, Spencer, Schneider, and Zernicke 1993.

Changes With Practice

Even after many months of practicing reaching, infants retained both their individually characteristic speeds and the close relationship between nonreaching and reaching movement speeds (Thelen et al. 1996). Figure 12.4 plots average reach speeds as a function of developmental period: the *early*, variable period, the later, *stable* period, and a middle period characterized by faster movements for Nathan, Gabriel, and Hannah but not for Justin. Note, however, that Gabriel and Nathan moved considerably faster than Hannah and Justin throughout the year, despite rather dramatic differences within infants in the developmental epochs. Most importantly, infants retained these preferred patterns throughout the first year, even in nonreaching movements. Figure 12.5 compares the average speeds of reaches with those of nonreaching movements, which were recorded before and after the reach itself during a particular week. Eleven of the 12 correlations are significantly positive. They suggest that despite fluctuating average speeds in different weeks, when infants reached fast, they were generally moving fast overall.

Why are these individual speed styles important in the developmental story? Movement speed—reflecting the amount of energy delivered to the limbs—is a critical parameter in many aspects of motor control. Faster movements are generally less accurate, probably because there is less time to make fine adjustments (Fitts 1954). Reach trajectories may require different control strategies and different

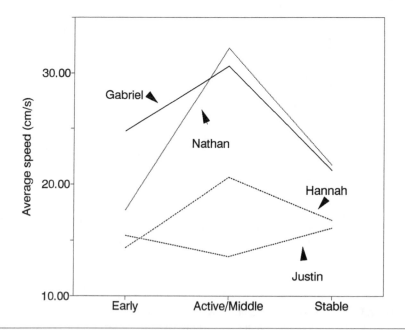

Figure 12.4   Average speed of reaches for the early, active/middle, and stable periods for the four infants.

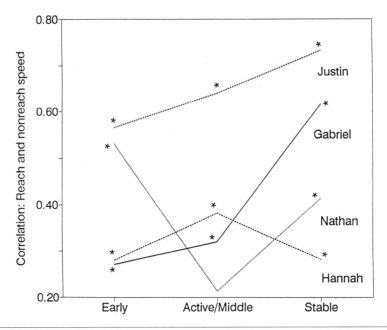

Figure 12.5   Average correlations between the speed of reaching movements and all nonreaching movements for each week for the three developmental periods.

patterns of muscle activation depending on whether they are performed slowly or rapidly (Flanders and Herrmann 1992; Gottlieb et al. 1989). Similarly, very fast movements produce greater motion-related passive forces than slow ones and thus pose different problems for neural control (Latash and Gottlieb 1991; Schneider et al. 1989). Interlimb coordination in cyclic movements is also affected by speed, because certain patterns of coordination become unstable as speed increases (Kelso et al. 1986).

Infants Find Individual Solutions

When the infants first learned to reach, they had individual and very different kinds of control issues to solve. Babies whose spontaneous movements were fast and forceful had to learn to slow down in order to be accurate and not just bat at the toy in front of them. Their fast movements were likely to be not only inaccurate but also wasteful of energy. Conversely, the slower-moving infants needed to generate sufficient forces to move their arms up and out toward the toy and, eventually, to move in an appropriately scaled speed. Fast infants needed to deal with motion-related forces, slower infants had to generate strong antigravity control. The solutions to these control problems cannot be built in. They must be discovered by each infant in relation to his or her given muscles, energetic levels, and the tasks at hand.

The persistence of the individual, speed-related movement styles over the transition to reaching throughout the first year are clues that trajectory control is intimately tied to the level of the amplitude and timing of forces. Since infants retain characteristic speed signatures throughout the year and the associated, characteristic control problems, the trajectory for reaching is not isolated from infants' other movements but is carved from them.

Other developmental scenarios are theoretically possible. For instance, reach trajectories could be planned and generated autonomously, that is, as a purely kinematic form to match the path of the hand to the target. Indeed, in both classic and recent work about infant reaching, an unquestioned assumption was that the baby's job was to match the seen (and felt) hand to the seen target. The separate starts and stops represent deliberate corrections to maintain a good trajectory (Berthier 1994; Fetters and Todd 1987; White, Castle, and Held 1964). If infants were learning visual motor mapping alone, the researchers would not necessarily expect such a pervasive influence of the force parameters—reflected in speed—on reach kinematics or such individually distinctive profiles. However, this was clearly not the case.

Levels of Control

The developmental picture is likely much more complex. A recent theoretical model by Gregor Schöner (1995) provides an insightful entry into the developmental process. The model addresses the classic Bernsteinian issue of what variable, given the rampant degrees of freedom, the nervous system controls. The fundamental assumption, from a dynamic pattern perspective, is that movement results from coordination of the multiple components and is governed by abstract coordination dynamics. From this view, Schöner offers the following solution to the control question. "Those measures of behavior that are stable against perturbations define what the nervous system controls" (Schöner 1995, page 300). Thus, in well-practiced reaching, the end point target posture is well buffered from transient mechanical perturbations (Polit and Bizzi 1978). In bimanual rhythmic tasks, the relative phase between the limbs is stable within a range of mechanical and perceptual perturbations (Scholz and Kelso 1989). Some variables remain uncontrolled. In both cases, for example, the set of relative joint configurations that accompany the movements might vary even in succeeding movements within the same individual.

An important consequence of this perspective is that variables may be controlled as a function of task; no one set of variables is privileged for all movement tasks. For example, in some forms of dance, maintaining joint configurations may be critical, while in other forms, the timing of forces and phasing between the feet must remain stable. Since humans can intentionally and rapidly shift control— imagine the dancer holding a posture and then beginning to tap—means that the manifold of uncontrolled and potentially controlled variables must itself be highly flexible and dynamic. This allows movements to be organized as the occasion demands. Additionally, this model takes an important step in erasing the boundaries between movement planning and execution, but that is yet another story.

This means that the components comprising the motor pattern must themselves be tightly interactive, allowing the same effector to be flexibly recruited as a controlled or uncontrolled variable as the occasion demands. Nonetheless, under specific circumstances, we can create conditions where the coupling between components is weaker, and we can analyze the role of stabilizing neural control. The best-known example is the now-classic equilibrium point experiment where the participant is told to reach to a target but "not to intervene" when the limb is mechanically perturbed (Feldman 1966). The person maintains the target goal despite being pulled off course. This is evidence that spatial end point is a controlled variable that can be separated from the parts of the system recovering from the tug. However, remember that these experiments deliberately isolate the level of control; they identify what can be protected during this experiment but not necessarily what occurs during natural movements.

These sorts of experiments led Schöner to propose a model for analyzing trajectory formation during goal-directed arm movements. He designated the levels as *goal*, *timing*, and *load* levels. It is important to emphasize that these are neither hierarchical nor anatomical levels. In reality, they are not separate because they are mutually and reciprocally interactive. Each is parameterized by the environment. Rather, they are heuristic because they are potentially separable by special manipulations. For trajectories, consider the following perturbations: phasic applications of mechanical load, perturbations in external timing as in instructions to move quickly or slowly, and global changes in movement goal, such as switching the target. Thus, the load level is the set of variables that stabilize the trajectory in the face of load level perturbations, and so on. Again, it is dangerous to interpret these levels too stringently or literally because in reality, the strict identification of a variable in this way is sometimes problematic.[2]

Levels of Control in Trajectory Development

Despite these difficulties, the model is very useful for thinking about the development of trajectory control. Two points seem apparent. First, in the initial state, the levels and components are tightly coupled. What develops is precisely this ability to control whatever is appropriate for the task flexibly and independently. Second, these levels, while not functionally hierarchical, may be developmentally sequential. That is, infants must gain a measure of control over the load level in order for stable timing dynamics to emerge. Likewise, they must gain control over load and timing level to be able to protect their goals from perturbation.

---

[2]In the spirit of the DDA (Disciplinary Day of Atonement), Bernstein put forth a very similar idea in the newly translated book *Dexterity and its Development* (Bernstein 1996). Bernstein's "levels of construction of movements" are perhaps more anatomical than envisioned by Schöner's abstract dynamics, but many parallels also exist.

The suggestion that load level control is primitive and other levels emergent is highly consistent with the evidence. First, infants move their arms around spontaneously for three to four months before they successfully reach out and grab an object. Some of these movements may put the hands into the visual field, but many do not. During these movements, infants learn the force pulses associated with the feel of particular movement velocities and amplitudes so that they may adjust these parameters. The first task for learning to reach is to get the arms near reachable objects or if hand-eye coordination is needed, into the visual field.

Principle of Linear Synergy

What are infants actually modulating at the load level as they wave their arms around in the air? Recall that they have the inverse dynamics problem: how to stabilize joints against motion-dependent forces in the linked system. These are pretty complicated equations for a one-month-old brain!

My colleagues and I are grateful to Dr. Gerald Gottlieb (and so are the babies) for suggesting that they look for a simplifying principle. In recent work, Gottlieb and his colleagues found a remarkable control principle for natural, unconstrained, multijoint reaching movements in adults (Gottlieb et al. 1996a). In brief, for reaching movements in much of the accessible work space, the torque generated at the shoulder was linearly proportional to the torque generated at the elbow. This is when the torque is considered a dynamic torque, defined as the portion of muscle torque beyond what is necessary to counteract gravity. This invariance held despite changes in speed or load on the limb. The slope of the linear regression line describing this relation was a function of the target position in the workspace (Gottlieb et al. 1996b, c).

It is well known that when adults perform such reaching and pointing movements, their hands trace a nearly straight path in Cartesian space from start to end, and the corresponding tangential velocity profiles are bell shaped. One possibility is that individuals plan their movements purely in this kinematic form and then calculate the muscle torques and muscle activation patterns to produce the desired kinematic effect. The implication of the linear synergy principle, however, is that the system can avoid this complex transformation by working at the load level. Straight trajectories may be a by-product, so to speak, of adjustments of limb force generation. However, in well-practiced adults, these possibilities are difficult to disentangle.

Infants, as we saw, do not have straight trajectories and bell-shaped velocity profiles, however. Thus, if the primary level of control is the path of the hand in relation to the visual target and the torque invariance principle is learned in order to produce a straight, smooth hand path, then the linear correlation of joint torques should increase as a function of trajectory control. If, on the other hand, good trajectories emerge from modulations of already coordinated joint torque dynamics, then the torque invariance should precede improvement in trajectories and, perhaps, the onset of reaching itself.

The second possibility is indeed what we (Thelen et al. 1997) found. We examined 115 reaches from the four infants, divided into the early and stable periods mentioned earlier, as well as 22 segments of pre-reaching movements. Remarkably, infants strictly adhered to the linear synergy principle throughout, whether their trajectories were poor or improved and, indeed, whether they were reaching at all. Figure 12.6 shows exemplar plots from one infant, Justin, at these three periods. Despite convoluted hand paths and poorly coordinated shoulder-elbow rotations, the synchrony and proportionality of the dynamic torques were maintained. Regression coefficients were uniformly high, and the slopes did not vary much, either, as indicated in table 12.1. The similarity of slopes suggests that infants are not apportioning torques differentially to the two joints. Shoulder torque is nearly always about two-and-a-half times that of the elbow. Neither hand path straightness nor coordination between joint rotations showed any regularity. Yet, at the level of the dynamic torque, a dramatic reduction occurred in the degrees of freedom.

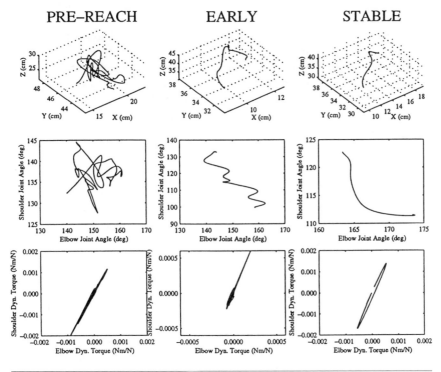

Figure 12.6   Exemplar trials of arm movement data for each of the three developmental periods for a single infant. The first row of three-dimensional trajectories depicts the infant's hand trajectory in pre-reaching, and early and stable reaching movements (with the infant facing along the positive y-axis). The second row shows the coordination between the displacement of shoulder versus elbow joint angles. The shoulder angle is the internal angle between the upper arm and a vertical line directly above the shoulder. The elbow joint angle is the internal angle between the upper and lower arm. The third row depicts the shoulder versus elbow dynamic torques during the same movement. Torques have been normalized to the infant's body weight at the time of data acquisition.

Table 12.1   Regression of Shoulder and Elbow Dynamic Torque

	Pre-reaching N = 22	Early N = 66	Stable N = 49
Slope	2.68 ± 0.70	2.45 ± 0.41	2.83 ± 0.40
R^2	0.94	0.94	0.94

This invariance held across the transition from pre-reaching to reaching despite evidence that infants recruited different sets of muscles to reach to the same place. John Spencer and I looked at patterns of coactivity in four muscle groups, biceps, triceps, trapezius, and deltoid, in reaching and nonreaching movements (Spencer and Thelen 1997). We found a rather dramatic shift between reaching and nonreaching movements in the proportion of particular muscle group activity. Large decreases occurred in triceps and biceps alone and in combination while corresponding increases occurred in deltoid activity and deltoid and trapezius activity with reaching. These differences were not a function simply of infants holding their arms in different postures. Rather, during the transition to reaching, infants were actually using different muscles (primarily those to elevate the shoulder and stabilize the neck) to get to the same places in space. Bernstein was correct once again. The invariance was clearly not at the level of particular muscles.

These results suggest that the linear synergy principle is a fundamental property of the human neuromotor system from early in life. Likely, it is not learned as a means to keep a visually monitored hand on a straight path. Rather, reaching for a target must be sculpted from this preference of the system to apportion torque proportionately and synchronously between shoulder and elbow. Even as the hand wandered from the path to the toy or as infants moved in a seemingly non-goal-corrected fashion, dynamic coordination was maintained. This constraint simplifies the problem of force control enormously. It suggests that mechanisms are already in place to compensate for complex motion-dependent and interactive torques that must arise from the linked system. Babies do not need to compute a long page of inverse dynamics equations. Rather, movement distance and speed may be modulated by scalar changes in the muscle activation that generates torques around the joints. Relatively subtle adjustments in the relative apportionment and timing of joint torque, learned by trial and error (Konczak et al. 1995), may be sufficient to specify trajectory direction. Natural reaching movements capitalize on the intrinsic, coupled force dynamics of the arm.

Load Level Comes First

Thus, we can reconsider Gabriel's transition from nonreaching to reaching as timing level emerging from load level dynamics. Consider Gabriel's arm movements before he learned to reach. These cyclical movements, shown in figure 12.7 on the phase plane, may be considered purely load level dynamics. The characteristic bursts of muscle activity appear to act like a forcing function on a damped, mass spring. The trajectory of the movement is emergent, much like that of a physical spring, and not controlled in Schöner's sense. Gabriel's first successful reaches appear to partake of the same dynamic regime. Consider the two reaches shown in the phase planes of figure 12.8, where the portion of the movement directed toward the object is dynamically configured in a similar topography as the preceding non-goal-directed part of the movement. The trajectory toward the toy is continuous with the springlike trajectory of the ongoing movements.

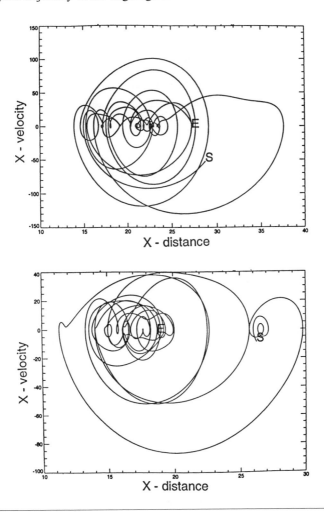

Figure 12.7   Two examples of Gabriel's spontaneous movements before he learned to reach, depicted on distance-velocity phase plane. (S) start of movement, (E) end of movement. Reprinted from Port and van Gelder 1995.

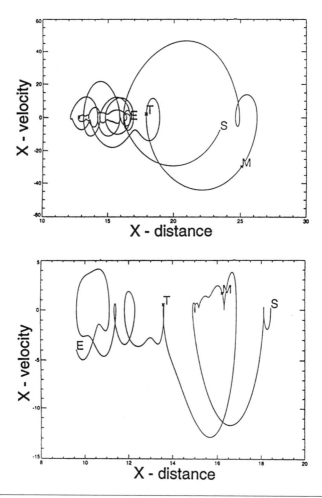

Figure 12.8 Two examples of Gabriel's early reaching movements, also on the phase plane. (S) start of movement, (M) start of reach, (T) contact with toy, and (E) end of movement. Reprinted from Port and van Gelder 1995.

Taken together, these data lead the reader to believe that straight, smooth hand paths are derived from underlying force dynamics and not vice versa. They also help explain why movement speed plays such an important role during the first year. Adults can produce good reaches at a wide range of movement speeds, but babies cannot. In Schöner's terms, infants cannot protect their timing level from load level perturbations. Thus, before the infants' trajectories stabilized, faster movements were highly disruptive, causing an increase in movement units and loss of straightness. During the stable period, faster movements continued to be less straight. Reaches were pretty good, but not yet fully controlled (Thelen et al. 1996).

Goal Level Control

Just as timing level control emerges only after load level, so too does goal level control depend on lower level stability. This point helps reinterpret some of the persistent goal level errors infants make during the last half of the first year. The classic one is Piaget's so-called A-not-B error, one of the most well-studied phenomena in developmental psychology (Piaget 1954; Wellman et al. 1987). The experimenter hides a toy repeatedly in one of two identically covered wells that are spaced several inches apart. After the infant recovers the hidden toy several times from this *A* location, the experimenter, in full view of the baby, then hides the toy at the second location, *B*. If a short 3–10 s delay occurs between hiding the toy and allowing the infant to reach, babies between 8 and 10 months robustly reach back to the original location. The classic interpretation of this task is that infants do not know that the hidden object exists when it has been transferred from the place they acted on it; they do not have a fully developed object concept independent of their own actions. Because other types of experiments have shown that infants do seem to remember the location of hidden objects, the classic interpretation has been questioned. Over the years, developmentalists have studied hundreds of variants of this task without reaching much agreement about what is occurring.

Linda Smith and I reinterpreted these results differently (Thelen and Smith 1994). We argued from a dynamic systems perspective that the task emerges because infants are only moderately good reachers. Their goals are not yet well controlled and become perturbed, so to speak, by their load and timing level dynamics. We suggested that in this perceptually confusing experiment (infants do not normally have two identical things to reach for) during repeated reaches to one target, a perceptual motor memory is built up that captures subsequent reaches. That is, reaches get increasingly stuck in the trajectory memory of previous reaches, as depicted in figure 12.9. Each reach to one goal location acts to deepen the attractor to that location. When the experimenter switches the goal, the babies cannot escape. Just as the trajectory was disrupted by goal level dynamics, so is the goal level tightly coupled to the lower levels. This is the essence of infants' being unskilled reachers; they lack adaptability and flexibility.

We confirmed our model in a series of experiments that showed three results. First, hidden objects were not necessary to produce the error. Perseveration was a function of the history of looking and reaching. Second, error could be created by manipulating the number of reaches to the *A* side—the more reaches, the more likely the infant would not switch. Third, the deepening attractor could be perturbed by a variety of distracting visual and postural disruptions (Smith et al. 1997).

Most dramatically, we also demonstrated kinematically the emergence of the stable attractor for the trained location (Diedrich et al. 1996). Figures 12.10 and 12.11 give two examples of trajectories and three-dimensional velocities of successive reaches in 9-month-old infants who made the A-not-B error. In figures 12.10 and 12.11, babies made the error on both *B* trials. With successive reaches, the trajectories look somewhat more alike, the speed profiles are remarkable in their

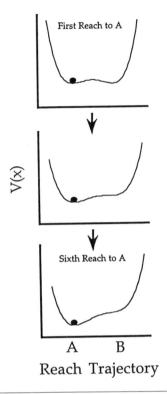

Reach Trajectory

Figure 12.9 Hypothetical potential function $V(x)$ illustrating the formation of a stable attractor for the A reach, where the ball represents the state of the system. At the first reach to A, two states are equally probable (the ball is likely to rest in either well). At the sixth reach, the stable attractor (deep well) for A acts to capture the system's behavior (the ball will likely roll into A).

increasing convergence. Infants whose velocity profiles showed strong convergence always perseverated (although a few infants who perseverated did not show strong convergence).

The implications of this work are profound. This suggests that the parameters of a performed reach are remembered by the system and influence the parameters generated to control the succeeding reach. (Even though the time between reaches in this task can be anywhere from 15 to 45 seconds or longer). The more reaches, the more the influence builds from the history of the system's activity.

Even more remarkable is that this memory encodes not just the spatial form of the trajectory but the precise speed changes, which are highly variable and idiosyncratic among infants. These changes—which are load and timing level characteristics—actually become part of the plan for the next movement. Indeed, in these not yet skilled reachers, these changes eventually overwhelm the changed spatial goal. The infant looks at the *B* location and for all an adult knows, intends to go there, but

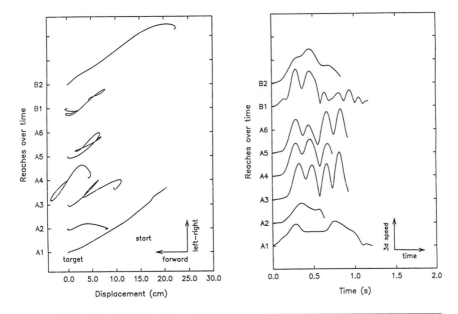

Figure 12.10   Reach trajectories viewed from above (left) and the corresponding three-dimensional speeds (right) for an infant who perseverated in both *B* trials of the A-not-B task. Trials begin with the *A* side (A1) and are repeated for 6 times, then the *B* side is cued twice.

the plan becomes swamped by the motor memory. Many additional experiments are needed to determine just what parameterizes the movement in this task, how these different stability levels can be more or less tightly coupled, and what changes with further development. I, however, feel this phenomenon provides an unprecedented window on motor memory as a function of the history of the system.

Conclusion

This view of trajectory development is testimony to Bernstein's insight about what drives motor skill development. The data confirm the growing belief that the plan cannot be separated from the execution. Perception, action, and cognition are so tightly coupled as to make useless any attempt to draw a line between them. However, this developmental window also allows the opportunity of assessing the relative stability of the component levels with the increasing ability of infants to control their limbs and bodies. Even in a visual task such as reaching and grasping, infants are both exploiting and modulating the force characteristics of the limbs. In this sense, the periphery is indeed setting the agenda for central nervous system changes.

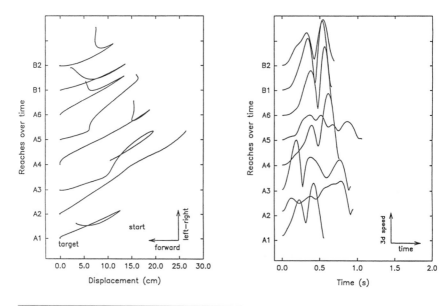

Figure 12.11   Reaching trajectories viewed from above (left) and the corresponding three-dimensional speeds (right) for a second infant who perseverated in both *B* trials. Note that the convergence is apparent only when viewing the speed profiles. Trial A6 is not plotted due to missing kinematic data.

Acknowlegments

I want to thank the following for their major contributions to the work discussed in this chapter: Daniela Corbetta, Kristin Daigle, Fred Diedrich, Dexter Gormley, Gerry Gottlieb, Gregor Schöner, Linda Smith, John Spencer, and Frank Zaal. This research was supported by NIH R01 HD22830 and a Research Scientist Award from the NIMH.

References

Bernstein, N.A. (1967) *The Coordination and Regulation of Movements.* New York: Pergamon Press.

Bernstein, N.A. (1996) *Dexterity and its Development.* Hillsdale, NJ: Erlbaum (Translated and edited by M.L. Latash and M.T. Turvey).

Berthier, N. (1994) Infant reaching strategies: Theoretical considerations. *Infant Behavior and Development,* 17: 521.

Bizzi, E., Hogan, N., Mussa-Ivaldi, F.A., Gitzer, S.F. (1992) Does the nervous system use equilibrium point control to guide single and multiple joint movements. *Behavioral and Brain Sciences,* 15: 603–613.

Corbetta, D., Thelen, E. (1995) A method for identifying the initiation of reaching movements in natural prehension. *Journal of Motor Behavior,* 27: 285–293.

Diedrich, F.J., Thelen, E., Smith, L.B., Corbetta, D. (1997) Motor memory is a factor in infant perseverative errors. (Submitted).

Feldman, A.G. (1966) Functional tuning of the nervous system with control of movement or maintenance of a steady posture. III. Mechanographic analysis of execution by man of the simplest motor tasks. *Biophysics*, 11: 766–775.

Feldman, A.G., Levin, M.F. (1995) The origin and use of positional frames of reference in motor control. *Behavioral and Brain Sciences*, 18: 723–744.

Fetters, L., Todd, J. (1987) Quantitative assessment of infant reaching movements. *Journal of Motor Behavior*, 19: 147–166.

Fitts, P.M. (1954) The information capacity of the human motor system in controlling the amplitude of movements. *Journal of Experimental Psychology*, 47: 381–391.

Flanders, M., Herrmann, U. (1992) Two components of muscle activation: Scaling with the speed of arm movement. *Journal of Neurophysiology*, 67: 931–943.

Gesell, A. (1945) *The Embryology of Behavior.* New York: Harper.

Gottlieb, G.L., Corcos, D.M., Agarwal, G.C. (1989) Strategies for the control of voluntary movements with one mechanical degree of freedom. *Behavioral and Brain Sciences*, 12: 189–250.

Gottlieb, G.L., Corcos, D.M., Agarwal, G.C., Latash, M.L. (1990) Principles underlying single-joint movement strategies. In *Multiple Muscle Systems: Biomechanics and Movement Organization*. J.M. Winters, S.L-Y Woo (Eds.), New York: Springer Verlag, 236–250.

Gottlieb, G.L., Song, Q., Hong, D., Almeida, G.L., Corcos, D.M. (1996a) Coordinating two joints: A principle of dynamic invariance. *Journal of Neurophysiology*, 75: 1760–1764.

Gottlieb, G.L., Song, Q., Hong, D., Corcos, D.M. (1996b) Coordinating two degrees of freedom during human arm movement: Load and speed invariance of relative joint torques. *Journal of Neurophysiology*, 76: 3196-3206.

Gottlieb, G.L., Song, Q., Hong, D., Corcos, D.M. (1996c) Directional control of human planar arm movement. (In preparation).

Guo, S., Roche, A.F., Fomon, S, Nelson, S.E., Chumlea, W.C., Rogers, R.R., Maumbartner, R.N., Ziegler, E.E., Siervogal, R.M. (1991) Reference data on gains in weight and length during the first two years of life. *Journal of Pediatrics*, 119: 355–362.

Halverson, H.M. (1931) An experimental study of prehension in infants by means of systematic cinema records. *Genetic Psychology Monographs*, 10: 107–286.

Hofsten, C. von (1991) Structuring of early reaching movements: A longitudinal study. *Journal of Motor Behavior*, 23: 280–292.

Hogan, N. (1984) An organizing principle for a class of voluntary movements. *Journal of Neuroscience*, 4: 2745–2754.

Kelso, J.A.S., Scholz, J.P., Schöner, G. (1986) Non-equilibrium phase transitions in coordinated biological motion: Critical fluctuations. *Physics Letters A*, 118: 279–284.

Konczak, J., Borutta, M, Topka, H., Dichgans, J. (1995) The development of goal-directing reaching in infants: Hand trajectory formation and joint torque control. *Experimental Brain Research*, 106: 156–168.

Latash, M.L., Gottlieb, G.L. (1991) Reconstruction of shifting elbow joint compliant characteristics during fast and slow movements. *Neuroscience*, 43: 697–712.

McGraw, M.B. (1943) *The Neuromuscular Maturation of the Human Infant.* New York: Columbia University Press.

Morasso, P. (1981) Spatial control of arm movements. *Experimental Brain Research*, 42: 223–227.

Piaget, J. (1954) *The Construction of Reality in the Child.* New York: Basic Books.

Polit, A., Bizzi, E. (1978) Processes controlling arm movements in monkeys. *Science*, 201: 1235–1237.

Schöner, G. (1995) Recent developments and problems in human movement science and their conceptual implications. *Ecological Psychology*, 7: 291–314.

Schneider, K., Zernicke, R.A., Schmidt, R.A., Hart, T.J. (1989) Modulation of limb dynamics during the learning of rapid arm movements. *Journal of Biomechanics*, 22: 805–817.

Scholz, J.P., Kelso, J.A.S. (1989) A quantitative approach to understanding the formation and change of coordinated movement patterns. *Journal of Motor Behavior*, 21: 122–144.

Smith, L.B., Thelen, E., Titzer, B., McLin, D. (1997) Knowing in the context of acting: The dynamics of the A-not-B error. (Submitted).

Spencer, J.P. and Thelen E. (1997) Changes in muscle activity during learning. II. A longitudinal study of infant reaching. (Submitted).

Thelen, E., Corbetta, D., Kamm, K., Spencer, J., Schneider, K., Zernicke, R.F. (1993) The transition to reaching: Matching intention and intrinsic dynamics. *Child Development,* 64: 1058–1098.

Thelen, E., Corbetta, D., Spencer, J.P. (1996) The development of reaching during the first year: The role of movement speed. *Journal of Experimental Psychology: Human Perception and Performance,* 22, 1059-1076.

Thelen, E., Daigle, K. Zaal, F.T.J.M., Gottlieb, G.L. (1997) An unlearned principle for coordinating natural movements. (Submitted).

Thelen, E., Smith, L.B. (1994) *A Dynamic Systems Approach to the Development of Cognition and Action.* Cambridge, MA: MIT Press.

Wellman, H.M., Cross, D., Bartsch, K. (1987) Infant search and object permanence: A meta-analysis of the A-not-B error. *Monographs of the Society for Research in Child Development,* 51: (3).

White, B.L., Castle, P., Held, R. (1964) Observations on the development of visually directed reaching. *Child Development,* 35: 349-364.

Whiting, H.T.A. (Ed.) (1984) *Human Motor Actions: Bernstein Reassessed.* Amsterdam: North-Holland.

Zaal, F.T.J.M., Bootsma, R.J. (1995). The typology of limb deceleration in prehension tasks. *Journal of Motor Behavior,* 27: 193–207.

13

Spatial Frames of Reference for Motor Control

Anatol G. Feldman

Research Center, Rehabilitation Institute of Montreal,
University of Montreal, Montreal, Canada

The most compelling message from Bernstein's teachings is that understanding the nervous system's function presupposes the elaboration of a theoretical foundation of movement science, which he called the physiology of activity. Such a foundation may not only allow researchers to organize the enormous amount of existing empirical data about motor control but also provide the predictive power of a scientific theory resulting in new conceptual and experimental paradigms. In other words, Bernstein understood that the scientific method so efficiently used, for example in physics, may also be efficient in investigating behavior. Bernstein was not the only person to teach this. Einstein expressed a similar view in 1922, "The object of all science, whether natural science or psychology, is to coordinate our experiences and to bring them into a logical system."

In spite of his significant efforts to integrate knowledge creatively from different fields (biomechanics, physiology, psychology, philosophy, physics, and mathematics), Bernstein approached but did not achieve this goal. However, the importance of Bernstein's work cannot be overstated. In particular, he clearly formulated fundamental problems in movement science. Bernstein triggered a process that caused movement science to begin releasing itself from the tenets of strict empirical studies. His work had an important impact on early thinking in the field of motor control and served as a departure point for many contemporary studies devoted to understanding the organization and control of motor activity.

The Necessity of a Critical Reassessment of Bernstein's Ideas

The potential of Bernstein's work may be fully realized only with critical reassessment of his concepts in the light of current experimental data and theoretical approaches. This may lead to a deeper elaboration of his theoretical framework even though, in the process, some of his ideas may be rejected. In this way, Bernstein's creative process of searching for a scientific theory of the physiology of activity can be continued and integrated with modern developments in the field. The fact that Bernstein's work is sometimes used, as Requin et al. (1984) noticed, as "a sort of bible" to which one may refer to find "support for whatever point of view one chooses" is not related to this creative process. This chapter, using data from Feldman and colleagues, will illustrate some examples of current ideas in motor control stemming from a critical analysis of Bernstein's ideas.

Bernstein stressed that the relationship between the pattern of central commands and motor output is ambiguous (Bernstein 1935, 1947, 1967; see also Requin et al. 1984). Bernstein knew that stretching a muscle that was simultaneously stimulated electrically at a constant strength in neuromuscular preparations results in a force dependency on length in that muscle (force-length curve). Bernstein emphasized, however, that muscle elastic properties were characterized not by a single force-length curve but by a family of such curves discriminated by the strength of stimulation. As mentioned by Partridge (1995), such families had been described for neuromuscular preparations as early as 1846 by Weber. A similar family was used later in the formulation of the α model for motor control (Bizzi et al. 1992). Bernstein also stressed that because of proprioceptive feedback to motoneurons (MNs), the level of muscle activation in the intact systems would continuously depend on muscle length and velocity. He took into account both muscle elastic properties and proprioceptive feedback in his theoretical analysis of the processes controlling the muscle interacting with a load.

Assume, argued Bernstein (1935), that the nervous system tries to select a single muscle curve of the family of force-length curves by issuing appropriate descending commands to MNs. These commands predetermine the relationship between muscle force and length but not a specific combination associated with a single point on the curve. This combination is determined by how the muscle interacts with the external force (e.g., created by gravity or by the counteraction of antagonist muscles). This interaction, dynamic in nature, may play a decisive role in specifying a final combination. The operational muscle length will be reached at the point where the muscle and external forces balance each other. The final combination of muscle force and length may not correspond to the desired ones, which necessitates a movement correction mechanism.

Proprioceptive feedback from the periphery to MNs was considered by Bernstein as an additional source of ambiguity in the relationship between the control input and the motor output. One type of such feedback called the stretch reflex (SR) is associated with an increase or a decrease in EMG activity in response to muscle lengthening or shortening, respectively. The term tonic SR refers to length-dependent changes in EMG activity at zero velocity. The term phasic SR is used when velocity is a dominant factor generating such activity (e.g., in the knee jerk).

Consistent with contemporary representations, Bernstein assumed that at given descending commands, proprioceptive feedback contributes to a continuous dependency of the muscle activation level on muscle length and velocity. As a result, the muscle force-length curve defined by the initial magnitude of muscle activation would change during movement regardless of the initial central commands. Some combination of muscle force and length would finally be reached as a result of the muscle-load interaction but not what had been anticipated. Thus, Bernstein concluded that the level of muscle activation, not to mention the final muscle force and length, could not be entirely predetermined by a unique central process. He postulated that, in addition, the central commands had to be periodically updated using proprioceptive signals about the motor output. The principle of proprioception-mediated corrections was thus fundamental to Bernstein's idea of the physiology of activity.

Note that Bernstein distinguished two levels at which proprioceptive signals play a role in motor regulation. The level of MNs is where signals are used continuously in muscle activation and force regulation. The level of central commands is where signals are used noncontinuously to make discrete corrections.

The Central Command Concept

A critical analysis of Bernstein's line of thought may be pursued. Bernstein assumed that proprioception-based corrections were discrete, i.e., not continuous. This implies that central commands may remain invariant for some time regardless of the events in the periphery. Simultaneously, the level of muscle activation may still vary continuously with muscle length and velocity provided by proprioceptive feedback. Therefore, static force-length curves associated with fixed levels of muscle activation postulated by Bernstein should be replaced with force-length curves defined by the length-dependent recruitment of motor units provided by continuous proprioceptive feedback. Then, the invariant central command may be associated with the selection of a specific curve from this family. Thus, this arrives at the notion of invariant force-length curves that characterize the behavior of the whole system rather than the muscle itself (Feldman 1986; Feldman and Levin 1995; Latash 1993). The interaction with the load will bring the system to an equilibrium state associated with a specific point on the invariant curve—the combination of static muscle force and length at which the load will be balanced. Note that the definition of the equilibrium point (EP) is based on the use of the force-length curve, which is only invariant in terms of central commands. In contrast, the strength of muscle activation is not invariant (length-dependent) for each curve. Using a force-length curve associated with a fixed level of muscle activation to find the final EP would result in an error (Feldman 1986). By changing the central command and thus specifying a new invariant characteristic, the system may become unbalanced and will be forced to move to a new EP. This critical analysis of Bernstein's reasoning brings the reader to the EP hypothesis formulated in terms of the λ model (Feldman 1966, 1986; Feldman and Levin 1993, 1995; Latash 1993).

This critical analysis strengthens the Bernsteinian idea that the relationship between the central process and the motor output is ambiguous. However, in this interpretation, the central command may unambiguously predetermine a static force-length curve (although not the original one postulated by Bernstein). Consequently, proprioception-based corrections of central commands may not be required to make a precise movement. The movement will be accurate in those cases when the system correctly anticipates the value of the load and appropriately specifies the central command establishing the force-length characteristic leading to the required final muscle length. An error is inevitable if the load is incorrectly estimated or suddenly changed, making adjustments of the central commands necessary during or after the end of the movement (Weeks et al. 1996).

The concept of the equilibrium state of a dynamic system used in the λ model and implicitly used by Bernstein is a basic one originating in classical physics and its branch called dynamic systems theory. The concept, however, is not trivial. The fact that two versions (α and λ) of the EP hypothesis exist (Feldman 1986; Bizzi et al. 1992; Latash 1993) mirrors the discrepancies in the understanding of the concept and its application to motor control.

Bernstein generally believed that actions may result from an imbalance in the interaction between the organism and the environment. These may be essentially described as processes involved in reaching an equilibrium state. Using, however, inappropriate (nonphysiological) muscle force-length characteristics, Bernstein was unable to indicate specific determinants of this state even in relatively simple tasks like making movements against a load. As has been argued (Feldman and Levin 1995), the same problem exists in the α version of the EP hypothesis. Moreover, such determinants could be elaborated only based on a hypothesis of the integration of central commands, proprioceptive feedback, and muscle properties in the production of muscle activation, force, and movement. A form of sensorimotor-control integration is partly reflected in the concepts of invariant characteristics and central commands in the λ model.

The concept of central commands is regularly used although not rigorously defined in the physiological literature. Usually, central commands or control variables are associated with voluntary actions. According to Bernstein, central commands may be independent and may even remain invariant regardless of the current muscle force and length. The experimental recording of invariant characteristics in intact humans (Asatryan and Feldman 1965; Feldman 1986; for more recent references see Latash 1993) gave rise to the notion that central commands produce shifts in their position defined by the threshold length (λ) at which MNs starts their recruitment. Several consequences of this notion can lead to a specific definition of the concept of control variables and improve the understanding of how voluntary movements are produced. First, it implies that a major control process underlying movement production is associated with a modification of SR parameters—thresholds. In other words, the relationship between the control and reflex systems in this model is hierarchical—control systems use reflexes to produce movements rather than directly producing MN activity (Adamovich et al. 1997). In this scheme, EMG signals always result from the action of both systems. Consequently, individual effects of control and reflex systems cannot be identified as separate components of EMG signals. This contrasts with the traditional view that the central contribution to EMG signals may be separated from the reflex one in deafferentation experiments.

The second consequence is consistent with Bernstein's view about the ambiguity of the relationship between the central control processes and the motor output (see above). The λ model is associated with the notion that kinematic, electromyographic, and force patterns are not predetermined by control signals. Instead, they emerge from the dynamic interaction of the system's components, including external forces.

Third, muscle activity is controlled in a one-dimensional frame of reference (FR) with muscle length as its only coordinate. The threshold length λ represents the origin point of this FR. Shifts in the origin point of the FR by central commands result in a change in the state of the muscle load system. For example, a shift in λ by central commands will result in a change in the position at which the load is balanced (figure 13.1). In other words, the activity of MNs and muscle forces are frame dependent. The nervous system may use this property of the system to produce motor actions by shifting the origin point of the FR.

This chapter will focus on the model that motor control is produced by shifting spatial FRs. The model will be described in detail and applied to multimuscle and multijoint motor tasks including locomotion (Feldman and Levin 1995). This will give the opportunity to discuss other ideas of Bernstein.

In what follows, any variable measurable in terms of length, angle, or distance defined in an FR will be called spatio-dimensional. The threshold λ is spatio-dimensional; its dimension coincides with that of actual muscle length *(x)*.

FRs for a Single Motoneuron and a Motoneuronal Pool

To clarify the concept of FR in the λ model, a physiological measure of central commands must first be defined. With this purpose, consider the effects of muscle stretch or a voluntary movement on the MN membrane potential.

In figure 13.2 (left panel), V is the membrane potential of the MN (vertical axis) at a muscle length x (horizontal axis) when descending control signals are fixed. Because of proprioceptive feedback, the state of the MN depends on muscle length. A quasi-static stretch of the muscle results in an increasing depolarization (δV_r) of the MN as a function of x (diagonal line) because of predominant facilitation from length-sensitive afferents (i.e., muscle spindle afferents in most muscles). The slope of this line

$$s = \partial V / \partial x \qquad (13.1)$$

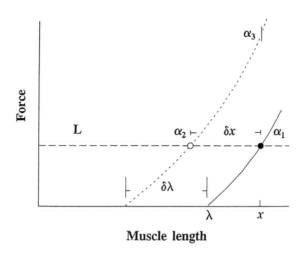

Muscle length

Figure 13.1   Movement production in the λ model. Diagonal lines rising to the right are the initial (solid) and final (dotted) invariant characteristics (ICs). Tonic levels of EMG *(α)* are different at different points on the IC. Initial variables: λ: recruitment threshold length; *x:* muscle length; α_1: EMG level. The initial equilibrium point (filled circle) is the point at which the solid line (i.e., initial IC) and the load characteristic (dashed horizontal line L) intersect. Change in the threshold length results in the displacement of the IC to the left and a shift of the EP (● → ○) leading to changes in muscle activation $\alpha_1 \to \alpha_3$ → α_2 bringing the system to the new EP. Note that the new EP remains the same regardless of the changes in muscle activation.
Adapted from Feldman and Levin 1995.

represents the positional sensitivity of the SR at the subthreshold level. The threshold membrane potential, V_+ and, consequently, the recruitment of the MN, will be reached at muscle length λ, the threshold of MN recruitment. Assuming, for simplicity, that the increase in membrane potential is a linear function of muscle length (figure 13.2, left panel) results in

$$V - V_+ = a(x - \lambda) \qquad (13.2)$$

The condition of MN recruitment $(V - V_+ \geq 0)$ can be represented in terms of variables having the dimension of length. The MN is recruited if

$$x - \lambda \geq 0 \qquad (13.3)$$

Presumably, a change in the tonic control signal may be independent of muscle afferent feedback. It can be measured by a decrement (δV_c) in the membrane potential at the initial muscle length (figure 13.2, right panel). The term *may be independent* should be understood in the sense that the δV_c may result from a planned action based on appropriate information, e.g., on errors in the previous trials (Weeks et al. 1996), as well as on the anticipated or initial muscle length and load. Once the

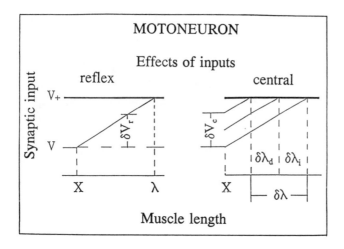

Figure 13.2   Identification of control variables at the level of the α MN in the λ model. Left panel: Effects of reflex inputs on the MN membrane potential. Slow muscle stretch from initial position *(x)* results in a length-dependent *(δV$_r$)* increase in membrane potential *(V)*. Threshold membrane potential *(V$_+$)* is reached at length λ, and tonic MN firing begins. Right panel: Two measures of control variables produced by descending systems—the length-independent increment of MN membrane potential *(δV$_c$)* and threshold muscle length (δλ). The latter consists of two components. One (δλ$_d$) is associated with direct synaptic inputs to the α MN and the other (δλ$_i$) with indirect inputs mediated by γ MNs and muscle spindle afferents.
Adapted from Feldman and Levin 1995.

decision has been made, changes in *V* are produced independently of the current muscle length until if necessary, a discrete correction is made. Starting from a higher level of membrane potential specified by the new central command (figure 13.2, right panel), muscle stretch will result in MN recruitment at a shorter threshold length. Thus, the change in the control signal can be measured as a decrement (δλ) of the threshold muscle length at which the MN is recruited. Like its electrical analog (δV$_c$), the decrement of the threshold length may be independent of the actual muscle length, *x*.

Several physiological mechanisms to regulate λ can be suggested. First, the independent change in the MN membrane potential and, as a consequence, the threshold length results from both direct influences of the central commands on the α MN and indirect influences mediated by γ MNs, muscle spindle afferents, and interneurons. These component changes in λ are denoted by δλ$_d$ and δλ$_i$ (figure 13.2, right panel). By combining direct and indirect inputs to MNs, the system may maximize the range of threshold length regulation and thus the range of actual muscle length and force regulation (Feldman 1986).

Second, tonic electrical stimulation of the lateral vestibular nuclei, pyramidal tract, red nuclei, or medial reticular formation results in a change of the SR threshold

(Capaday 1995; Feldman and Orlovsky 1972; Nichols 1994). This implies that the threshold may be controlled by most descending systems.

Third, the elevation of the membrane potential (figure 13.2, left panel) would be steeper if the muscle were stretched at a rate eliciting velocity-dependent changes in the activity of muscle spindle afferents. A velocity-dependent decrease in the threshold muscle length would be observed (figure 13.3, right panel). This implies that in dynamic conditions, the threshold is a decreasing function of velocity. This assumption is consistent with experimental observations such as a silent period in agonist EMG activity after unloading and tendon jerks reflexes.

Fourth, in figures 13.2 and 13.3, only the length-dependent feedback that originates from the same muscle (autogenous SR) was considered. In natural conditions, changes in the length of a muscle is accompanied by changes in the joint angle and lengths of the other muscles spanning the same and other joints. This may affect the activity of interneurons mediating reflex interactions between muscles such as Ia interneurons of reciprocal inhibition between agonists and antagonists. The elevation membrane potential during the muscle stretch and, as a result, the threshold length may thus be affected by reflex interaction.

Fifth, additional force-dependent feedback to α MNs is mediated by tendon organ afferents and Ib interneurons (which also receive inputs from Ia muscle spindle and cutaneous afferents). Recurrent inhibition of MNs mediated by Renshaw cells is another type of feedback to MNs dependent on their output activity. Tendon organ

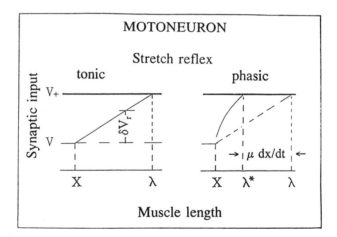

Figure 13.3 Proprioceptive feedback in the λ model. Left panel (as in figure 13.2): Slow muscle stretch from current position (x) results in a length-dependent (δV_r) increase in membrane potential (V). Threshold membrane potential (V_+) is reached at length λ, and tonic MN firing begins. Right panel: Threshold muscle length decreases $(\lambda \rightarrow \lambda^*)$ as a result of rapid muscle stretch. The decrease is assumed proportional to velocity $(dx/dt,$ positive for muscle stretch). Coefficient μ represents the dynamic sensitivity of the threshold to velocity.
Adapted from Feldman and Levin 1995.

afferents and likely Renshaw cells, however, become active after recruitment of appropriate MNs. Thus, the threshold length for the first recruited MN during muscle stretch is defined without their participation. However, the tendon afferents and Renshaw cells that become active following recruitment of the first MN may influence the thresholds of subsequently recruited MNs. These types of feedback may, however, be functional in initially active muscle. Therefore, they may contribute to the threshold length changes in response to shortening or lengthening.

Based on the above description, the dynamic threshold length (λ^*) for activation of a muscle may likely be presented, to a first approximation, as the sum of several components

$$\lambda^* = \lambda^c_d + \lambda^c_i - \mu \cdot dx / dt + \rho + \sigma + \tau \qquad (13.4)$$

The first component with subscript d ("direct") is specified by independent, control inputs to α MNs not involving muscle spindle afferents (superscript c stands for the control component). These inputs may be monosynaptic or polysynaptic. The second component with subscript i ("indirect") represents the effect of control inputs to γ MNs. The other components describe the influence of, respectively, velocity (dx/dt), reflex intermuscular interaction, recurrent inhibition, and force-dependent afferent feedback on the dynamic threshold. Time dimension parameter μ characterizes the dynamic sensitivity of muscle spindle afferents.

The model introduces a measure of muscle activation as an increasing nonnegative function of variable A

$$A = b \ [x - \lambda^*]^+ \qquad (13.5)$$

where $[u]^+ = u$ if $u \geq 0$ and $u = 0$ otherwise. Active muscle force may be defined based on this function, velocity-dependent and force-dependent properties of sarcomeres consistent with the shape of the experimental invariant characteristics as well as with the kinematic and EMG patterns of single- and double-joint movements (St-Onge et al. 1997; Flanagan et al. 1993). Positive coefficient b is constant if the differences between the recruitment thresholds of MNs remain invariant regardless of changes in the thresholds themselves. The experimental, invariant torque/angle or force/length characteristics become somewhat steeper with a decrease of the threshold (Feldman 1986; Capaday 1995). This implies that the b may increase when the threshold length decreases. Alternatively, the increase in the slope (stiffness) of the characteristics may be explained by the dependency of the elasticity of sarcomeres on the muscle length (Gordon et al. 1966). Therefore, the shape of the invariant characteristics may be different even though parameter b is constant.

The above illustrates that MN threshold properties and proprioceptive feedback may be essential components of the mechanism that relates basic electrical parameters of the MN to spatio-dimensional variables. This mechanism causes current and threshold membrane potentials to be associated with the current and threshold muscle lengths, respectively. As a result, MN functioning becomes associated with external, one-dimensional space represented by the muscle length. The association of the threshold membrane potential with the threshold length is mutable. The threshold length may be shifted by changing the membrane potentials of interneurons and MNs by independent control inputs.

The λ model implies that the nervous system deals with the muscle in a one-dimensional FR with muscle length as its only coordinate. This FR can be called natural since it is similar to what one would choose for a mathematical description of muscle behavior. For this purpose, the origin is selected, i.e., a specific point associated with a particular muscle length. Using a scale in this FR, any other muscle length is quantified, for example, in centimeters. Similarly, in the FR for muscle activity regulation, the threshold length appears to be the origin point. The magnitude of muscle activation is defined by the distance between the point associated with the actual muscle length and the origin point of the FR.

In the FR used by the nervous system, muscle force is frame dependent in the sense that shifts of the FR origin produced by central commands provoke changes in the muscle activation as defined by equation 13.5. FRs in which forces are frame dependent may be called physical. For example, a hermetically sealed vessel may be considered as a physical FR for the gas molecules that it contains since the movement of molecules is constrained by the vessel. As a planet, earth may also be considered as a natural, physical FR for everything gravitating to its surface. These examples also illustrate the notion that a shift of a physical FR in space affects the movement of all bodies in it. Thus, according to the λ model, the nervous system utilizes a simple physical principle for movement production or, if movement is prevented, isometric torque generation. This principle implies the ability to produce shifts in the origin of a physical FR based on the sensori-control integration described above.

Single-Joint Level

Now consider a one-dimensional FR with the coordinate representing joint angle θ for a single degree of freedom of the body. The degree of freedom is subserved by multiple muscles. In this angular FR, the SR threshold for activation of each muscle is measurable in terms of the threshold angle at which MNs of this muscle begin to be recruited. Taking into account that the influence of descending systems is typically mediated by interneurons terminating on MNs of different muscles, it seems physiologically plausible to suggest that the angular thresholds for different muscles are not entirely independent. In particular, situations are likely when the threshold angles for different muscles are identical. This common threshold angle is denoted by R (figure 13.4B). At this angle, all muscles are silent. However, one group or the remaining, antagonistic group of muscles may be activated depending on the direction of a passive deflection from the threshold angle. The angle R may be considered as a referent angle for activation/deactivation of the joint muscles. It also represents the origin point of the united angular FR. Control inputs to MNs may shift the origin point (R) of this FR. These control influences are called the reciprocal or R command. Similar to the FR for MNs of a single muscle, the scales for changes in the

A MOTONEURONAL LEVEL

Threshold length   *EMG activity*

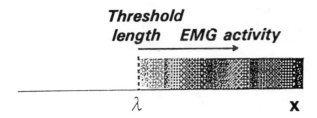

λ **x**

Muscle length

B JOINT LEVEL

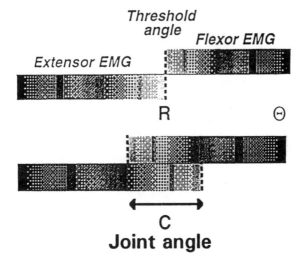

Threshold angle   *Flexor EMG*

Extensor EMG

R Θ

C

Joint angle

(continued)

Figure 13.4 Spatial frames of reference at different levels of motor control. *(A)* Motoneuronal recruitment starts at a threshold muscle length (λ) and increases (as shown by shaded area) as the actual length *(x)* increases. *(B)* Two spatial frames of reference for activation of flexor and extensor muscles of a joint defined by a transitional angle *(R)* and the coactivation zone *C*.

Adapted from Feldman and Levin 1995.

C BODY LEVEL

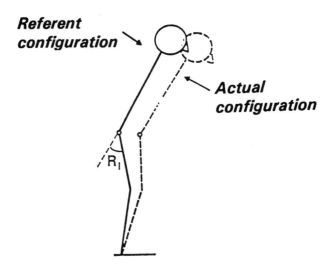

Referent configuration

Actual configuration

R_1

D EXTRAPERSONAL LEVEL

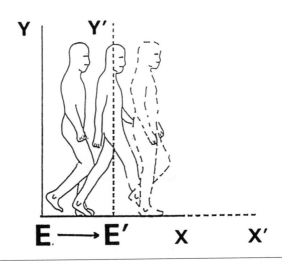

Y   Y'

E ⟶ E'   X   X'

Figure 13.4   *(continued) (C)* Referent body configuration (solid diagram) defined by the threshold angles *(R₁)* for each degree of freedom of the body. Actual and, in particular, equilibrium configuration (dashed diagram) result from neuronal and mechanical interactions of the system with external forces in the frame of reference designated by control signals. *(D)* Extrapersonal frame of reference associated with external three-dimensional space. Shifts in the origin point of the frame *(**E** → **E'**)* may be used to elicit a step or train of steps.
Adapted from Feldman and Levin 1995.

muscle activity in the angular FR are defined individually for each muscle by the sensitivity of the MNs to changes in the joint angle when the R command is constant.

Indeed, the situation when the angular thresholds are identical is not universal. The cases when agonist and antagonist muscles are simultaneously active indicate that the system may specify different thresholds for the agonist and antagonist muscles. The transition angle R will be surrounded with an angular zone (C) in which these groups of muscles may be simultaneously active (if $C > 0$) or silent (if $C < 0$). Control inputs to MNs may change the width (C) of this zone. These control influences are called the C (coactivation) command.

When requested to coactivate muscles of the arm, people usually do this while preserving the initial arm position. Therefore, the C command, by definition, does not affect the equilibrium position of a joint whether or not this position coincides with the angle R. Indeed, in the presence of a C command, the R command no longer represents the transition angle or switching point for agonist/antagonist muscles. When the C command is positive, angle R represents the one at which the net joint torque is zero. However, any passive deflection from this angle elicits active resistance. When the C command is negative, the R command is a referent angle specifying the center of the coactivation zone. A change in the R command is thus associated with a shift in the coactivation zone in the angular FR (Levin and Dimov 1997).

Most essentially, R and C commands are functionally different. The R command defines an origin point of the angular FR for all muscles of the joint regardless of their biomechanical functions or anatomical arrangements. In contrast, the C command distinguishes between agonist and antagonist muscles and provides different origin points for these muscle groups in the angular FR. Additional functional differences between the two commands are also important. Voluntary movement from one position to another cannot be produced unless the system shifts the equilibrium state of the system (Feldman and Levin 1995). Of the two commands, only the R command is able to produce such shifts by changing the origin point of the angular FR. A positive C command may contribute to the net joint stiffness defined as the slope of the invariant torque/angle characteristic specified by the R and C commands. Applied in combination with an R command, the C command contributes to movement speed (Feldman and Levin 1995; Latash 1993). Changes in stiffness and speed also depend on the R command. However, the C command is able to produce these changes without modifying the timing of shifts in the equilibrium state.

The R and C commands are independent of each other. The nervous system may use them in various combinations to produce different single-joint movements, including the fastest ones. This has been explained in recent publications in which other predictions of the model are also tested (Adamovich et al. 1993, 1997; Feldman et al. 1995; Feldman and Levin 1995; Levin et al. 1992; Levin et al. 1995; Weeks et al. 1996). These tests deal with the timing of the commands, strategies utilized by the nervous system in correcting movement errors, as well as the effects of perturbations on the EMG and kinematic patterns in the fastest elbow and wrist movements.

The difference between the traditional approach and the author's approach will be emphasized. In the former, control processes are considered in terms of reciprocal and coactivation electromyographic patterns for agonist and antagonist muscles during different motor tasks. In the latter, the reciprocal and coactivation commands are position-dimensional determinants (origins) of FRs for muscle activation. These commands are, in essence, independent of EMG patterns (but not vice versa). For example, the R command in the author's approach indicates the joint angle, R, at which the transition of activity from agonist to antagonist may reciprocally occur (if $C = 0$). Whether or not the transition actually occurs depends on the spatial relationship between the actual joint position and R. In addition, muscle activity also depends on the C command and, because of proprioceptive feedback, on movement speed. Similarly, although the coactivation zone may be present $(C > 0)$, the load may be balanced outside this zone so that only agonist muscles will be active. The distribution of activity among different agonist muscles (synergists) may also be affected by the C command (Levin and Feldman 1995).

The definition of the R command as a threshold angle at which the transition of activity from agonist muscles to antagonists occurs when $C = 0$ requires some reservation concerning biarticular muscles. The length of such muscles is a function of two joint angles. As a result, the threshold angle, R, for muscle activation is reached only when an appropriate angle at the second joint is established. This problem is solved naturally in the definition of the R command generalized to all skeletal muscles of the body (next section).

Referent Configuration of the Body: The Origin of the United FR for all Skeletal Muscles

Now consider an n-dimensional FR (configurational FR) representing all possible configurations of the body of a living organism (n is the number of degrees of freedom of the body). Each configuration resembles the actual body configuration perceived by an onlooker (figure 13.4C, dashed diagram). In a mathematical sense, it may be defined as a point in this FR or a vector θ composed of individual angles for each degree of freedom

$$\theta = (\theta_1, \theta_2, \ldots, \theta_n) \tag{13.6}$$

In this configurational FR, motor actions produced by control levels of the nervous system may be described as R and C commands generalized to all skeletal muscles and degrees of freedom of the body. In particular, the generalized R command is a vector, R, composed of individual R commands for each degree of freedom

$$R = (R_1, R_2, \ldots, R_n) \tag{13.7}$$

The R command may be considered a referent body configuration (figure 13.4C, solid diagram) in the configurational space. EMG activity and muscle forces depend on $\theta - R$, i.e., on the deflection of the actual configuration from this referent one.

When the C commands for each joint are zero, R represents a threshold configuration for recruitment/derecruitment of each skeletal muscle of the body. In other words at this configuration, EMG activity of all skeletal muscles is zero regardless of their biomechanical functions. However, any deflection of the body from configuration R whether elicited by external forces or changes in R itself is counteracted by recruitment and increasing activity of agonist muscles while antagonists remain silent.

The R command may also be considered as the origin point of the configurational FR. Since EMG activity depends on θ–R, the nervous system may produce active movements by shifting the origin of the FR.

Although all skeletal muscles are controlled in the united FR, their activation patterns may differ. In particular when taking into account the differences in the anatomical arrangement of muscles, their activation also depends on the direction and moment arm of muscle action. As a consequence, the individual EMG levels and the division of muscles into the agonist/antagonist groups depend on the direction of the difference θ–R Similar to threshold lengths (see equation 13.4), threshold angles and, as a consequence, muscle activation depend on position-, velocity-, and force-dependent homonymous and heteronymous reflexes mediated by proprioceptive afferents.

External forces (e.g., gravity) deflect the body from the referent configuration. Muscles resist the deflection. Finally, they balance these forces at some body configuration called the equilibrium configuration. In this process, the nervous system, by specifying the referent configuration, may only restrict the set of possible equilibrium body configurations. In contrast, a specific equilibrium configuration emerges from the interaction of the body with external forces.

The difference between the referent and equilibrium configurations is not considered an error necessitating correction. This contrasts with the servo-system theory would suggest. The nervous system may decide whether the equilibrium body configuration is suitable for the motor task (e.g., grasping an object in space). If it is not, the nervous system modifies the referent configuration until the final configuration becomes adequate.

Extrinsic FRs for Actions in a Three-Dimensional Environment

At the top of the spatial FR hierarchy may be an extrinsic FR utilized for body actions in external three-dimensional space. Consider a Cartesian three-dimensional FR associated with a person, localized and oriented in some way in extrapersonal space (figure 13.4D). Such an extrinsic FR will be called the relative FR to distinguish it from an absolute, motionless Cartesian FR, for example, associated with the ground surface. Localization of the relative FR is defined by a three-dimensional vector, L, representing the coordinates of the relative FR origin in the absolute FR. The origin may be associated with a point of the body, for example, with the heel of a foot. The previously defined referent body configuration R is imbedded in this FR with the attachment of its one point (the heel) at the origin point.

Orientation of the relative FR may be described by a variable M composed of three rotations of the relative FR in the absolute FR. These rotations may be characterized by three Euler's angles or three other parameters (Branetz and Schmiglevsky 1973). One may assume that in standing, the orientation of the relative FR axes corresponds to vertical, frontal, and sagittal directions for the body.

Two variables, L and M, define the localization and orientation of the relative FR in the absolute FR. One may assume that by combining independent, control inputs with inputs from vestibular, visual, and proprioceptive receptors on neurons of descending systems, the brain may change the referent values of these variables. As in other FRs, the generation of muscle activity is frame dependent. In other words, by influencing the referent values, the nervous system may shift the relative FR and change its orientation. This provokes an active movement of the body in external space. Although specific details have not yet been elaborated, the united FR dealing with the body configuration is likely subordinated to this FR dealing with the localization and orientation of the body in external space. Specifically, consider the case when the system only controls the localization of the body in external space by changing the referent value, E, of the localization. As a working hypothesis, one can assume that R may be a function of the deflection of the actual localization of the body (L) from the referent one (E) and velocity dL/dt

$$R = R_o + f(L + v \cdot dL/dt - E) \qquad (13.8)$$

where R_o is the initial referent body configuration and v is a time-dimensional damping coefficient.

Different Actions: Explanations in Terms of the FR Hypothesis

In previous experimental and theoretical studies, the λ model has been applied to single and double-joint arm movements (Flanagan et al. 1993) and four–degrees of freedom jaw-hyoid motion during mastication and speech production (Laboissière et al. 1996). Different components of the system were described by differential equations of motion. These dynamic equations produce kinematic and EMG patterns of movement in response to comparatively simple shift patterns in the appropriate FRs (direct dynamic solutions). The notions of the united configurational and extrinsic relative FRs substantially extended the applicability of the model. This chapter briefly reviews the explanatory capacity of the model.

Single-Joint Movements

Fast, single-joint movements from one position to another are characterized by an initial agonist EMG burst followed by an antagonist burst and then a secondary agonist burst (the tri-phasic EMG pattern). The origin of this pattern can be explained by the λ model in qualitative and quantitative terms using basic knowledge of static and dynamic properties of muscles and proprioceptive reflexes. In addition, one must assume that a change in the joint position results from a monotonic, ramp-shaped shift of the spatial FRs for activation of agonist/antagonist muscles. Such a shift results in a modification of the equilibrium state of the system causing EMG bursts. A three-burst EMG pattern has also been obtained in a mathematical formulation of the model (St-Onge et al. 1997). In general, this pattern is a trivial consequence of the suggested model of motor control. The model has explained such detailed, nonlinear effects of perturbations as the prevention or suppression of the antagonist burst when long and short elbow movements, respectively, are suddenly arrested (Feldman et al. 1995). Both experimental and theoretical studies are consistent with the notion that in fast movements, shifts in the EP end before the movement peak velocity, i.e., long before the end of movement (Feldman et al. 1995; cf. Latash 1993; Gomi and Kawato 1996). The long held belief that the onset time of the first agonist burst is preprogrammed and is not modified by the stretch reflex (Wadman et al. 1979) has recently been challenged by opposite findings consistent with the model (Adamovich et al. 1997).

A Single Step, Locomotion, Jumping, Pointing, and Somersault

Except for somersault, all these actions are presumably produced by shifts in the referent value of the origin *(E)* of the relative FR. A sagittal shift in the relative FR will elicit changes in neuronal and muscle activity resulting in the transition of the body from the initial to a final equilibrium configuration reached in another part of external space. In terms of locomotion, a step forward or backward will be made depending on the direction of the shift. To further elaborate this idea, assume that an initial standing configuration is established when a referent configuration of the body, R_o, is specified. Assume that the body weight is shifted to one leg to prepare the other leg for the swing phase, thus making locomotion possible. To initiate a step, the nervous system makes a change in the *E* to shift the relative FR in a sagittal direction. According to equation 13.8, this results in changes in the referent body configuration, *R.* These changes are generally assumed to diminish the distance between the actual and referent location of the body in external space. This process is defined by function *f* in equation 13.8. The translation of the body may mainly be reached by changes in the *R* resulting in hip flexion of the unloaded leg and simultaneous ankle flexion of the loaded leg. These actions may be associated with the swing and stance phases of locomotion. The first phase contributes to the forward translation of the unloaded leg while the second to the translation of both the legs and trunk. The step will be completed when an equilibrium body configuration is restored at a new location in external space.

Indeed, a continuous sagittal change in *E* will elicit locomotion. Perfection of the gait cycle may be achieved by combining the sagittal shift in *E* with a small vertical shift to elevate the feet above the ground during the swing phase. By adding rhythmical frontal shifts in *E,* the necessary transfers of body weight from leg to leg are produced. The trunk is preserved in a vertical orientation if function f in equation 13.8 provides an ankle flexion in combination with a hip extension during the stance phase. The gait direction may also be controlled by shifts in E. One may also assume that with increasing speed of the shift, walking will be transformed into running characterized by the presence of a short flight phase (cf. Kelso 1995).

The shift rate of the relative FR as well as other FRs considered here is, indeed, physically limited. Empirical and simulation studies show that the rate of changes in the R command in fast, single-joint movements is less than 700 deg/s; the shift rate in the E command for pointing is typically less than 4 m/s (Flanagan et al. 1993). Sprinters may likely increase this limit up to 10 m/s when running for a short distance.

A jump up may be produced by a rapid vertical shift of the extrinsic FR. For a forward jump, a vertical shift should be combined with a horizontal shift of the FR.

Shifts of a relative extrinsic FR may also be used to produce arm pointing movements. The origin of such an FR may be associated with a point in actual space (e.g., the corner of the table above which the subject moves his or her arm) or with the main effector (e.g., with the arm end point that moves to the target). The suggestion that the system select a shoulder-centered FR for pointing (Soechting and Flanders 1992) would be inconsistent with the main idea of the model presented in this chapter—motor actions (including pointing) are produced by shifting the origin point of a spatial FR. Pointing may be produced by shifting a relative extrinsic FR involving either only arm joints or also other joints, e.g., those of the trunk (Ma and Feldman 1995), depending on the task requirement (instruction, external constraints, and so forth). In the actual movement, shifts in the FR may be depicted as a displacement of the arm end point. The possible discrepancy between the referent shift in the origin of the FR and actual movement directions of the end point is not considered an error. An error exists when the actual movement direction of the arm end point differs from the intended one. Correcting the error is accomplished by modifying the direction of the shift in the FR (Flanagan et al. 1993).

Somersaults and other similar movements (e.g., tumbling) imply that the nervous system may rotate the relative FR (change in the referent value of the variable M) to generate them.

Motor Learning and Development

The FR hypothesis may be used to suggest that motor learning and development (Thelen and Smith 1994) are associated with the formation and management of appropriate FRs. For example, a toddler's ability to stand may appear only after the formation of the united configurational FR is complete. The ability to produce a step or gait arises after the formation of a relative extrinsic FR. Interestingly, a two year old is able to walk and jump upward but not to jump while propelling forward. This implies that the child is at the stage of development when he or she can produce separate but not simultaneous horizontal and vertical shifts in the relative extrinsic FR.

The origin point of the configurational FR may be associated with a minimal reflex interaction between different muscles (Feldman and Levin 1995). This implies that the principle of minimal interaction formulated by Gelfand and Tsetlin (1971) may underlie the formation of the united FR during development and continue to function in adults.

Discussion

The idea of FRs will be compared with other approaches to motor control, in particular, with Bernstein's idea of hierarchical functional organization of movement production.

Natural and Physical Frames of Reference for Motor Control

According to the model, different FRs used for movement production have common attributes despite their broad variety. First, an FR is created by the convergence of sensory information (visual, vestibular, auditory, proprioceptive, and so forth) and independent control inputs on common neurons or MNs. Sensory systems are sensitive to variables associated with the environment. Therefore, they relate the FR to the physical world. These FRs are naturally occurring ones in which muscle length, joint angle, body configuration, localization, and orientation of the body in extrapersonal space is measured. Motor actions are described as shifts in the origin and changes in the localization and orientation of FRs. Finally and most importantly, the generation of neuronal and muscle activity and forces is frame dependent. Thus, the nervous system uses physical FRs for motor control. Shifting the FRs by independent control inputs compels the system to modify its activity. The referent body configuration may also be a key element in the perception of body posture. Therefore, the model may be extended to explain kinesthesia. The model may also be helpful in understanding the basic mechanism that orients a single neuron in space with maximal sensitivity to a specific direction (Georgopoulos et al. 1992; Kalaska and Crammond 1992).

Comparison With Bernsteinian Ideas and Other Approaches to Motor Control

As illustrated, physical FRs are hierarchically organized such that FRs at higher levels subordinate those at lower levels. The λ model accepts Bernstein's notion about the hierarchical functional organization of motor control. This notion does not conflict with modern data showing that control functions may be widely distributed across different brain structures. In fact, Bernstein himself did not exclude the possibility of distributed control processes (Bernstein 1935).

Bernstein systematically postulated the existence of higher levels (called the levels of spatial fields; see Bernstein 1947) used to plan movement in terms of desired spatial characteristics. These levels did not directly participate in the awkward job of generating muscle forces. This was left to executive levels consisting of, in particular, muscles and joints interacting with the environment (Kalaska and Crammond 1992). In the FR hypothesis. such a division of labor is realized. Control levels may be preoccupied with spatial shifts in the origin points of the FRs and allow the executive levels to bring the system to a final state.

Bernstein (1967) suggested that control functions may be transferred from one neuronal level to another, causing the latter to become the leading level. He also assumed that this process is associated with a change in the type of leading sensory information required for functioning at the new level. These ideas are actually presented in the λ model in a very specific form. Different FRs are hierarchically organized. However, the nervous system may transfer control functions, for example, from the joint-oriented, configurational FR to an FR associated with external space appropriate for reaching arm movements and locomotion.

A fundamental for the λ model is that MN recruitment may be described in terms of spatio-dimensional variables (threshold lengths). Since threshold lengths in an MN pool are interdependent, only two independent variables (the actual muscle length and the threshold length for a single MN of the pool) define all possible states of the MN pool in static conditions. In particular, these two independent variables are sufficient to determine which MNs are recruited and which are silent. The two variables define all points of an invariant force/length characteristic. Using this characteristic and the value of the threshold length, one can find the threshold force at which the MN is recruited.

In some studies, the threshold and actual forces are considered the only determinants of MN recruitment. Although this may be convenient in some cases, e.g., in slow generation of isometric force, it is not convenient in ballistic force generation, where most motor units are recruited before force generation (Desmedt and Godaux 1977). In general, the description of MN recruitment in terms of force thresholds cannot replace the description in terms of spatio-dimensional variables and may actually be logically inconsistent. In particular, one can say that the smallest motor unit has a zero force threshold and, therefore, is recruited first in all cases when active muscle force is greater than zero. However, motor unit recruitment precedes the generation of active force by this unit and muscle. In other words, recruitment is not related to the active muscle force exceeding the zero force threshold. The actual cause of recruitment thus remains hidden in models based on the notion that the threshold force is a determinant of MN recruitment.

Some experimental data are considered as conflicting with the λ model. Before discussing these data, note that the model is based on the well-established threshold properties of motoneurons, reflexes, and motor control processes. For example, many spinal and descending pathways are able to modify muscle activation thresholds (Matthews 1959; Feldman and Orlovsky 1972; Nichols and Steeves 1986; Capaday 1995). In addition, a threshold mechanism underlies movement initiation

(Hanes and Shall 1996; Adamovich et al. 1997). Mathematically, threshold functions are singular. In other words, motor systems are fundamentally non-linear and cannot be considered linear, even locally, for small changes in variables.

In contrast, not only local but global linearity of the system was assumed in the computation of the equilibrium trajectory in the recent study by Gomi and Kawato (1996). It is not surprising therefore that they obtained a paradoxical result: Figure 3 in their study shows movements ceasing about 250 ms before the end of the shift in the equilibrium state. Physically, a shift in the equilibrium state provokes movement and, consequently, the movement can only end after, not before, the shift. The paradoxical finding of Gomi and Kawato was a consequence of their false assumption of the linearity of the system and brings into question the validity of their conclusion that the pattern of shift in the equilibrium position is complex, with multiple velocity peaks. Indeed, the experimental data on which Gomi and Kawato based their computations has been reproduced using a λ model incorporating a simple, ramp-shaped shift in the equilibrium position (Gribble and Ostry 1997). Finally, it has also been demonstrated experimentally that the EP shift in fast uncorrected arm movements is a short-duration monotonic process ending long before the end of actual movement (Feldman et al. 1995a). Thus, claims that Gomi and Kawato's findings invalidate the λ model are unfounded.

The report of positional errors in arm pointing movements perturbed by Coriolis forces in a rotating room (Lackner and Dizio 1994) is another source of controversy associated with the EP hypothesis. It was assumed in these studies that since the Coriolis force is velocity dependent, it could not modify the final EP ("movement equifinality"). The term equifinality was originally used to describe a feature of the system when the pattern of central commands remained unchanged (Asatryan and Feldman 1965; Feldman and Levin 1995). It implies that the arm movement would arrive at the same final EP after transient perturbations as when they are made without perturbations. We do not think, however, that the finding of inequifinality justifies the rejection of the EP hypothesis. According to the EP hypothesis, an essential requirement for equifinality is that the pattern of central commands underlying the movement remains the same regardless of perturbation. Equifinality may thus not occur if perturbations cause changes in the initial pattern of central commands.

Note that the Coriolis force deflects the arm from the equilibrium position in proportion to the movement velocity, in contrast to muscle forces resisting the deflection. Coriolis forces thus belong to the family of anti-damping, destabilizing perturbations. Control systems may be forced to actively react to such perturbations to preserve movement stability. From this point of view, positional errors resulting from Coriolis forces can be explained in the framework of the EP hypothesis as has recently been demonstrated (Feldman et al. 1995b).

The author's approach concerning the nature of the FRs used in motor production differs in several aspects from that developed in other studies dealing mainly with reaching movements. In those studies, the FR origin is associated with different anatomical sites defined within the body architecture: head, shoulder, and trunk centered (Blouin et al. 1993; Jeannerod 1988; Soechting and Flanders 1992). In the author's formulation, no motionless site of the body can be considered as the origin point of an FR used for movement production. This would be inconsistent with the idea that actions are produced by shifting the origin of FRs. In pointing movement, a movable body site may be the end point of the arm. As a consequence in this approach, pointing movement may be associated with a hand vector representing the shift of the appropriate extrinsic FR (Flanagan et al. 1993). In general, the idea of physical FRs conflict with the idea of stationary, motionless FRs used by the nervous system for movement planning, as suggested, for example, by the shoulder-centered FR hypothesis of Soechting and Flanders (1992). In addition, computing torques based on the desired movement kinematics cannot be reconciled with this approach.

Experimental data about the neurons involved in producing the united, configurational FR and about their sensory and control inputs and output projections to MNs are practically absent. Possible candidates may be propriospinal neurons and neurons of descending tracts with their nonlocal projections to MNs of multiple body muscles (Lemon 1990; Shapovalov 1975). To unify the body frame, control systems should project widely to these neurons and so should proprioceptive muscle afferents. Reflex intermuscular interaction may also play a significant role in the organization of the united body FR (Lacquaniti 1992; Nichols 1994). The extrinsic FR controlling the orientation of the whole body in space may be provided by vestibulospinal neurons. The existence of efferent projections to the vestibular system may be a part of the mechanism involved in controlling shifts in the FR.

The referent body configuration is unique because in the absence of coactivation, this configuration is the threshold for all skeletal muscles when active muscle forces are zero (minimal interaction). At this configuration, the threshold length coincides with the actual muscle length for each muscle ($\lambda_i = x_i$ for each muscle i). In other words, the relationship between λs mirror the anatomical relationship between actual muscle lengths (the principle of anatomical correspondence in the control of muscles). In essence, this solves the problem of muscle redundancy: the system may be controlled at a high level by changing commands not differentiating between muscles (like the R and E commands). The reflex interaction between muscles likely obeying the principle of minimal interaction (Gelfand and Tsetlin 1971) may specify λs according to the anatomical arrangement of muscles.

Conclusion

Bernstein's creative process in which he searched for a scientific theory of the physiology of activity was examined in relation to some recent developments in the field. Some hypotheses in motor control stemming from a critical analysis of Bernstein's ideas were illustrated. One of them is that the relationship between the pattern of central commands and motor output is ambiguous, an idea inherent in the λ model of motor control. In the most advanced formulation of the model, the nervous system uses sensory information to organize spatial FRs for the sensori-control apparatus. These FRs are physical in the sense that the generation of relevant neuronal and muscle activity and forces is frame-dependent. Active movements may be produced by shifting the origin points or rotating the FRs. In the framework of this FR hypothesis, different FRs were described starting from the level of a single neuron and ending at the levels of the configuration of the body and its orientation in external space. The FR hypothesis offers a solution to the Bernsteinian problem of "division of labour" between different levels of motor regulation: Control levels may be pre-occupied with shifts of the spatial FRs and relegate movement of the system to a final state to executive levels interacting with the environment. These ideas were used in the explanation of basic control mechanisms underlying multi-muscle co-ordination, single-joint movements, pointing, single steps, locomotion and jumping. The FR hypothesis was also extended to motor learning and development.

This chapter illustrated that a critical approach to Bernstein's ideas may be productive. Further development of the model may be produced based on a critical analysis of its present formulation. The concepts of the referent body configuration and relative extrinsic FRs were defined in a way that permitted, after cautious consideration, the formulation of experimentally testable predictions concerning multimuscle and multijoint coordination (Feldman and Levin 1995).

Acknowledgments

I would like to acknowledge my deep obligation to Bernstein, whose views stimulated the development of my work as well as that of others making up a strong group of motor-control scientists in the former Soviet Union. I would also like to thank my numerous colleagues who, throughout the years, have acted as mentors, opponents, and supporters of the development of the model. Particular thanks are extended to Mindy Levin for her help developing and expressing the ideas represented in this article. I would also like to acknowledge financial support from the Natural Science and Engineering Research Council of Canada and the Medical Research Council of Canada.

References

Adamovich, S.V., Levin, M.F., Feldman, A.G. (1993) Merging different motor patterns: Coordination between rhythmical and discrete single-joint movements. *Experimental Brain Research*, 99: 325–337.
Adamovich, S.V., Levin, M.F., Feldman, A.G. (1997) Central modifications of reflex parameters may underlie the fastest arm movements. *Journal of Neurophysiology*, 77: 1460-1469.
Asatryan, D.G., Feldman, A.G. (1965) Functional tuning of the nervous system with control of movement or maintenance of a steady posture: I. Mechanographic analysis of the work of the limb on execution of a postural task. *Biophysics*, 10: 925–935.
Bernstein, N.A. (1935) The problem of interrelation between coordination and localization. *Archives of Biological Science*, 38: 1–35 (In Russian).
Bernstein, N.A. (1947) *On the Construction of Movements*. Moscow: Medgiz (In Russian).
Bernstein, N.A. (1967) *The Coordination and Regulation of Movements*. Oxford, UK: Pergamon.

Bizzi, E., Hogan, N., Mussa-Ivaldi, F.A., Giszter, S. (1992) Does the nervous system use equilibrium-point control to guide single and multiple joint movements? *Behavioral and Brain Sciences,* 15: 603–613.

Blouin, J., Bard, C., Teasdale, N., Paillard, J., Fleury, M., Forget, R. Lamarre Y. (1993) Reference systems for coding spatial information in normal subjects and a deafferented patient. *Experimental Brain Research,* 93: 324–331.

Branetz, V.N., Schmiglevsky, I.P. (1973) *The Use of Quaternions in the Task of Rigid-Body Orientation.* Moscow: Nauka, Moscow (In Russian).

Capaday, C. (1995) The effects of baclofen on the stretch reflex parameters of the cat. *Experimental Brain Research,* 104: 287-296.

Desmedt, J.E., Godaux, E. (1977) Ballistic contractions in man: Characteristic recruitment pattern of single motor units of the tibialis anterior muscle. *Journal of Physiology (London),* 184: 170–173.

Einstein, A. (1922) *Meaning of Relativity.* London, Methuen.

Feldman, A.G. (1966) Functional tuning of the nervous system with control of movement or maintenance of a steady posture. II. Controllable parameters of the muscles. *Biophysics,* 11: 565–578.

Feldman, A.G. (1986) Once more on the equilibrium-point (λ model) for motor control. *Journal of Motor Behavior,* 18: 17–54.

Feldman, A.G., Adamovich, S.V., Levin, M.F. (1995a) The relationship between control, kinematic and electromyographic variables in fast single-joint movements in humans. *Experimental Brain Research,* 103: 440–450.

Feldman, A.G., Levin, M.L. (1993) Control variables and related concepts in motor control. *Concepts in Neuroscience,* 4: 25–51.

Feldman, A.G., Levin, M.F. (1995) The origin and use of positional frames of reference in motor control. *Behavioral and Brain Sciences,* 18: 727–804.

Feldman, A.G., Orlovsky, G.N. (1972) The influence of different descending systems on the tonic stretch reflex in the cat. *Experimental Neurology,* 37: 481–494.

Feldman, A.G., Ostry, D.J., Levin, M.F. (1995b) Velocity-dependent Coriolis force perturbations: an explanation of the positional errors and adaptation. *Society for Neurosciences. Abstracts.* 21: 681.

Flanagan, J.R., Ostry, D.J., Feldman, A.G. (1993) Control of trajectory modifications in target-directed reaching. *Journal of Motor Behavior,* 25:140-152.

Gelfand, I.M., Tsetlin, M.L. (1971) On mathematical modeling of mechanisms of central nervous system. In *Models of the Structural-Functional Organization of Certain Biological Systems.* I.M. Gelfand, V.S. Gurfinkel, S.V. Fomin, M.L. Tsetlin (Eds.), Cambridge, MA: MIT Press, 1-22.

Georgopoulos, A.P., Ashe, J., Smyrnis, N., Taira, M. (1992) The motor cortex and the coding of force. *Science,* 233: 1416–1419.

Gomi, H., Kawato, M. (1996) Equilibrium-point control hypothesis examined by measured arm stiffness during multijoint movement. *Science,* 272: 117-120.

Gordon, A.M., Huxley, A.F., Julian, F.J. (1966) The variations in isometric tension with sarcomere length in vertebrate muscle fibres. *Journal of Physiology (London),* 184: 170–173.

Gribble, P.L. and Ostry, D.J. (1997) Are complex computations required for the control of arm movement? *Neural Control of Movement,* 2: 27.

Hanes, D.P., Shall, J.D.(1996) Neural control of voluntary movement initiation. *Science,* 274: 427-430.

Jeannerod, M. (1988) *The Neuronal and Behavioral Organization of Goal-Directed Movements.* Oxford, UK: Clarendon Press.

Kalaska, J.F., Crammond, D.J. (1992) Cerebral cortical mechanisms of reaching movements. *Science,* 255: 1517–1523.

Kelso, J.A.S. (1995) *Dynamic Patterns of the Self-Organization of the Brain and Behavior.* Cambridge, MA: MIT Press.

Laboissière, R., Ostry, D.J., Feldman, A.G. (1996) The control of multi-muscle systems: Human jaw and hyoid movements. *Biological Cybernetics,* 74: 373-384.

Lackner, J. R., Dizio, P. (1994) Rapid adaptation to Coriolis force perturbations of arm trajectory. *Journal of Neurophysiology,* 72: 1-15.

Lacquaniti, F. (1992) Automatic control of limb movement and posture. *Current Opinion in Neurobiology,* 2: 807–814.

Latash, M.L. (1993) *Control of Human Movement.* Champaign, IL: Human Kinetics.

Lemon, R.N. (1990) Mapping the output functions of the motor cortex. In *Signal and Sense.* G.M. Edelman, W.E. Gall, W.M. Cowan (Eds.), New York: Wiley-Liss, 315–355.

Levin, M.F., Dimov, M. (1997) Spatial zones for muscle coactivation and the control of postural stability. *Brain Research*, 757: 43–59.

Levin, M.F, Feldman, A.G. (1995) The λ model for motor control: more than meets the eye. *Behavioral and Brain Sciences*, 18: 786–798.

Levin, M.F., Feldman, A.G., Milner, T.E., Lamarre, Y. (1992) Reciprocal and coactivation commands for fast wrist movements. *Experimental Brain Research*, 89: 669–677.

Levin, M.F., Lamarre, Y., Feldman, A.G. (1995) Control variables and proprioceptive feedback in fast single-joint movement. *Canadian Journal of Physiology and Pharmacology*, 73: 316–330.

Ma, S., Feldman, A.G. (1995) Two functionally different synergies during arm reaching movements involving the trunk. *Journal of Neurophysiology*, 73: 2120–2122.

Matthews, P.B.C. (1959) The dependence of tension upon extension in the stretch reflex of the soleus in the decerebrated cat. *Journal of Physiology* 47: 521–546.

Nichols, T.R. (1994) A biomechanical perspective on spinal mechanisms of coordinated muscular action: An architectural principle. *Acta Anatomica*, 151: 1–13.

Nichols, T.R., Steeves, J.D. (1986) Resetting of resultant stiffness in ankle flexor and extensor muscles in the decerebrate cat. *Experimental Brain Research* 62: 401–410.

Partridge, L.D. (1995) Let us accept a "controlled trade-off" model of motor control. *Behavioral and Brain Sciences*, 18: 773–775.

Requin, J., Semjen, A., Bonnet, M. (1984) Bernstein's purposeful brain. In *Human Motor Actions— Bernstein Reassessed.* H.T.A. Whiting (Ed.), Elsevier Science, Amsterdam, 467–504.

Shapovalov, A.I. (1975) Neuronal organization and synaptic mechanisms of supraspinal motor control in vertebrates. *Review of Physiology, Biochemistry and Pharmacology*, 72: 1–54.

Soechting, J.F., Flanders, M. (1992) Moving in three-dimensional space: Frames of reference, vectors, and coordinate systems. *Annual Review of Neuroscience*, 15: 167–191.

St-Onge, N., Adamovich, S.V., Feldman, A.G. (1997) Control processes underlying elbow flexion movements may be independent of kinematic and electromyographic patterns: Experimental study and modeling. *Neuroscience* 79: 295-376.

Thelen, E., Smith, A. (1994) *A Dynamic Systems Approach to the Development of Cognition and Action.* Cambridge, MA: MIT Press.

Wadman, W.J., Denier van der Gon, J.J., Geuze, R.H., Mol, C.R. (1979) Control of fast goal-directed arm movements. *Journal of Human Movement Studies*, 5: 3–17.

Weber, E. (1846) Muskelbewegung. *Wagner's Handwirterbuch der Physiologie*, 3: 1–122.

Weeks, D.L., Aubert, M.-P., Feldman, A.G., Levin, M.F. (1996) One-trial adaptation of movement to changes in load. *Journal of Neurophysiology*, 75: 60–74.

14

CHAPTER

Control of Multijoint Reaching Movement: The Elastic Membrane Metaphor

Mark L. Latash

Pennsylvania State University, University Park, PA, U.S.A.

I am going to suggest a metaphor for equilibrium point control of human, multijoint, limb-reaching movement based on moving a ball across an elastic membrane by changing the membrane shape. This approach circumvents some of the notorious problems of motor control such as muscle redundancy, inverse kinematics, and inverse dynamics. Controlling endpoint movement is based on distorting the physical properties of an adjacent area of space with the help of changes in reflex properties of all limb muscles and creating three-dimensional manifolds in the six-dimensional state space of the endpoint. The manifolds are assumed to be direct reflections of independently controlled variables generated by the central nervous system. Muscle forces and EMGs are viewed as emergent properties of the movement. This process can be described with a vector (equilibrium point vector) and a matrix describing the properties of the potential well in which the endpoint resides. Specific features and advantages of this mechanism of control will be discussed.

The Problem

Nikolai Aleksandrowitsch Bernstein formulated the major problem of multijoint motor control as the elimination of redundant degrees of freedom (DOFs) (Bernstein 1935, 1967). His original formulation referred to the apparently redundant kinematics of human effectors during most natural movements. On the other hand, Bernstein also emphasized that the human brain is unable to control performance variables (including individual joint rotations) because they depend on external force fields and reactive forces that may be poorly predictable. This apparent contradiction between formulating a basic problem of control using peripheral variables and simultaneously emphasizing the uncontrollability of the same peripheral variables leads the author to believe that Bernstein actually implied redundancy at a control level when he formulated the principle of elimination of redundant DOFs (Latash 1996). Bernstein might have used the language of individual joint rotations because of its intuitive clarity and the actual possibility to count DOFs. At the time of Bernstein's vigorous activity, no hypotheses existed about the nature of control variables that might be used by the brain to coordinate multijoint movements. During the last 30 years, the situation has improved somewhat. However, a clear, accepted control language applicable to processes involved in voluntary motor control is still lacking. This chapter will suggest that problems of peripheral redundancy (e.g., muscle redundancy) are not solved explicitly by the central nervous system (CNS) but may be ignored as irrelevant to the central mechanism of control.

When a person repeats one and the same voluntary reaching movement several times in reproducible external conditions, some of the movement characteristics demonstrate considerable variability across trials while others may demonstrate relatively high reproducibility. If the external force field changes and the subject does not try to correct the movement, all the commonly measured movement characteristics (kinematic, dynamic, and electromyographic) change, including those describing the behavior of the endpoint. I am going to assume, however, that there exist internal variables, generated by the central nervous system and associated with the process of control of a movement, whose time patterns can be preserved or reproduced by the CNS despite possible changes in the external conditions of movement execution. These variables will be called independently controlled variables (Latash 1992, 1993; Feldman and Levin 1995).

Contemporary motor control studies are based on explicit or implicit assumptions concerning the nature of independently controlled variables used by the brain to plan and execute movements (for reviews see Feldman and Levin 1995; Gottlieb et al. 1989; Stein 1982). For example, attempts at solving problems of inverse kinematics and inverse dynamics with respect to human movements (Atkeson 1989; Hollerbach and Atkeson 1987) imply that somewhere in the central nervous system, individual joint rotations or patterns of individual muscle forces are calculated to bring about a desired movement. Other studies investigated patterns of activation of individual muscle and considered them as candidates for or direct reflections of variables used by the brain for the purposes of control (Gottlieb et al. 1995; Karst and Hasan 1991). Other studies consider the kinematic trajectory of the limb's endpoint (working point) in the external, Cartesian space as the variable controlled by the brain (Hogan 1984; Rosenbaum et al. 1993). This variable looks like a better candidate when compared to individual joint rotations, muscle forces, or muscle activation patterns because of its smaller variability across trials. However, even the trajectory of the working point changes immediately following a change in the external force field if the subject is not undertaking corrective actions, for example, when the limb hits an invisible (glass) wall while moving. Obviously, all the other mentioned variables will change too. Some of them (e.g., joint positions and torques) will change instantaneously while others (muscle activation patterns) will change at a relatively short reflex delay.

The equilibrium point (EP) hypothesis of motor control has reached considerable success on the way to identifying independently controlled variables for single-joint movements (Feldman 1986; Feldman and Levin 1995; Latash 1993; Latash and Gottlieb 1991). It has also led to the formulation of promising ideas about controlling multijoint movements (Bizzi et al. 1992; Feldman and Levin 1995; Flash 1987; Mussa-Ivaldi and Giszter 1992; Shadmehr et al. 1993). These ideas formed the foundation of the present chapter which suggests a physical metaphor for the process of controlling voluntary multijoint movements. It leads to a reconsideration of some of the basic problems of motor control such as the muscle redundancy problem (Bernstein 1947, 1967; Turvey 1990). This physical metaphor follows a hypothesis concerning the nature of the independently controlled variables used by the central nervous system to control voluntary movements.

The Metaphor

The Demon-of-the-Endpoint

Consider what happens during a simple reaching or pointing movement from the point of view of a tiny demon sitting on the limb's endpoint. The demon does not know the nature of the forces that act at the endpoint. It knows, however, that in any static position, the endpoint resists attempts by external forces to move it as if attached to the position by springs (cf. Flash 1987; Giszter et al. 1993; Shadmehr et al. 1993). Thus, a static position is seen as the bottom of a potential well (figure 14.1). There are two basic ways of moving to a new position in space. The first one is to calculate and apply certain force patterns to the endpoint. The second one is to change the potential field, i.e., to modify the physical properties of the surrounding space replacing the original potential well with another one, which would induce movement of the endpoint with the demon in a desired direction.

In the second case, the space around the endpoint becomes distorted, allowing the endpoint to move more easily in some directions as compared to others. If the demon had an opportunity to control the subject's brain and if the brain had an opportunity to change the physical properties of an area of space near the working

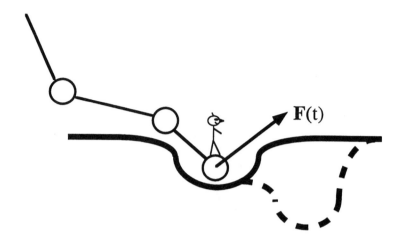

Figure 14.1   The endpoint of a multijoint limb is in an equilibrium in an elastic force field (in a potential well). From the point of view of a demon sitting on the endpoint, there are two basic ways of moving to a new position. First, calculate and apply a pattern of external force, *F(t)*. Second, change the shape of the original force field, creating a new potential well (broken line).

point, the demon would be able to prescribe any trajectory of the endpoint in terms of desired kinematics in the external space. I am going to suggest that *the human brain can change the physical properties of the space around a working point and that, when someone wants to move a finger from one point to another, an appropriate demon-of-the-finger issues a motor command in terms of actual coordinates in external space.*

The Ball on the Membrane

Consider a light metal ball on a wide horizontal elastic membrane in the field of gravity (figure 14.2). The ball will create a relatively shallow potential well on the membrane with its weight and will be there in an equilibrium. You can press with a finger on the membrane and create another potential well on the bottom of which the ball will be in a new equilibrium. Since the ball is light, its weight is a minor factor as compared to changes in the membrane geometry induced by active pressing on it with the finger. If you want to move the ball, you can press with the finger at a nearby spot thus distorting the original potential field and then move the finger to a new desired position. The ball will move in the direction the finger specifies. The finger can be moved along any trajectory, and the ball will follow it, although with deviations depending on the particular dynamics of the system (e.g., ball inertia, membrane elasticity, and friction).

The forces acting on the ball will be created by the field of gravity and by the elastic properties of the membrane. (Note that the finger is not touching the ball

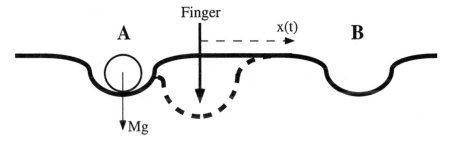

Figure 14.2   A ball on an elastic membrane is in an equilibrium at a point A. To move the ball to another point (B), press with a finger near the ball and move the finger toward point B. The trajectory of the finger (the virtual trajectory, *x(t))*, will differ from the actual ball trajectory.

directly.) However, these are only seeming reasons for the ball's movement that are likely to be confused with the actual reasons by an external observer who can see only the ball's movement but neither the membrane nor the moving finger. The observer can measure the ball kinematics, solve inverse problems, and find out time patterns of forces that act on the ball and induce its movement. The observer may go a step further and ask a question: "How does the controlling system calculate patterns of forces to move the ball from a starting to a final position?" This approach is quite legitimate. However, it ignores the specificity of control in this particular system and, as such, is likely to be inefficient if one wants to understand how ball movement is actually controlled.

The primary reason for ball movement is pressing on the membrane with the finger, i.e., distorting the force field in a certain area of space in a close proximity to the ball. Note that the forces acting on the ball in each moment of time are secondary to changes in the force field (which are direct reflections of control signals) and actual movement of the ball. Note also that if the mechanical properties of the ball (i.e., its inertia) are not exactly known to the hypothetical controller or if small, transient perturbations occur during the movement, direct control of forces acting on the ball is likely to lead to errors that need to be monitored and corrected. Control by pressing on the membrane with a finger is more universal and effective in reaching the same final position even in cases of misassessments of the ball properties or transient perturbations. Actual ball movement may deviate from the *virtual* or *equilibrium trajectory* specified by the finger (cf. Flash and Hogan 1985), but the goal will ultimately be reached.

The Physiological Mechanisms

The elastic membrane metaphor is based on assumptions regarding certain properties of force fields created by muscles. First, the metaphor implies that muscles create elastic force fields that keep a limb's endpoint in an equilibrium or violate an existing equilibrium, leading to movement. Second, they imply that the CNS controls muscles by changing their contribution to the overall elastic field, thus shaping it. These assumptions form the core of the EP-hypothesis (Bizzi et al. 1992; Feldman 1986; Feldman and Levin, 1995; Latash 1993).

Muscles as Force Field Generators

The fact that individual muscles demonstrate elastic as well as viscous properties has been known for a long time (for a review see Zajac 1989). Tendons and other passive tissues also contribute to the viscoelastic behavior, but their contribution will not be discussed separately. Muscle elasticity has two major components. The first is defined by the mechanical properties of muscle fibers and other peripheral tissues. The second is defined by the properties of reflexes originating from proprioceptors within the muscles, mostly from muscle spindles sensitive to muscle length and velocity. Both peripheral and reflex components of muscle elasticity are velocity dependent.

Muscle elastic properties create a force field around the endpoint, which can be tested with local mechanical perturbations. These force fields were originally studied for individual muscles, later for joints, and more recently for multijoint limbs (Feldman 1966; Giszter et al. 1993; Matthews 1959; Shadmehr et al. 1993). Thus, *a muscle does not simply apply forces at the points of its insertion but introduces a local force field making the space surrounding the joints and the endpoint of the limb uneven.* From the point of view of the demon-of-the-endpoint

$$F_m(x,y,z) = f(x - x_0, \ y - y_0, \ z - z_0) \tag{14.1}$$

where F_m is a force field created by a muscle in the vicinity of an equilibrium point with coordinates (x_0, y_0, z_0); x, y, and z are spatial coordinates; f is a function.

Within a limb, all the joints are mechanically coupled so that a change in muscle force acting on one of the segments generally leads to a change in torques acting at all limb segments with respect to all the joints. The presence of biarticular and polyarticular muscles that cross virtually every joint in the human body adds to this effect. Thus, all the muscles of a limb contribute, albeit to a different extent, to the total force field around the limb endpoint (figure 14.3).

$$F_t(x,y,z) = \Sigma f_i(x - x_0, \ y - y_0, \ z - z_0) \tag{14.2}$$

where F_t is total force field, and f_i are functions for individual muscles.

Basic Principles of Equilibrium Point Control

Controlling movements by controlling equilibrium states of the neuromuscular apparatus and external force field form the basis of EP control. Assume that an

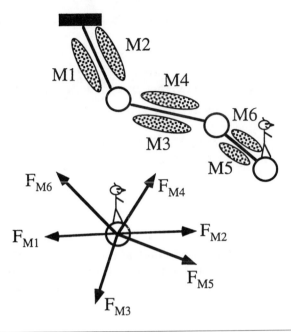

Figure 14.3   All muscles of a multijoint limb act on the endpoint, creating an elastic field. The more muscles with different lines of force action available, the more uniform and controllable is the created elastic field.

external force field does not dramatically restrict endpoint movements (e.g., the field of gravity). Then, a unique active force vector at the endpoint will be required to keep it at an equilibrium in a given location. In general depending on specific characteristics of the force field, its effect may be described with three equations

$$\phi_1(x_1, x_2, x_3, f_1, f_2, f_3) = 0$$
$$\phi_2(x_1, x_2, x_3, f_1, f_2, f_3) = 0 \qquad (14.3)$$
$$\phi_3(x_1, x_2, x_3, f_1, f_2, f_3) = 0$$

where ϕ_{1-3} are functions, and x_{1-3} and f_{1-3} are spatial and force coordinates of the endpoint state space. If these equations are independent, they cut a three-dimensional manifold in the six-dimensional state space of the endpoint.

Imagine now that a subject of a mental experiment has been asked to occupy a position in space against an external force. The subject is instructed not to intervene voluntarily, and the experimenter slowly moves the endpoint by changing the externally applied force. It is possible to move the endpoint to any point of accessible space, which means that three dimensions are still available under a "fixed central command". However, keeping the endpoint in an equilibrium at a chosen location requires a strictly defined external force vector equal in magnitude and opposite in direction to the active force at the endpoint. Thus, central command defines three more equations

$$\psi_1(x_1, x_2, x_3, f_1, f_2, f_3) = 0$$
$$\psi_2(x_1, x_2, x_3, f_1, f_2, f_3) = 0 \qquad\qquad (14.4)$$
$$\psi_3(x_1, x_2, x_3, f_1, f_2, f_3) = 0$$

that cut another three-dimensional manifold in the state space of the endpoint. If all the 14.3 equations and 14.4 equations are independent, their combination defines an equilibrium point for the limb endpoint.

Within this framework, *voluntary movements are assumed to be controlled by hypothetical signals which distort the physical properties of the space around the working point and create three-dimensional manifolds (force fields) in the six-dimensional state space of the working point. The manifolds are direct reflections of independently controlled variables while muscle forces and EMGs are emergent properties of the movement.* One can visualize this method of control as forming and manipulating landscapes with ridges and valleys across which the working point travels.

The Generalized Displacement Reflex

The λ-model of the EP-hypothesis considers the tonic stretch reflex (TSR) as the basic mechanism defining viscoelastic properties of muscles (Feldman 1986). The reflex arc of the TSR is unknown. It is assumed to originate from the activity of muscle spindles. However, it also depends on the activity of other receptors and demonstrates intermuscle, interjoint, and interlimb components (Lundberg 1979; Nichols 1989). As such, TSR can be defined as a hypothetical mechanism giving rise to the observed dependence between muscle length and active force.

Recall the mental experiment with a person balancing an external force in a point of the external space (e.g., pushing against the palm of the experimenter). Remember that the subject of this mental experiment is instructed not to intervene voluntarily. The experimenter can push the fingertip in different directions by changing the original force and bring it into new equilibrium states. Apparently, the location of the fingertip, force, joint angles, torques, and levels of muscle activation will all change while the central command of the subject presumably stays the same. The pattern of a dependence between external force and endpoint coordinates is defined by an involuntary mechanism that may be considered reflex.

Let me introduce a notion of generalized displacement reflex (GDR) as a hypothetical mechanism that brings about the relationship between external force and endpoint location in space. As such, it is going to be a multijoint, three-dimensional expansion of the notion of TSR. GDR may receive contributions from all the receptors whose firing levels change as a result of displacement. It does not control a single muscle but, instead, a muscle group involved in an action (cf. the notions of coordinative structures or action units, Bernstein 1967; Greene 1971; Kelso et al. 1979; Whiting 1984). GDR is poly-dimensional. It may be described with two characteristics, an equilibrium vector and a gain (cf. ellipses of stiffness in Flash 1987). The author assumes GDR to be the basic involuntary mechanism used by the hypothetical control system to define manifolds in order to assure a required movement or position of a working point.

Recent Criticisms of the Equilibrium Point Hypothesis

In the recent impressive series of studies, Lackner and DiZio (1994) have shown reproducible errors in final position during arm reaching movements in the presence of transient perturbations due to Coriolis forces when the subjects were slowly rotated in a dark room. These violations of equifinality have been presented as disproving the EP-hypothesis. However, equifinality and the EP-hypothesis are not synonyms. In particular, the EP-hypothesis predicts equifinality if no changes occur in the central command in response to transient external perturbations. In the experiments by Lackner and DiZio, the subjects reported that they had not perceived the rotation. This does not mean, however, that the signals from the vestibular apparatus were not used by the hypothetical controller to correct the ongoing command. If the command was changed, the EP-hypothesis predicts violations of equifinality and, thus, is likely to survive this latest threat. Actually, a recent series of simulations have shown that Lackner and DiZio's findings can be accounted for within the framework of the EP-hypothesis (Shapiro 1996).

Another recent study by Gomi and Kawato (1996) studied reconstructed equilibrium trajectories during voluntary arm movements in human subjects and reported nonmonotonic trajectories during fast movements. This finding was also presented as disproving the EP-hypothesis. Note, however, that the basic principles of EP-control do not assume a particular shape of the control signal. They are compatible with both monotonic and nonmonotonic equilibrium trajectories. Actually, nonmonotonic equilibrium trajectories were reconstructed earlier during single-joint movements within the framework of the EP-hypothesis (Latash, 1993; Latash and Gottlieb, 1991).

Specific Features and Advantages of the Elastic Membrane Control

The Role of Muscle Forces and Reflexes

Within the suggested approach, *muscle forces and activation patterns (EMGs) are not primary causes of movement but its emergent properties.* They may be compared to tension patterns generated by segments of the metaphorical elastic membrane when the ball rolls over it following movement of the metaphorical finger (cf.

figures 14.2 and 14.3). Thus, muscle forces are not directly controlled by the CNS and depends on both the central command and actual motion, which, in turn, depends on the external force field.

The whole variety of muscle reflexes (compare the earlier-introduced notion of GDR) create a manifold (an elastic field) that plays the central role in the suggested scheme. Until now, this chapter has not discussed viscous properties of muscles and their reflexes that are likely to be important in movement stabilization and termination. Whether these properties are sufficient by themselves to brake a fast movement or a nonmonotonic equilibrium trajectory is required is still an open question (Latash and Gottlieb 1991; Latash 1993; Feldman and Levin 1995; Gomi and Kawato 1996). I have also not discussed the possible role of peripheral muscle elasticity which depends on the level of muscle activation and is likely to contribute to the overall elastic properties of the force fields. Since the level of muscle activation is assumed to be controlled through the hypothetical mechanism of GDR, the contribution of muscle peripheral elasticity is automatically taken into account.

The Problem of Muscle Redundancy

One of the most attractive features of the suggested type of control is the elimination of the problem of muscle redundancy. If each muscle plays the role of an elastic force generator acting at a working point in a certain direction, the more muscles that are available, the more uniform and controllable is the created elastic field (figure 14.3). Thus, *muscle redundancy becomes not a problem but an essential part of the solution.* Remember that within this scheme, the hypothetical controller specifies a force field (a manifold in the state space), while actual forces emerge depending on both the field and the dynamics of the effector.

Recent studies by Nichols (1994) have demonstrated that the lines of muscle action differ from those expected from pure flexors-extensors or abductors-adductors and cover a wide range of directions. Muscles that have commonly been considered as agonists (e.g., the heads of triceps surae) may demonstrate noticeably different directions of force action. Such design may look awkward from the computational view. However, it is natural if individual muscles are not independently controlled force generators but elements of a larger, distributed system which is used to change elastic force fields around the working point.

Preparation for a Movement

The suggested scheme allows active subthreshold preparation for a movement without visible changes in muscle activity if a person knows in which direction to move. Rosenbaum (1980) and Anson et al. (1994) have shown that if a subject is presented with two targets, the reaction time is shorter if the targets are in the same direction

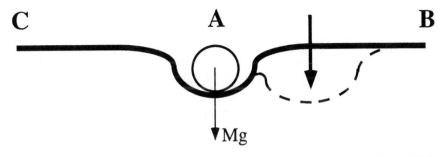

Figure 14.4 If a subject knows in which direction to move (toward target B), the subject can prepare for a movement with a subthreshold change in the original force field (the broken line), thus decreasing reaction time.

but at different distances than if they are in the opposite directions but at the same distance. If the demon-of-the-endpoint knows in which direction to move, it can start pressing slightly on the membrane at a point in the desired direction (figure 14.4), thus decreasing the total time of movement initiation. This initial distortion of the force field should be small so that it does not violate the existing equilibrium. If the direction of a future movement is unknown, this possibility is unavailable.

Neuronal Population Coding

A number of studies have demonstrated relations between movement direction and a characteristic of the activity of a population of neurons (neuronal population vector) in different brain structures such as cortical areas and the cerebellum (Fortier et al. 1989; Georgopoulos 1986). The hypothesis that groups of neurons in brain structures code movement direction for the endpoint in the external space rather than individual joint rotations or individual muscle activation patterns corresponds to the basic idea in this chapter.

Obstacle Avoidance

Within the suggested scheme, avoiding obstacles during a reaching movement is straightforward. Getting back to the metaphor discussed earlier, if one needs to get from point A to point B and avoid obstacle C, an explicit equilibrium (finger) trajectory may be used. It will clear the obstacle so that the actual ball trajectory does not touch it. The important question of how an equilibrium trajectory is chosen has not been addressed. This question is beyond the scope of this chapter. A number of promising hypotheses have been formulated in this area (Flash 1987; Rosenbaum et al. 1993).

Implications for Control of Multilimb Movements

Creating a manifold to control movement of a limb's endpoint may be a demanding and nontrivial task for the hypothetical central controller. As a result, independently controlling a number of manifolds may be next to impossible. One solution that the control system may accept when faced with a task of coordinating movements of two or more limbs is creating for all the endpoints manifolds whose changes are coordinated according to a certain law. This situation may be viewed as a synergy, emerging at a rather low functional level and coordinating manifolds rather than individual joints or muscles. This type of control is expected to lead to reproducible, stable phase relations among movement trajectories or force patterns of individual endpoints which will also depend on the dynamics of the effectors (inertia, external loading, and so forth) and were described in a number of studies (Kelso 1995; Schoner 1990; Schoner and Kelso 1988).

Conclusion

My task has been not to introduce a new theory but rather to describe a metaphor which makes application of the EP-hypothesis to control of mutli-joint movements intuitively more clear, suggests interpretations of some of the recently published data, and creates a potentially fruitful environment for future studies. There is a sharp contrast between the ideas of control of movements through direct computation and application of forces to working points or intermediate segments or through control of muscle activation patterns and the ideas of EP-control when movements are consequences of changes of an elastic force field which is controlled by the brain making use of both peripheral and reflex-mediated viscoelastic muscle properties. Similar ideas, expressed as control of endpoint movement by shifting positional reference frames, have recently been developed by Feldman and Levin (1995).

Acknowledgments

Preparation of this chapter was supported in part by grant HD-30128 from the National Center for Medical Rehabilitation Research, NIH.

References

Anson, J.G., Wickens, J.R., Hyland, B.I., Kötter, R. (1994) Inequivalence of direction and extent precue effects on reaction time. *Soc. Neurosci. Abstr.,* 20: 1200.
Atkeson, C.G. (1989) Learning arm kinematics and dynamics. *Ann. Rev. Neurosci.,* 12: 157–183.
Bernstein, N.A. (1935) The problem of interrelation between coordination and localization. *Arch. Biol. Sci.,* 38: 1–35 (In Russian).
Bernstein, N.A. (1947) *On the Construction of Movements.* Moscow: Medgiz.

Bernstein, N.A. (1967) *The Co-Ordination and Regulation of Movements.* Oxford, UK: Pergamon Press.

Bizzi, E., Hogan, N,, Mussa-Ivaldi, F.A., Giszter, S. (1992) Does the nervous system use equilibrium-point control to guide single and multiple joint movements? *Behav. Brain. Sci.* 15: 603–613.

Feldman, A.G. (1966) Functional tuning of the nervous system with control of movement or maintenance of a steady posture. II. Controllable parameters of the muscle. *Biophysics,* 11: 565–578.

Feldman, A.G. (1986) Once more on the equilibrium-point hypothesis (λ model) for motor control. *J. Mot. Behav.,* 18: 17–54.

Feldman, A.G., Levin, M.F. (1995) The origin and use of positional frames of reference in motor control. *Behav. Brain Sci.,* 18: 723–804.

Flash, T. (1987) The control of hand equilibrium trajectories in multi-joint arm movements. *Biol. Cybern.,* 57: 257–274.

Flash, T., Hogan, N. (1985) The coordination of arm movements: An experimentally confirmed mathematical model. *Journal of Neuroscience,* 5: 1688–1703.

Fortier, P.A., Kalaska, J.F., Smith, A.M. (1989) Cerebellar neuronal activity related to whole-arm reaching movement in the monkey. *Journal of Neurophysiology,* 62: 198–211.

Georgopoulos, A.P. (1986) On reaching. *Ann. Rev. Neurosci.,* 9: 147–170.

Giszter, S.F., Mussa-Ivaldi, F.A., Bizzi, E. (1993) Converging force fields organized in the frog's spinal cord. *Journal of Neuroscience,* 13: 467–491.

Gomi, H., Kawato, M. (1996) Equilibrium-point control hypothesis examined by measured arm stiffness during multijoint movement. *Science,* 272: 117–120.

Gottlieb, G.L., Chen, C.-H., Corcos, D.M. (1995) Relations between joint torque, motion, and electromyographic patterns at the human elbow. *Exp. Brain Res.,* 103: 164–167.

Gottlieb, G.L., Corcos, D.M., Agarwal, G.C. (1989) Strategies for the control of voluntary movements with one mechanical degree of freedom. *Behav. Brain Sci.,* 12: 189–250.

Greene, P.H. (1971) Introduction. In *Models of the Structural-Functional Organization of Certain Biological Systems.* I.M. Gelfand, V.S. Gurfinkel, S.V. Fomin, M.L. Tsetlin (Eds.), Cambridge, MA: MIT Press.

Hogan, N. (1984) An organizational principle for a class of voluntary movements. *Journal of Neuroscience,* 4: 2745–2754.

Hollerbach, J.M., Atkeson, C.G. (1987) Deducing planning variables from experimental arm trajectories: pitfalls and possibilities. *Biol. Cybern.,* 56: 279–292.

Karst, G.M., Hasan, Z. (1991) Timing and magnitude of electromyographic activity for two-joint arm movements in different directions. *Journal of Neurophysiology,* 66: 1594–1604.

Kelso, J.A.S. (1995) *The Self-Organization of Brain and Behavior.* Cambridge, MA: MIT Press.

Kelso, J.A.S., Southard, D.L., Goodman, D. (1979) On the nature of human interlimb coordination. *Science,* 203: 1029–1031.

Lackner, J.R., DiZio, P. (1994) Rapid adaptation to Coriolis force perturbations of arm trajectory. *Journal of Neurophysiology,* 72: 299–313.

Latash, M.L. (1992) Are we able to preserve a motor command in the changing environment? *Behav. Brain Sci.,* 15: 771–773.

Latash, M.L. (1993) *Control of Human Movement.* Champaign, IL: Human Kinetics.

Latash, M.L. (1996) How does our brain make its choices? In *Dexterity and Its Development.* M.L. Latash, M.T. Turvey (Eds.), Mahwah, NJ: Erlbaum, 277–304.

Latash, M.L., Gottlieb, G.L. (1991) Reconstruction of elbow joint compliant characteristics during fast and slow voluntary movements. *Neuroscience,* 43: 697–712.

Lundberg, A. (1979) Multisensory control of spinal reflex pathways. In *Reflex Control of Posture and Movement.* R. Granit, O. Pompeiano (Eds.), Amsterdam: Elsevier, 11–28.

Matthews, P.B.C. (1959) The dependence of tension upon extension in the stretch reflex of the soleus of the decerebrate cat. *Journal of Physiology,* 47: 521–546.

Mussa-Ivaldi, F.A., Giszter, S.F. (1992) Vector field approximation: A computational paradigm for motor control and learning. *Biol. Cybern.,* 67: 491–500.

Nichols, T.R. (1989) The organization of heterogenic reflexes among muscles crossing the ankle joint in the decerebrate cat. *Journal of Physiology,* 410: 463–477.

Nichols, T.R. (1994) A biomechanical perspective on spinal mechanisms of coordinated muscular action: An architecture principle. *Acta Anatomica,* 151: 1–13.

Rosenbaum, D.A. (1980) Human movement initiation: Specification of arm, direction and extent. *Journal of Experimental Psychology (General),* 109: 444–474.

Rosenbaum, D.A., Engelbrecht, S.E., Busje, M.M., Loukopoulos, L.D. (1993) Knowledge model for selecting and producing reaching movements. *J. Mot. Behav.* 25: 217–227.

Schöner, G. (1990) A dynamic theory of coordination of discrete movement. *Biol. Cybern.,* 63: 257–270.

Schöner, G., Kelso, J.A.S. (1988) Dynamic pattern generation in behavioral and neural systems. *Science,* 239: 1513–1520.

Shadmehr, R., Mussa-Ivaldi, F.A., Bizzi, E. (1993) Postural force fields of the human arm and their role in generating multijoint movements. *Journal of Neuroscience,* 13: 45–62.

Shapiro, M.B. (1996) Control of forward reaching movements made during body rotation: A simulation study. In *Abstr. Conf. "Bernstein's Traditions in Motor Control."* University Park, PA: Penn State University, 112.

Stein, R.B. (1982) What muscle variable(s) does the nervous system control in limb movements? *Behav. Brain Sci.,* 5: 535–577

Turvey, M.T. (1990) Coordination. *Amer. Psychol.,* 45: 938–953.

Whiting, H.T.A. (1984) *Human Motor Actions: Bernstein Reassessed.* Amsterdam: Elsevier.

Zajac, F.E. (1989) Muscle and tendon: Properties, models, scaling, and application to biomechanics and motor control. *CRC Crit. Rev. Biomed. Eng.,* 17: 359–411.

15

CHAPTER

Generalized Motor Programs and Units of Action in Bimanual Coordination

Richard A. Schmidt
Human Performance Group, Failure Analysis Associates, Inc.,
and Department of Psychology, University of California,
Los Angeles, CA, U.S.A.

Herbert Heuer
Abteilung Arbeitsphysiologie, Institut für Arbeitsphysiologie
an der Universität Dortmund, Dortmund, Germany

Dina Ghodsian
Department of Psychology, University of California,
Los Angeles, CA, U.S.A.

Douglas E. Young
Human Performance Group, Failure Analysis Associates, Inc.,
and Department of Kinesiology, California State University,
Long Beach, CA, U.S.A.

Bernstein (1967) taught that a full understanding of human skilled performance awaits an account of how the motor system controls coordination among the many degrees of freedom necessary for smooth, efficient, and effective actions. Toward this end, issues in coordination have been studied in a wide variety of ways. This chapter focuses on the intriguing results from laboratory studies where the subject is requested to produce different, simultaneous actions with two effector systems (e.g., vocal-manual or bimanual tasks). The research literature in this area reveals findings that seem remarkably discrepant to everyday observations about coordination. This discrepancy is a major motivation for the work reported in this chapter.

Fundamental Problems in Coordination

At one level, people seem able to produce coordinated actions easily, almost trivially. In countless activities, effectors produce different actions at the same time, seemingly without any interference among the limbs (e.g., using a knife and fork, playing piano, walking while chewing gum, and so forth). In tasks like throwing and kicking, the limbs perform very different functions and movements, all without difficulty. Yet, at another level, the laboratory data (and a few everyday examples) suggest remarkable interference between the effectors under some conditions. Well-known cases involve patting one's head while rubbing one's tummy. Both hands appear to produce either rubbing or patting actions but not different actions. Other examples come from Peters (1977, experiment 3), who showed that not one of 100 subjects was able to recite a nursery rhyme with proper rhythmic intonation while tapping a 1–3–123 rhythm. Summers et al. (1993) provide evidence about difficulties in tapping different rhythms with two hands (producing so-called polyrhythms). These and many other observations suggest the existence of barriers to interlimb coordination. These barriers likely support genetically defined and biologically important activities such as locomotion. However, they tend to impede more arbitrary yet culturally important (e.g., piano playing, throwing) or unimportant (e.g., rubbing the head while patting the tummy) skills that require other coordination patterns.

For several decades (Klapp 1979, 1981), a recognized major source of interlimb coordination interference seemed related to the temporal structure(s) of the actions being coordinated. Actions with the same temporal organization are easily coordinated with tight temporal coupling between the limbs. In contrast, activities with different temporal organizations are not easily produced, if they can be done at all. There is clearly more to coordination than this, however. Recently, the authors have been studying rapid bimanual actions where the temporal organizations in the two limbs are deliberately very different from each other. Seemingly in contrast with Klapp's simple generalization, undeniable evidence of remarkably tight temporal coupling has been found between the two hands in these tasks (Heuer et al. 1995) even though the hands perform different actions.

This chapter briefly describes a theory and a simple mathematical model the authors have developed that account for these findings. The model is based on three interrelated ideas. The first is the two-level control concepts from Wing and Kristofferson (1973). The second is multiplicative rate control (the action is temporally scaled with invariant relative timing) from the notion of generalized motor programs. The third is the concept of units of action. Together, these ideas capture the notion of common timing control for two hands doing different things simultaneously. Theoretical concepts and experimental data are reviewed first. Then, the concept itself is briefly described and followed by a discussion of various ways of testing it. The test procedures are then illustrated with laboratory data from human subjects and data from computer simulations. Finally, the results of several new experiments that evaluate some additional predictions of the model are reported.

Units of Action

At the level of observable motor behavior, many have argued that at least very brief movements (200–300 ms in duration) are controlled by a pattern of neuromuscular activity largely organized in advance. This motor behavior may involve coordination among several limbs for as long as a few seconds. Of course, this is the essential feature of the theory of motor programs, of which many separate versions have been brought forth over the years (Grillner 1975; Henry and Rogers 1960; Keele 1968, 1986; Lashley 1917; Rosenbaum 1991; Schmidt 1976, 1988b; Weiss 1941) with an early statement dating back to James (1890). Bernstein (1967) also referred to motor programs although he did not use that term (Schmidt 1988a). In all these proposals once this program is initiated, it provides certain features of the action's structure and controls it for several hundred milliseconds or perhaps longer (Shapiro 1977).

Motor Programs

The motor program does not provide *all* the details, of course. Many modern viewpoints contain not only the open-loop feature of the earlier views but also extensive sensorimotor integration. Current evidence shows that the centrally produced, programmed activities are blended with inputs from a variety of sources (vision, touch, and so forth) to provide the final movement output (Grillner 1975; Rosenbaum 1991; Smith 1978). Many have debated the relative proportion of central versus peripheral contributions. However, in all these views, *at least some* fundamental features of the action are specified in advance. This chapter makes that assumption. If it can be accepted, this assumption justifies research into how such a structure is organized, and how and under what conditions it operates.

Units as Programs

Next, consider an action somewhat longer in duration and complexity (involving more than one limb), such as serving a tennis ball or shifting gears in a car. Most workers in motor behavior have been unwilling to claim that such longer sequences are controlled by single programs, and the authors agree. This chapter proposes that such actions are composed of a *string* of programs (each lasting 200–300 ms or more). Furthermore, each of these programs can be thought of as a *unit* of action— a piece of behavior that can be utilized repeatedly in various actions, producing essentially the same movements (but scaled to the environment) each time. For example, shifting gears in a car might be thought of as composed of three units: gas up/clutch down; shift lever up-over-up; and gas down/clutch up. (A race car driver might perform this task as one unit.) Furthermore, each unit must be objectively identifiable so that it could be observed in other actions. For example, a clutch down/gas up unit could be involved in both a gear change from first to second and from second to third, while another unit (shift lever movements) would differ in the two actions.

The Discovery and Measurement of Units

Theorizing that actions might be composed of sequences of programs raises several problems. First, what evidence shows that such units actually exist? Second, how can they be described, measured, and identified? Third, how can they be defined in a way that allows one action to be identified as unit A and another as unit B (Schmidt 1985a)? This chapter has addressed all of these problems using methods and rationale from the theory of generalized motor programs (GMPs).

Generalized Motor Programs

About 20 years ago, Schmidt (1975, 1976) proposed that a programmed action could be generalizable across certain dimensions. For Schmidt, the initial idea came from Pew's (1974) observations about the patterns of action shown in a thesis by Armstrong (1970), Pew's student. Subjects learned a pattern of elbow movements lasting about 4 s (figure 15.1, solid trace). Sometimes, the subject made this action too quickly, as seen in the dotted trace. Pew noticed that the *whole movement* was made too quickly, as if the dotted trace were a temporally compressed version of the solid trace. This suggested that actions could not only be scaled temporally but be scaled in other analogous ways as well, based on a generalizable motor program. These ideas led to what is now known as the theory of generalized motor programs (GMPs) (Schmidt 1985b, 1988b; Heuer 1991).

Pew's notion of generalizable movement representation can be formalized in many ways. In the original, and simplest, version of this theory, the GMP is a program specifying the sequence among muscles, relative timing, and relative forces

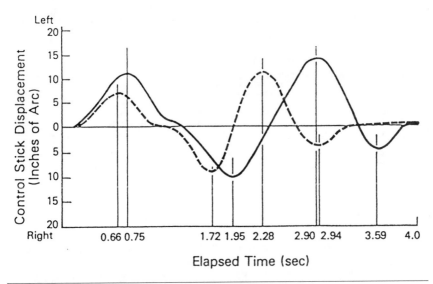

Figure 15.1   Position-time traces from Armstrong's (1970) experiment. The solid trace is the goal pattern, and the dotted trace is a movement performed too quickly.
Reprinted from Schmidt, 1988, *Motor Control and Learning,* Champaign: Human Kinetics, 247.

among the muscle contractions (sometimes called structural features). For a given GMP, the sequence, relative timing, and relative force are assumed to be essentially invariant even when superficial features such as overall movement duration and movement size are altered. These latter aspects (often called metrical features) are thought to be controlled by parameters that adjust the surface expression of the GMP's output yet allow it to retain its invariant structure or pattern. This GMP model allows the same pattern of activity with different speeds (a fast or slow signature), different sizes (large or small writing), or even with different limbs (right or left hand). However, relative timing and relative force are invariant.

Considerable converging support now exists for the view that rapid, skilled actions are based on a generalizable fundamental structure that is then parameterized to match the environment. In reaction time paradigms, Rosenbaum (1991) cited much evidence that the time to specify an action is faster if choices are among parameter (scale) variations than among pattern (program) variations (Quinn and Sherwood 1983). In a completely different paradigm, Wulf et al. (1993) showed that reducing feedback frequency during practice facilitates the learning of dependent variables attributed to GMP (pattern) accuracy. However, it also degrades the learning of dependent variables indexing parameterization (Wulf and Schmidt 1996). This supports a theoretical separation of GMP- and parameterization-processes as suggested here.

This initial version of the GMP assumed that the central representation could be scaled *linearly* in both duration and amplitude. Other types of scaling could have been considered, but linear scaling was favored because of its simplicity. Most importantly, the notion of linear scaling seemed to fit at least some of the existing data. This suggests an invariance in relative timing.

Relative Timing Invariances

On the basis of existing evidence concerning the temporal organization of actions that are generally rapid, discrete, and operate in stable environmental contexts, Schmidt and others have suggested that a quantity termed *relative timing* might be an invariance common to a set of actions governed by a GMP (Schmidt 1975, 1985b, 1988b). Relative timing is thus one way to define the temporal structure of the pattern. Given a set of temporal landmarks in an action (its initiation, time of peak acceleration, and so forth) and the durations of intervals between the landmarks, relative timing can be represented as the set of ratios among these intervals. For relative timing to be invariant, these ratios must remain constant as other features (e.g., overall movement duration, movement amplitude, loads on the limb, and so forth) are allowed to change from attempt to attempt.

Over the past decade, considerable attention has been devoted to whether or not movements scale in this way. It has been studied in such tasks as typing and writing (Terzuolo and Viviani 1979, 1980), hand and arm movements (Carter and Shapiro 1984; Schmidt et al. 1985; Shapiro 1978), and tapping (Summers 1977). For reviews, see Gentner (1987, 1988), Heuer (1988, 1991), Heuer et al. (1995), or Schmidt (1985b, 1988b). Generally, most of these actions—particularly if they are relatively rapid—seem to approximate this kind of invariance.

However, the concept of a GMP with invariant relative timing has had several difficulties when confronted with the experimental data. In an extensive review, Gentner (1987) has shown that relative timing is almost never *perfectly* invariant. Sometimes, the deviations from invariance are relatively severe. A summary of this evidence suggests that these deviations are generally of two types.

First, many skills contain clear, environmentally determined modifications during the action. In reaching, for example, researchers have long known (Woodworth 1899) that such movements are controlled by an open-loop, distance-covering phase (governed by the GMP, presumably) followed by a closed-loop, homing-in phase responsive to sensory information during the action. These and other tasks of relatively long duration (juggling, stair climbing, and so forth) typically fail to show invariances of the type described here. This largely occurs because at least a portion of the action would not be expected to be controlled by features defined in the GMP (Beek 1992). Evidence for this kind of organization has been produced by Young and Schmidt (1990, 1991) and Schneider and Schmidt (1995).

Second, a number of other actions that are faster and presumably more completely open loop tend, in fact, to show approximate invariance in relative timing. As a test of invariance, Gentner has advocated using the slope of the relationship between, for example, movement time and the proportion of movement time occupied between a pair of landmarks, computed across trials and within subjects. Of course, a relative timing invariance would require that such slopes be zero. Such slope tests are extremely sensitive to small deviations from invariance, particularly if the number of trials used is very large. Many instances have been found where the slope was significantly different from zero. However, the numerical departure from a zero slope was very small and perhaps even biologically trivial. In many such cases, the violations of invariance can probably be considered as acceptable deviations between nature and human conceptualizations of it (Heuer and Schmidt 1988). Even if relative timing is not invariant in the periphery, it may be invariant at a central level. This is Heuer's (1988) main reaction to Gentner's (1987) concerns about relative timing. As a result of such actions, we are fascinated by the conspicuous tendency toward relative-timing invariance that gives important clues to the system's operation. The authors prefer to regard the pitcher as partly-full rather than partly-empty.

Finally, no strong reason actually exists to assume that all actions are scaled with an invariance in relative timing as opposed to some other type of invariant feature(s) (Heuer 1991). For example, actions could scale nonlinearly, with certain portions of the actions changing their durations differently than others. Thus, while it would be overly bold to expect that a model proposing such a simple (i.e., linear) basis for invariant relative timing would be predictive for all GMPs, beginning with this invariance is relatively tractable mathematically. In addition, it does not seriously violate the empirical data for at least rapid actions, which are the main focus of these investigations. These procedures should be capable of providing tests that can detect systematic deviations from relative timing invariance, perhaps leading to the discovery of other forms of invariance.

Relative Timing and Units of Action

In a number of experiments, the authors considered an invariance in relative timing for a behavioral sequence as evidence at least consistent with the view that a GMP produced that behavior. To measure these relative timing invariances, Young and Schmidt (1990, 1991; also Schneider and Schmidt 1995) initially used a correlation criterion (logically similar to Gentner's method). They examined the within-subject (across trials) correlations among the intervals between landmarks in an action. The hypothesis of invariance in relative timing predicts that these correlations should approach 1, limited, of course, by errors in measurement, variability, and other statistical factors.

Now, consider another case with a longer movement time, again with landmarks scattered across the movement's duration. Suppose that durations of the first several intervals in the movement correlate highly among themselves but that the duration of a later interval does not correlate with any of these earlier ones. One interpretation would be that the first set of intervals was governed by a GMP because relative timing was approximately invariant. However, the later interval was not part of the unit because the invariance was no longer present. This, then, provides the essential idea for identifying units.

A *unit* is defined as a sequence of behavior with essentially invariant relative timing. That is, it has high correlations among the durations of the intervals in it. When later intervals of the action no longer share this invariance (i.e., intervals do not correlate with intervals from earlier portions), the first unit has ended or some sort of boundary exists between units. Under this view while such a unit is assumed to occur in a single trial, the existence of these units can be measured only across a *group* of trials. Next, a concrete example is provided in which units and boundaries are clearly demonstrated.

Methods in Unit Identification

Young and Schmidt (1990, 1991; Schneider and Schmidt 1995) asked subjects to learn a coincident-timing task, which involved a moving object simulated by a series of lights that had to be struck with a handheld lever. The seated subject (figure 15.2) began with the lever at 75 deg (the frontal plane was 0 deg). After the light illumination began (duration = 1.176 s; speed = 3 mph), the subject made a backswing to the left

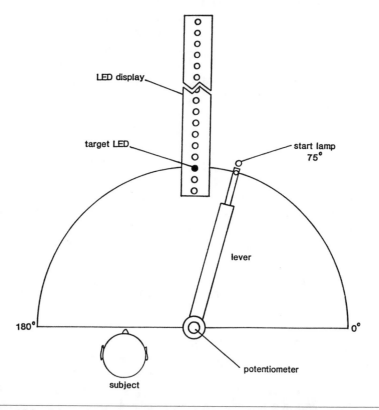

Figure 15.2   Schematic diagram of the lever apparatus used by Young and Schmidt.
Reprinted from Schneider and Schmidt 1995.

(between 90 and 180 deg) then a movement back to the right (with a follow through) with the goal of striking the passing target light with the lever. The subject could start at any time, backswing to any place beyond 90 deg, and move forward at any time and velocity. The goal was to propel the imaginary object as far as possible to the right. Scoring considered both velocity at the target and spatial error. Feedback (when given) was defined in terms of how far the ball was propelled. This was a combination of the lever's momentary velocity at the coincidence point and the spatial accuracy in striking the ball (Schmidt and Young 1991; Young and Schmidt 1991, 1992). Position time data (sampled at 500 Hz) were filtered (12-Hz cutoff frequency) and differentiated to obtain the velocity—and acceleration—time functions.

Various kinematic landmarks (shown as points a, b, c, ... , g in figure 15.3) were defined as maxima, minima, or zero crossings (except point g which was when the lever crossed the light path, indicated by the arrow). The *times* that each of these events occurred were measured with the start of the light travel beginning at 0 s. The amplitudes were not considered. Each landmark was present in every trial for every subject and was easily and objectively identified. The within-subject (over trials) correlations among all possible pairs of landmark times were computed. A useful examination involved the correlation of the first landmark with the second, with the third, and so on (a–b, a–c, a–d, ... , a–g). They also examined the correlations with the last landmark (g) and all possible earlier landmarks (a–g, b–g, ... , f–g). If the entire action was governed by a single unit, these correlations should all be relatively high, because all intervals would be scaled proportionally in time.[1]

These correlations are shown in figure 15.4, averaged after Z' transformations across 10 subjects in each of two groups (one with knowledge of the results (KR) about the distance the ball traveled, one without KR). For the no-KR subjects (left) and using landmark a as the common point (top), a weak decreasing trend from a–b through a–g was seen, with the correlations near 0.70. Using landmark g as the common point (lower), the correlations were again roughly constant but slightly larger. This suggested a pattern of action controlled as a single unit based on a correlation criterion.

The KR subjects—who were far more proficient in this task than the no-KR subjects—showed a different pattern (figure 15.4, right). When landmark a was the common point (top), the correlations with later landmarks (a–b, a–c, a–d) were high (above 0.80). However, correlations a–e, a–f, and a–g were abruptly lower (about 0.20), $p < .05$. The interpretation was that landmarks a through d were contained in a single unit, and this unit did not contain landmarks e, f, and g. When landmark g was used as the common point (lower), the correlations among e, f, and g were all very high (around 0.85), but the correlations with earlier landmarks a to d were low (around 0.40). These differences were reliable, $p < .05$. Landmarks e, f, and g were

---

[1]These analyses involved overlapping intervals. For example, the intervals from a to b and from a to c both contain the interval from a to b, forcing the correlations among intervals to be somewhat too large. In subsequent work with coordination (presented later), the authors have used nonoverlapping intervals (e.g., from landmarks a to b versus c to d), which lowers the correlations slightly. However, it does not alter the form of the functions nor the resulting conclusions.

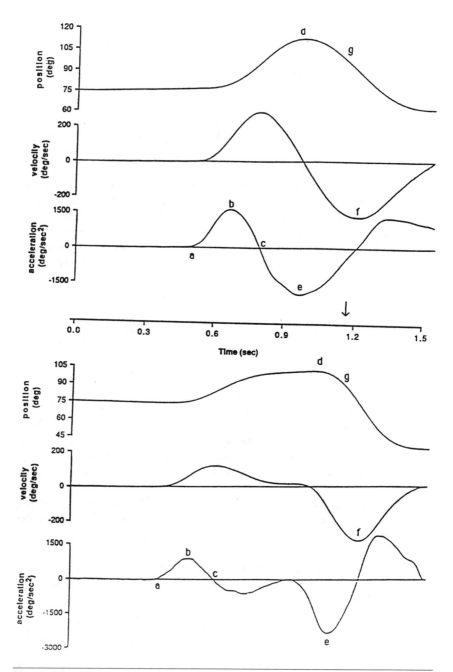

Figure 15.3   Kinematic traces from a subject in the no-KR group (upper traces) and the KR group (lower traces). Landmarks (a–g) are defined as the times of distinctive features.

Reprinted from Young and Schmidt 1990.

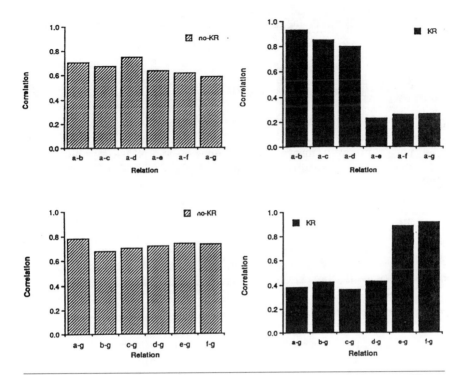

Figure 15.4   Mean within-subject correlations among landmark times for the no-KR subjects (left panels) and the KR subjects (right panels). The first landmark (a) is correlated with each later landmark in the top figures, and the last landmark (g) is correlated with each earlier landmark in the lower figures.
Reprinted from Young and Schmidt 1990.

contained in a unit, and this unit did not contain landmarks a to d. The important feature here was the *abrupt change* in the pattern of correlations rather than their absolute values.

Finally, a unit, defined in this way, does *not* correspond merely to a single movement direction (e.g., a backswing, a forward swing). Schneider and Schmidt (1995) added a small rightward segment to the beginning of the backswing. Subjects organized the actions so that this rightward movement and the backswing were combined into a single unit (correlations of about 0.90). The forward swing was again seen as a separate unit. Also, the overall duration of the entire action was markedly shortened by speeding up the target light. However, the fundamental organization of the units and their location in the sequence of behavior was almost completely unaffected.

By now, several experiments have revealed that these units are identifiable, robust and repeatable, modifiable with practice and feedback, and sensitive to several task variables. Importantly, their existence is principled. It emerges from GMP theory and the tendency toward invariance in relative timing. Next the main goal of this chapter is presented—the extension of the notion of units of action to the problems inherent in bimanual coordination.

Units in Bimanual Coordination

If the notion of units of action has validity to real-world tasks, such units should also be identifiable in two-limb (or even multilimb) actions. In these actions, it might (or might not) be discovered that the involved limbs are controlled as a single unit, as defined in the previous section. This brings in issues of *coordination* that, despite its obvious importance, had surprisingly been largely ignored until the 1980s. Recently, it has attracted attention from researchers operating with several different theoretical perspectives. These include movement dynamics (Schöner and Kelso 1988; Turvey 1990; Wallace 1989), programming in RT paradigms (Kelso et al. 1979; Rosenbaum 1985, 1991), and the theory of motor programs (Heuer et al. 1995). A persistent problem in all this work has been how the individual organizes limb control when the limbs are to produce different patterns simultaneously, which of course was one of Bernstein's (1967) major concerns.

As mentioned earlier, the obvious discrepancies between some findings in the coordination literature versus the common, everyday observations of motor skills are fascinating. In one sense, doing different things with the limbs is relatively easy. Examples can be seen simply by observing people pressing the clutch and shifting gears, playing the piano, or typing. Yet, it has become clear that in some cases at least, such performances can be quite difficult. One common, well-studied result (Heuer 1996) is that when the left and right actions have different temporal structures—such as in tapping polyrhythms (Summers and Kennedy 1991)—at least some combinations are difficult or impossible to perform simultaneously. Even though these investigations have furthered the understanding greatly, an examination of continuous, simultaneous movement studies for which different spatiotemporal patterns have to be performed has been missing (an exception—Swinnen and Walter 1991). Perhaps this is understandable. Based on the earlier literature, the expectation was that such two-hand tasks would either be ". . . impossible or extremely difficult to produce simultaneously" (Heuer 1996 page 132).

In addition in work about rhythms (Klapp 1979), performance tends to break down as the ratio of left-hand and right-hand periods deviates from integer values. The performance of polyrhythms is much poorer than performance of simple rhythms. The exact form of the integrated timing control in tasks requiring polyrhythms is a matter of some debate lately (Klapp et al. 1985; Jagacinski et al. 1988; Summers et al. 1993). However, the fundamental problems for coordination appear to be temporal in nature.

Another line of evidence comes from tasks in which the left and right hands are to produce simultaneous, discrete aiming responses to targets. When the temporal aspects of the actions are the same (whether or not the spatial aspects, such as movement distance, are), these actions are easily done together, a general phenomenon recognized long ago (Woodworth 1903). However, if the two hands are to produce simultaneous actions with different movement times, under several conditions the actions are produced nearly simultaneously (Kelso et al. 1979), although a small part of the goal difference remains (Corcos 1984; Marteniuk and MacKenzie 1980). Along similar lines, Schmidt et al. (1979) found that simultaneous aiming movements with the two hands produced high between-hands correlations for movement times but not for movement amplitudes. This suggests that the two hands tend to be linked in time. All of this evidence is consistent with the view that the system tends to use a common rate parameter for the two hands, which interferes with attempts to produce bimanual actions with different rate parameters or even different temporal structures in the two hands.

Finally, it is often not difficult to produce bimanual actions with different spatial characteristics (e.g., differences in amplitude or direction) as long as the temporal structures in the two hands are the same. However, when the temporal structures of the actions differ—with or without similar spatial characteristics—producing the actions together becomes greatly more difficult (Konzem 1987). Several experimenters have examined such tasks. However, they showed that subjects can acquire the capability to produce such actions with practice although the subjects deal with the coordination problem in rather variable ways (Swinnen et al. 1988; Walter and Swinnen 1992).

These kinds of observations have recently led the authors to study movement performances where the two hands produce deliberately different patterns that are rapid and discrete, and have been considerably practiced. The results, described next, show a surprising deviation from Heuer's (1996) reading of the current views about coordination. Evidence will be shown supporting the view that the two hands can be controlled as a single unit (i.e., by a single GMP) even though the two limbs are producing completely different spatiotemporal patterns.

As a result of these findings, Heuer et al. (1995) have formalized the notion of the GMP with invariant relative timing and have generalized it to multiple-limb actions. This process began with the assumption of a continuous prototypical function as the GMP that can be scaleable in time and/or amplitude (Meyer et al. 1982; Schmidt et al. 1979, 1985). However, a slightly different approach—the one used in this chapter—assumes that the GMP produces discrete events distributed in time, where the focus is on the intervals among such events (Gentner 1987). The discrete version of the GMP model can be viewed as a simplification of the continuous version, because continuous movements can be described in terms of intervals between discrete events. Consistent with this view, the temporal occurrence of discrete values (landmarks) can be measured from the essentially continuous kinematic traces produced during the action. These values could be almost any feature of the traces, such as the time of peak acceleration, time of peak deceleration, time of maximum velocity, and so forth, forming a set of landmarks for each trial.

Units, Coordination, and Relative Timing

Recently, the authors have examined rapid tasks in which the two hands must simultaneously produce different spatiotemporal patterns. The seated subjects moved two horizontal levers with elbow flexion-extension movements (without the row of moving lights, however). Subjects attempted to produce particular space-time patterns that differed for each limb (see figure 15.5). The right arm (upper left panel) produced a flexion-extension-flexion pattern through about 70 deg, while the left arm produced a flexion-extension pattern through about 60 deg. These actions were simultaneously produced with movement times ranging from 400 to 600 ms. Subjects were to align the peak displacement of the left hand midway between the reversal points in the right to ensure that any portions of the left- and right-hand patterns (or of their acceleration profiles) were not simply mirror images of each other. Position-time functions were given as feedback on an oscilloscope after the action. Subjects were to initiate the two movements within 30 ms of each other. Rare trials (< 2%) with unequal starting times were rerun.

Subjects had much difficulty with this task at first; the two hands generated considerable mutual interference (see also Swinnen et al. 1991). After several hundred trials, the subjects were able to produce the actions very easily. After giving initial practice with the two-hand coordination task, an additional requirement was introduced. The movement times had to vary, in response to instructions, between

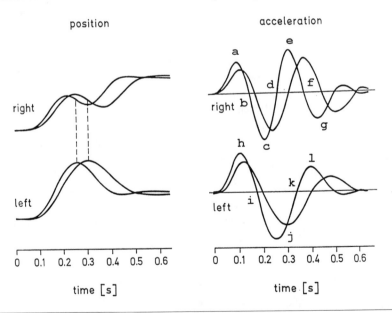

Figure 15.5   Position- and acceleration-time traces for the right and left hands in the bimanual coordination task. Two trials with different movement times are shown, with landmarks defined from the acceleration traces.
Reprinted from Heuer, Schmidt, and Ghodsian 1995.

400 and 600 ms on successive trials. Using a 36-trial test given on the third practice day, also with varied movement times, the two-hand coordination was examined using the unit analysis described earlier.

The acceleration-time traces for the two-hand actions are shown to the right of the respective position-time traces in figure 15.5. Several landmarks (minima, maxima, and zero crossings) are defined in the left- and right-hand acceleration traces, labeled a through g in the right hand and h through l in the left. As before when analyzing units, the temporal occurrence of these landmarks was measured. Next, the within-subject (across trials) correlations among all possible pairs of landmarks were computed. Landmarks a and h were not used, because the absolute time to peak acceleration tended to be invariant (Heuer 1991; Zelaznik et al. 1986).

Figure 15.6 contains the mean within-subject correlations for four subjects. For the hands separately, the correlations with the first point (landmarks B and I, respectively; left graphs) with successively later points decreased slightly (0.96 to 0.80 for the right hand; 0.84 to 0.72 for the left) as more distant landmarks were correlated and with no abrupt changes in magnitude. When viewed with the last point in the action as the common one (landmarks G and L, respectively; right graphs), the correlations increased (0.80 to 0.99 in the right hand; 0.72 to 0.99 in the left) as closer landmarks were considered, again with no abrupt changes. The relatively high correlations without abrupt changes suggested a one-unit pattern for each hand where the actions tended to be linearly scaled. This is essentially the same result seen in the no-KR subjects in figure 15.4. These analyses have also been performed with nonoverlapping intervals, e.g., intervals from landmarks A–B versus C–D, and so forth. Somewhat lower correlations, but essentially similar patterns, resulted.

The initial excitement, however, came from an analysis of mean correlations between the right and left hands. For simplicity, figure 15.6 (lower) shows the correlations only between those points that occur at roughly the same times in the movement—namely, B–I, C–I, D–J, E–J, F–K, and G–L (see figure 15.5). These correlations averaged about 0.82. The right-left correlations were almost the same size as the left-left correlations (mean = 0.827), suggesting that the left hand limited the size of the right-left correlations. The main finding was that the right and left hands were scaled together in time as the overall movement time intentionally varied. One interpretation of this result is that the right and left hands were a part of the same unit—that is, controlled by the same GMP.

While these results provided an interesting new direction for the study of two-hand coordination, they were limited by the use of correlations as the basic measures of the unit structures. Part of the problem is that correlations are subject to numerous sources of variability, such as errors in measurement, neuromuscular noise (Schmidt et al. 1979), and so forth, that make their absolute size difficult to evaluate. A second concern was the determination of how large a change in correlation was sufficient to conclude that one unit had ended and another had begun. As a result, the thinking moved in a different direction. The most promising concerned new ways of analyzing the patterns of covariances calculated among the intervals (after Heuer 1988). Predictions were also derived, based on patterns of covariances, for the hypothesis

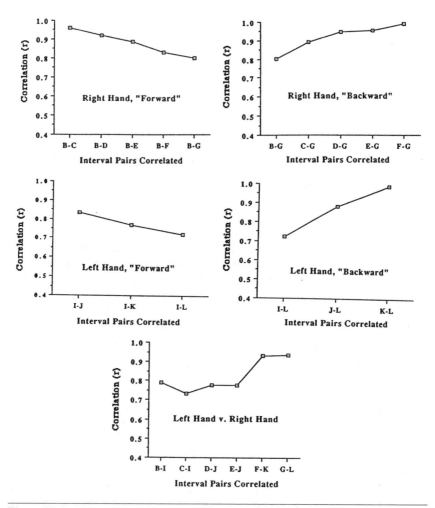

Figure 15.6   Mean within-subject correlations between landmarks for the right hand (top figures) and the left hand (center figures). Between-hand correlations for landmarks occurring at approximately the same time are shown in the lower figure.
Reprinted from Heuer, Schmidt, and Ghodsian 1995.

that the two hands were programmed with the same GMP and for several alternative possibilities (Heuer et al. 1995).

Models of Two-Hand Coordination Based on GMPs

The authors began with the assumption that the GMP for the two hands produces a continuous, prototypical function that can be scaled linearly in time and/or in amplitude (Meyer et al. 1988; Schmidt et al. 1979). This function generates intervals between kinematic landmarks such as those seen in the movement records (e.g., figure 15.5). Any interval between two landmarks is then modeled as the prototypical duration of that interval defined by the GMP's relative timing multiplied by the rate parameter that scales all of the intervals linearly. In addition, a motor delay was assumed (Norman and Komi 1979) between the central GMP output and its realization in the musculature, exactly as Wing and Kristofferson (1973) and Heuer (1988) did earlier (Vorberg and Wing 1996). This motor delay is assumed to have some variability (random noise). The motor delays and their variability can differ for different intervals. With this basic framework, predictions concerning the values of various combinations of covariances computed among these intervals were derived, as described next.

The Development of Covariance Tests

Assume that for each action such as shown in figure 15.5, the temporal occurrence of each of several (here, 12) landmarks is measured, with landmarks A through G, in the right hand and landmarks H through L in the left hand, just as described earlier. However, to avoid the problem of overlapping intervals, only the intervals between the landmarks, such as A to B, D to E, and so forth, are used as the basic data. These intervals are labeled in lowercase letters, where interval a is from the start of the movement to A, b is from A to B, and so forth, where the label indicates the landmark at the end of the interval (e.g., interval e is the interval from D to E). In addition, sets of between-hand intervals are defined, such as the interval from landmark A (right) to J (left) or from landmark H (left) to G (right). Then, within each subject over trials, the covariances are computed among all possible pairs of within- and between-hand intervals. The model makes specific predictions about particular combinations of these covariances, as discussed next. (See Heuer et al. 1995 for derivations and a fuller account.)

Covariance Ratios (COVRs)

First, consider various sets of four intervals, defined with the following restrictions. Two of the intervals are from the right hand and two are from the left. In addition, the within-hand intervals must be nonoverlapping in time and nonadjacent. They are nonoverlapping to avoid the problem of inflated covariances from overlapping variance (see footnote 1). They are nonadjacent to eliminate the tendency for a negative covariance component because the intervals have a common border. In this data set with seven right-hand and five left-hand landmarks, 90 sets of four intervals satisfy these restrictions. For any four such intervals, a covariance ratio (COVR) is the product of the covariances of the two between-hand intervals divided by the product of the covariances of the two within-hand intervals, as follows

$$COVR = \frac{\text{Product of Two Between-Hand Covariances}}{\text{Product of Two Within-Hand Covariances}}. \tag{15.1}$$

Because each of the 90 tetrads of allowable intervals can be combined in two ways, this actually results in 90 × 2 separate COVRs for a set of movements with 12 landmarks. (See Heuer et al. 1995 for details.) If the two hands are governed by a single GMP with a constant rate parameter throughout each movement (but this rate parameter varies across trials), each of the 180 COVRs described in equation 15.1 should be 1.0 for all qualifying intervals.

This COVR test for two-hand coordination can also evaluate whether each hand, treated separately, shows an approximation to invariant relative timing. This is essentially the method proposed by Heuer (1988, equation 6) using so-called tetrad differences. One of the difficulties in evaluating relative-timing invariances concerns the test procedures used. Heuer's method appears to address at least some of the problems raised by Gentner (1987). Essentially, the COVR test described for two-hand coordination is an extension of Heuer's original method for the one-hand case.

Computer Simulations

Computer simulations of this model have also been created. These allowed the authors to determine the extent to which collecting COVRs generated from a human subject conform to the predictions from the model. One example of this simulation is provided here. First, a hypothetical bimanual coordination task with the number of landmarks specified for each hand was created. In the present example, seven landmarks were selected for the right hand and five for the left to conform to the real-world task. The relative timing for this coordination, defined as the set of proportions describing the timing between each of the landmarks in the right hand and between landmarks in the left hand, was then established. The exact values of these proportions are arbitrary, but they must sum to 1.0 for each hand. The simulation has several parameters: the mean and SD of the overall rate parameter, analogous to the subject's overall movement time (and SD); the mean and SD of the motor delays for each landmark; and the number of trials. The simulator then generates the landmark times for each trial, calculates the intervals among landmarks, computes the covariances among all intervals, and computes the set of COVRs for each group of trials. These simulations can be run repeatedly, providing information about the expected variability in the COVRs from sample to sample.

Agreement of Simulated and Actual Data

Table 15.1 shows the simulated COVRs based on parameters that reasonably approximate those in the data set. In particular, the set of proportions for the within-hand intervals were 0.1, 0.2, 0.1, 0.2, 0.1, 0.2, and 0.1 for the right hand and 0.1, 0.2, 0.1, 0.2, and 0.4 for the left hand. The motor delays were assumed to be 50 ms with SDs of 5 ms. The mean rate parameter was 500 ms with an SD of 100 ms.[2] The various allowable right-hand intervals are listed down the left column (a–c, a–d, . . . , e–g). The allowable left-hand intervals are across the top row (h–j, . . . , j–l). For each combination of four intervals, two sets of COVRs, labeled 1 and 2 in the table, correspond to the two allowable ways that each tetrad of intervals can be combined. For example, the entries in the upper left corner (COVR = 0.87, 1.08) are the two COVRs computed using intervals a and c in the right hand and intervals h and j in the left hand.

If one examines the 180 values in this table, it is clear that the COVRs were generally close to 1.0 with some unsystematic scatter for the different intervals. This scatter represents sampling variability. Sampling variability increases as the motor delay variability increases, as the rate parameter variability decreases, and as the number of trials decreases. These are only initial attempts at simulation. However, they provide a general picture of the amount of instability in the COVRs that would be expected from an actual subject behaving according to the assumptions of the model. While some of the factors that influence this variability are known, future efforts will be directed at estimating the sampling distribution of the COVRs so that statistical tests can be run to contrast actual and simulated data.

Table 15.2 shows corresponding COVRs from one of several subjects tested with this task. The data are based on a 36-trial test given after five days of practice. Consider first the set of 90 COVRs in column 2 of each combination in the table. These COVR values agree relatively well with the prediction from the model, because they seem to hover around 1.0. The degree of scatter is roughly the same as with the simulated values. This analysis thus supports the hypothesis of a two-hand GMP. Certainly, more work is needed with simulations. However, that subjects seem to produce behavior at least consistent with the rather restrictive predictions from the model is encouraging.

Simulated data were deliberately created to violate the model and to check the method's sensitivity. In one simulation, the restriction that the rate parameter be the same for the two hands was relaxed. Nonidentical but correlated ($r = 0.40$) rate parameters were used instead. The other parameters were the same as in the earlier simulation. This represents just one version of a set of hypotheses that claim each hand is controlled somewhat separately but with constraints that provide loose coupling. For this version, though, rather marked deviations of the COVRs from 1.0 were found. The COVRs ranged from –0.02 to 0.23 (see table 15.3 here or table 4 in Heuer et al. 1995).

---

[2]The mean values of the motor delays and rate parameter do not affect the pattern of covariances, but the SDs do.

Table 15.1 Tetrad Ratios (COVRs) From a Simulated Data Set With Equal Rate Parameters for the Two Hands, Where the Two Hands Are Controlled by the Same GMP

Right-hand intervals	h–j 1	h–j 2	h–k 1	h–k 2	h–l 1	h–l 2	i–k 1	i–k 2	i–l 1	i–l 2	j–l 1	j–l 2
a–c	0.87	1.08	0.95	1.06	0.86	0.82	0.83	1.02	0.86	0.89	1.13	0.87
a–d	0.94	1.07	0.97	1.06	0.80	0.81	0.85	0.91	0.79	0.80	1.05	0.94
a–e	0.98	1.13	1.07	1.11	0.83	0.86	0.93	1.02	0.83	0.89	1.10	0.97
a–f	1.01	1.37	1.08	1.35	0.90	1.04	0.94	1.13	0.89	0.99	1.18	1.01
a–g	0.79	0.90	0.88	0.88	0.68	0.68	0.77	0.84	0.67	0.73	0.89	0.79
b–d	1.12	1.16	1.15	1.26	0.95	0.91	1.11	1.08	1.04	0.89	1.13	1.05
b–e	0.86	0.91	0.95	0.98	0.74	0.71	0.91	0.90	0.81	0.74	0.88	0.81
b–f	0.84	1.02	0.89	1.11	0.74	0.80	0.85	0.93	0.81	0.76	0.88	0.78
b–g	0.97	1.00	1.08	1.08	0.83	0.78	1.04	1.03	0.92	0.85	1.00	0.91
c–e	1.09	1.03	1.20	1.11	0.94	1.01	1.15	1.02	1.02	1.05	1.00	1.15
c–f	0.96	1.05	1.02	1.14	0.85	1.04	0.98	0.96	0.93	0.99	0.91	1.01
c–g	0.98	0.89	1.08	0.97	0.84	0.88	1.04	0.92	0.91	0.95	0.89	1.02
d–f	1.02	1.21	1.09	1.25	0.91	1.02	0.93	1.04	0.89	0.97	1.04	1.00
d–g	1.08	1.07	1.19	1.10	0.92	0.90	1.02	1.04	0.90	0.97	1.06	1.05
e–g	0.94	0.92	1.04	1.01	0.81	0.79	0.95	0.95	0.84	0.85	0.91	0.91

Note: Allowable (i.e., nonadjacent) right-hand intervals are shown along the left column, and the allowable left-hand intervals are shown across the top. The numerals 1 and 2 in the top margin indicate the two separate ways of grouping the between-hand intervals to form each tetrad ratio. The results using the second set of tetrad ratios agree reasonably well with the actual data from the human subject in table 15.2.

Table 15.2 Tetrad Ratios (COVRs) From a Single Subject for all Allowable Tetrads of Intervals in the Two Hands

LEFT-HAND INTERVALS

Right-hand intervals	h–j		h–k		h–l		i–k		i–l		j–l	
	1	2	1	2	1	2	1	2	1	2	1	2
a–c	1.45	0.90	1.42	0.97	1.51	1.05	1.07	1.05	1.02	1.03	0.99	1.07
a–d	1.35	0.86	1.68	0.93	1.67	1.00	1.26	1.01	1.13	0.98	1.09	1.03
a–e	1.40	1.00	1.72	1.08	1.72	1.16	1.29	1.03	1.16	1.00	1.13	1.07
a–f	1.17	1.07	1.67	1.15	1.65	1.24	1.25	0.94	1.12	0.92	1.08	0.89
a–g	1.33	1.25	1.79	1.34	1.86	1.45	1.34	1.05	1.26	1.03	1.23	1.02
b–d	1.02	0.88	1.28	0.90	1.26	0.93	1.22	0.98	1.09	0.91	1.11	0.95
b–e	1.05	1.01	1.29	1.03	1.29	1.06	1.23	0.99	1.11	0.92	1.13	0.98
b–f	–.96	1.17	1.36	1.20	1.35	1.24	1.30	0.98	1.16	0.91	1.19	0.89
b–g	1.01	1.26	1.35	1.30	1.41	1.34	1.29	1.02	1.21	0.95	1.24	0.94
c–e	0.91	1.04	1.11	1.02	1.11	1.08	1.21	0.97	1.09	0.93	1.17	1.00
c–f	0.85	1.26	1.22	1.23	1.20	1.30	1.32	1.00	1.18	0.96	1.27	0.94
c–g	0.93	1.39	1.24	1.36	1.29	1.45	1.35	1.07	1.27	1.02	1.37	1.02
d–f	0.67	0.96	0.96	1.20	0.95	1.19	1.04	0.98	0.93	0.88	0.98	0.86
d–g	0.75	1.10	1.01	1.37	1.05	1.36	1.10	1.08	1.03	0.96	1.08	0.95
e–g	0.86	1.13	1.15	1.38	1.20	1.38	1.10	1.09	1.04	0.98	1.11	0.97

Note: Allowable (i.e., nonadjacent) right-hand intervals are shown along the left column, and the allowable left-hand intervals are shown across the top. The numerals 1 and 2 in the top margin indicate the two separate ways of grouping the between-hand intervals to form each tetrad ratio.

Table 15.3 Tetrad Ratios (COVRs) From a Simulated Data Set With Unequal Rate Parameters for the Two Hands (Correlated 0.40), Where the Two Hands Are Controlled by Separate, Loosely Coupled GMPs

Right-hand intervals	h–j		h–k		h–l		i–k		i–l		j–l	
	1	2	1	2	1	2	1	2	1	2	1	2
a–c	0.13	0.06	0.08	0.07	0.13	0.03	−0.01	0.06	−0.02	0.03	0.08	0.04
a–d	0.06	0.08	0.11	0.09	0.10	0.04	−0.02	0.07	−0.02	0.04	0.06	0.02
a–e	0.23	0.09	0.13	0.10	0.16	0.05	−0.02	0.08	−0.02	0.04	0.10	0.07
a–f	0.19	0.13	0.13	0.14	0.08	0.06	−0.02	0.01	−0.01	0.00	0.05	0.05
a–g	0.09	0.02	0.04	0.02	0.11	0.01	−0.01	0.05	−0.02	0.02	0.07	0.03
b–d	0.07	0.11	0.12	0.13	0.12	0.13	0.09	0.11	0.09	0.11	0.08	0.06
b–e	0.19	0.09	0.11	0.11	0.13	0.11	0.08	0.09	0.10	0.09	0.06	0.15
b–f	0.19	0.15	0.13	0.18	0.08	0.18	0.09	0.01	0.06	0.01	0.06	0.15
b–g	0.17	0.04	0.07	0.04	0.22	0.04	0.05	0.11	0.16	0.12	0.15	0.14
c–e	0.13	0.10	0.07	0.07	0.09	0.10	0.06	0.05	0.08	0.09	0.10	0.15
c–f	0.14	0.19	0.10	0.13	0.06	0.19	0.08	0.01	0.06	0.01	0.07	0.17
c–g	0.06	0.02	0.03	0.02	0.08	0.02	0.02	0.04	0.08	0.06	0.10	0.07
d–f	0.19	0.10	0.13	0.17	0.08	0.16	0.10	0.01	0.07	0.01	0.04	0.14
d–g	0.10	0.01	0.05	0.03	0.13	0.02	0.04	0.06	0.12	0.07	0.06	0.08
e–g	0.08	0.03	0.03	0.02	0.10	0.02	0.03	0.05	0.08	0.06	0.14	0.08

LEFT-HAND INTERVALS

Note: Allowable (i.e., nonadjacent) right-hand intervals are shown along the left column, and the allowable left-hand intervals are shown across the top. The numerals 1 and 2 in the top margin indicate the two separate ways of grouping the between-hand intervals to form each tetrad ratio. These results differ markedly from the COVRs from the human subject and from its simulation (tables 15.1 and 15.2).

Violations of the Model

This mathematical modeling provides not only predictions for the values of the COVRs under the model's assumptions but also a way that specific kinds of model violations can be detected via the COVRs. (See Heuer et al. 1995 for details.) For example, one alternative model has a GMP for the right hand and another for the left so that each hand has its own rate parameter. This is somewhat analogous to the so-called half-center model of locomotion in which each limb has its own oscillator or GMP (Grillner 1981). In this case, COVRs are predicted to be less than 1.0. Another model concerns movements where no linear scaling exists at all, such as in very long duration actions where no GMP could control behavior. In such cases, the predicted values will scatter widely about 1.0, and the probability that all 180 COVRs will equal 1.0 is negligible.

Another violation of the model has both hands governed by one rate parameter at one time early in the action and then by another rate parameter later in the action. Here, both limbs would be controlled at all times by the same rate parameter. However, the rate parameter could change across the action, as if all the behavior sped up or slowed down as it progressed. For this case if intervals on *both* sides of the rate parameter shift are included in the computation of a COVR, the value will be greater than 1.0. However, if the intervals are chosen so that they all lie either after or before the shift in rate parameters, the COVRs will be 1.0.

This latter case is seen in some of the authors' data. Examine the series of column 1 values in table 15.2. Most of these values hover around 1.0 just as the values for column 2 did, except for those (italicized) in the upper left corner of the matrix where the values are all systematically greater than 1.0. These latter COVRs involve sets of intervals that include the first interval in the right hand (interval a) and the first interval in the left hand (interval h). These particular combinations involve the computation of the covariance of intervals a and h to generate the COVRs, whereas the combinations under column 2 did not. One interpretation of this violation is that the rate parameter for the first portion of the action (where intervals a and h are located) was different from the rate parameter in the later parts. This suggests a shift in control from one unit to another during the action. In fact, some earlier work showed that the initial time to peak acceleration was invariant when movement time was changed (providing a kind of start-up unit) even though the remainder of the movement scaled essentially proportionally with movement time once it got going (Zelaznik et al. 1986). These methods should allow the authors to determine where in an action and under what circumstances the initially organized unit structure might have been altered.

Additional Features of Bimanual Coordination

We started with the general idea that the hands, in rapid actions at least, can be controlled by the same GMP. Other features of program theory—which have been examined primarily in single-limb actions—may then be added to form additional hypotheses about the nature of two-hand coordination. These ideas provide interesting and sometimes counterintuitive predictions about the behavior of the hands under various instructions and tasks. They form the basis for some recent experimentation.

First, subjects can clearly alter the speed of the two-hand coordination task (once learned) without any difficulty, varying the movement times from 400 through 600 ms. The COVR tests indicate that, for the last portions of the movement at least, subjects were able to adjust the two hands together. All the individual elements were tightly correlated with each other in time. Subjectively, the subjects (and experimenters as well) found the changes in duration to be almost as easy as altering the speed of a one-hand action. Schmidt claims that this two-hand action is now (after practice) a part of his own movement repertoire. It can now be produced easily on demand almost anywhere and at various speeds. The argument is that he has learned a new two-hand GMP.

In recent work, subjects were asked to perform these actions as rapidly as possible in transfer tests. After two days of practice, overall movement times were reduced from about 450 ms in the last practice day to 397 ms under this maximum-speed instruction. When this occurred, correlations among the action intervals increased markedly. In this test, correlations for the right hand ranged from 0.78 to 0.99 (mean = 0.91) and for the left hand ranged from 0.91 to 0.99 (mean = 0.94). More importantly, the across-hands correlations were extremely high as well, ranging from 0.79 to 0.98 (mean = 0.91). The general finding, which was also seen in several earlier studies, is that as the coordination pattern is scaled toward maximum speed, a substantial increase occurs in the correlations among landmarks (both within and between hands). It would appear that with maximal speed instruction, the coordination pattern conforms maximally to the idealized model of a two-hand GMP with linear scaling. This occurs even though the range of movement times decreases under these maximal-speed instructions. From a statistical standpoint, it would tend to decrease—not increase—the absolute sizes of these correlations.

This capability to scale the actions in time has been, to this point, limited to when the two hands are scaled together. If, for example, a subject were instructed to modify the movement duration of one hand while leaving the other one constant, this modification should provide great difficulties—and might even be essentially impossible. This emerges directly from the theory that the bimanual GMP can have only a single rate parameter at any one instant. Alternatively, changing the movement duration of one hand but not the other can be conceptualized as requiring a different bimanual relative timing. The temporal structure of the task—viewed as a bimanual event with coordination defined between the hands—would be disrupted. In this sense, a new GMP for this new coordination would be required. It would have to be practiced just as the original bimanual task had to be when first attempted. The almost trivial task of changing the duration of both hands, coupled with the apparent impossibility of changing only one hand, provides an important prediction from the hypothesis that might allow it to contrast with other theoretical positions about bimanual coordination.

Experimentation about this issue has begun. Subjects were asked to learn the bimanual coordination task described earlier. Then, after two days of practice, the subjects were asked to produce the same pattern in the right hand but to slow the speed of only the left hand. This instruction was combined with visual information about how the pattern for the right hand should be unchanged and how the left hand's pattern should be modified. This resulted in a bimanual pattern where the left-hand peak amplitude occurred after the peak amplitude of the right—which was opposite to that in the earlier-practiced pattern. Subjects performed 30 additional trials under this instruction.

Differential-Speed Scaling

As expected from the earlier informal tests, subjects had a very difficult time with this transfer test. Whereas subjects did tend to follow the instruction for slowing the left-hand action (39%, or from about 500 to 700 ms in overall duration), they also violated the instruction and slowed the right hand as well (by 18%, or from about 500 to 600 ms in overall duration). Furthermore, while the left hand appeared to be slowed similarly in both the initial and final portions, the later portions of the right-hand action were lengthened more than the earlier portions (15% versus 6%). This, of course, implies that the transfer test created a disruption in the relative timing of the original right-hand pattern. It was not simply a slowed version of the task originally practiced.

This disruption was also manifested in the correlation matrices produced for the bimanual task. Whereas the matrices for the standard bimanual task had values that resembled those found earlier, these correlations were severely disrupted for the transfer test where only the left hand was to be slowed. The across-hands correlations were quite variable (range from –0.19 to 0.59), with a mean of 0.20. Note that in the earlier retention tests with the standard pattern, the across-hands correlations were far more stable (ranging from 0.25 to 0.90) and larger (mean of 0.59). Also, compare these across-hands correlations for the differential-speed scaling with those found in the maximum-speed condition discussed in the previous section (where the mean = 0.91).

Generally, this differential-speed instruction seemed to produce a new action quite different from the earlier-practiced action in several respects. First, it had a differently structured right-hand portion, which was slowed more in the later segments as compared with the earlier segments. Second, stable between-limb coordination was not strongly present. This was indicated by the more variable and smaller within- and across-hands correlations among the landmarks. Finally, subjects expressed a strong sense of subjective difficulty in complying with the differential-speed instructions, indicating that the new coordination was particularly troublesome to accomplish. This contrasts with the almost trivial task of speeding both hands together.

Amplitude Scaling

Based both on RT evidence (Rosenbaum 1991) and data about aiming tasks (Schmidt et al. 1979), it is tempting to suggest that amplitude scaling is done after the rate scaling and separately for each hand (unlike rate scaling, which is the same for both hands, as just discussed). This nonintuitive set of ideas implies that in the two-hand task, one could increase the *size* of one hand's action while holding the size of the other hand's action constant. Theoretically, the performer increases the size of the amplitude (i.e., force) parameter for one hand but not for the other. In fact with a little practice, one should be able to modulate the sizes of each hand's patterns more or less independently and hold the relative timing between hands constant.

This prediction was tested by using a second transfer test in which the subjects were instructed to hold the durations of the hands constant but to reduce the size of the left-hand movement from 90 to 75 deg, keeping the temporal structure of the one- and two-hand actions constant. Subjects found this task relatively easy, especially in contrast to the changed-speed instructions just discussed. Here, the right-arm movement was made with about the same duration as in the retention test (448 versus 438 ms), but the left-arm movement was speeded slightly by the amplitude instructions (451 versus 390 ms). Nevertheless, it was not clear that these instructions altered the structure of the bimanual coordination revealed in the correlation matrices. The values of the across-hand correlations (mean = 0.51) were reasonably similar to those from the retention tests (mean = 0.59). This suggests that the amplitude instruction did not appreciably influence the between-hands coordination pattern. These differential-amplitude instructions appeared to have a smaller effect on the coordination patterns than did the differential-speed instructions, where the between-hands correlations dropped to 0.45. This effect could be most clearly seen in the variability of the across-hands correlations. These ranged from 0.31 to 0.68 in the differential-amplitude test versus –0.19 to 0.59 in the differential-speed test. Also, the instructions for holding the right-hand task constant were more closely followed in this test than in the differential-speed test, suggesting far less interference with the amplitude alterations in the left-hand task than for speed alterations.

Load Scaling

Similar to amplitude scaling, subjects should be able to produce horizontal movements with an added inertial load for one of the two hands, maintaining both the temporal structure and the coordination trajectory. Under this view, changes in inertial load (with amplitude held constant) are generated by scaling a force parameter (Schmidt et al. 1979, 1985). These force parameters are differentially assigned to the hands to meet the inertial demands of each, while the common rate parameter controls the timing for both hands.

These ideas motivated a third transfer test in which amplitudes and durations of the two hands were held constant, but the inertial load on the left hand was increased (300 g added at the handle). Again, subjects were instructed to hold the right-hand task constant in the face of the left-hand loading. Subjects generally found this task to be relatively easy. The movement times of the left hand increased only slightly as the load increased (451 to 473 ms), while the right hand maintained the movement time very closely (448 to 456 ms). The between-hand correlations decreased slightly in relation to the earlier retention test with the standard task (from 0.59 to 0.45). This suggests some disruption in the patterning of the hands when the load was applied to one of them.

Effects of the Differential-Hand Instructions

It is perhaps too early to be certain about the relative level of disruption caused by the instructions to change the left hand slightly in various ways while keeping the right hand constant. Both the changed-amplitude and changed-load conditions caused some disruption of the coordination pattern and of the right hand's action. It was difficult to discriminate among them given the level of analysis so far used. Additional COVR tests asking about the extent and nature of the disruption are needed. One possibility is that the changes in amplitude and/or load may have interfered with the position-time trajectories, more or less as mass-spring approaches to movement control would predict. However, such changes may not have altered the underlying force characteristics of the action, which are thought to be somewhat more reflective of the GMP's final output to the muscles.

What seems very clear is that the changed-load and changed-amplitude instructions were far less disruptive than the changed-speed condition. This latter instruction produced much larger changes in the overall durations and in the patterns of bimanual coordination than did the amplitude and load changes. Subjectively, the speed change instructions seem very difficult to do. More work is needed on these predictions in terms of ways to study these changes. However, the initial impression is that the differential-speed instruction acts completely differently than either the differential-amplitude or differential-load instructions. This, at least, was predicted from the model.

Conclusion

This chapter has reviewed some of the authors' earlier work on bimanual coordination. This work studied the idea that complex actions involving many different limbs—even those tasks where the limbs do different things simultaneously—can be thought of as controlled by generalized motor programs (GMPs) with a single, scaleable rate parameter that governs both limbs at the same time. The authors' model (Heuer et al. 1995) of bimanual coordination results from the combination of these ideas with the mathematics of covariances to provide many testable predictions for tasks of this type. These ideas have just begun to be examined. However, initial findings summarized here suggest that under some conditions (chiefly rapid, well-practiced tasks in stable environments), human bimanual coordination can be modeled successfully in terms of these fundamental ideas.

The ideas presented here, if accepted, require some very different ways of thinking about the development of coordination than used in the past. One idea that has achieved some support is that in early practice, the limbs are constrained to act in a mirror image pattern. Overcoming the tendency to do the same thing with each hand (e.g., rubbing both the head and tummy instead of rubbing one and patting the other) becomes very difficult. With practice, the system becomes able to produce these nonsymmetrical patterns somewhat more easily. Swinnen et al. (1993) have argued that the hands become decoupled with practice so that the actions of the hands can be more or less independent. Their arguments are based on analyses showing that over trials, the between-hands correlations between the space-time trajectories decrease. They interpret the findings as a decreased influence of the mirror-image aspects of the two hands.

On the basis of the present data, the authors view this finding in a somewhat different light. First, on theoretical grounds, the idea that the hands might become more independent begs the question about what controls each hand. Suggesting a kind of decoupling means that people must have several agents to control actions, one for each of the decoupled hands. Second, empirically, the fact that the correlations between the hands' trajectories decrease with practice might simply mean that the hands are becoming more facile at doing different things at the same time. Swinnen et al. (1993) agree that these findings do not necessarily mean that the hands are becoming more independent in the sense of decoupled railroad cars.

However, the ideas presented here suggest that while the hands are becoming more adept at generating different trajectories together, the hands are doing it by an increased dependence between the hands. Evidence was generated by examining the correlations among the intervals between landmarks rather than by examining the correlations between right-left positions in time as Swinnen et al. (1993) had done. It suggests that the two hands come to be tightly linked to each other in a complex way that allows them to produce different trajectories. When the action speed is changed, the pattern of coordination is adjusted so as to maintain temporal structure of the bimanual action. This, of course, is seen as the relative timing described between the landmarks in the two hands. Furthermore, under the proper circumstances (rapid actions, stable environments, and so forth), this coupling is very tight. It allows the hands to produce actions that satisfy this coordination pattern and allows the performer to change the action speed very easily.

The authors argue that the phenomenon wherein the hands do become increasingly constrained in time is the result of learning a new bimanual GMP. The new bimanual action is very difficult at first. The performer is constrained by the tendency to make mirror image movements, and no other structure can be used to produce the desired action. With practice, however, the performer constructs a new bimanual GMP that allows the hands to satisfy the bimanual coordination problem. This program—apparently like unimanual motor programs studied in the past decade or so—can be scaled more or less linearly in time. This allows or requires that the two hands maintain their own trajectories with their own temporal structures. Other coordinations (such as slowing the left hand and keeping the right constant) are not easily done, because this would require two different rate parameters at the same time—one for each hand. Alternatively, this can be thought of as requiring a new bimanual coordination pattern with its own coordination goals, which the performer must now learn from scratch. This does not mean that the person should not be able to do the new coordination pattern. Instead, it means that the new pattern requires some practice in order to develop a new GMP.

These notions are certainly preliminary, and much work needs to be done to test the generality of these ideas. How such GMPs might interfere with each other, how they interact with other more fundamental or natural constraints, and whether many such limbs (as in pole vaulting) could also be constrained by a single GMP are examples of future directions. However, the work presented here provides a strong suggestion that open-loop processes are important contributors to at least some kinds of coordination.

Acknowledgment

Thanks to Sue Hillebrand of UCLA for help with collecting data for these experiments.

References

Armstrong, T.R. (1970) *Training for the Production of Memorized Movement Patterns.* Technical Report No. 26. Ann Arbor, MI: Human Performance Center, University of Michigan.

Beek, P.J. (1992) Inadequacies of the proportional duration model: Perspectives from a dynamical analysis of juggling. *Human Movement Science,* 11: 227–237.

Bernstein, N.A. (1967). *The Co-Ordination and Regulation of Movements.* Oxford: Pergamon Press.

Carter, M.C., Shapiro, D.C. (1984) Control of sequential movements: Evidence for generalized motor programs. *Journal of Neurophysiology,* 52: 787–796.

Corcos, D.M. (1984) Two-handed movement control. *Research Quarterly for Exercise and Sport,* 55: 291–298.

Gentner, D.R. (1987) Timing of skilled motor performance: Tests of the proportional duration model. *Psychological Review,* 94: 255–276.

Gentner, D.R. (1988) Observed movements reflect both central and peripheral mechanisms: Reply to Heuer. *Psychological Review,* 95: 558.

Grillner, S. (1975) Locomotion in vertebrates: Central mechanisms and reflex interaction. *Physiological Reviews,* 55: 247–304.

Grillner, S. (1981) Control of locomotion in bipeds, tetrapods, and fish. In *Handbook of Physiology: The Nervous System.* Vol. 2. V.B. Brooks (Ed.), Bethesda, MD: American Physiological Society, 1179–1236.

Henry, F.M., Rogers, D.E. (1960) Increased response latency for complicated movements and a "memory drum" theory of neuromotor reaction. *Research Quarterly,* 31: 448–458.

Heuer, H. (1988) Testing the invariance of relative timing: Comment on Gentner (1987). *Psychological Review,* 95: 552–557.

Heuer, H. (1991) Invariant relative timing in motor-program theory. In *The Development of Timing Control and Temporal Organization in Coordinated Action.* J. Fagard, P.H. Wolff (Eds.), Amsterdam: Elsevier Science, 37–68.

Heuer, H. (1996) Coordination. In *Handbook of Perception and Action: Motor Skills.* Vol. 2. H. Heuer, S.W. Keele (Eds.), London: Academic Press.

Heuer, H., Schmidt, R.A. (1988) Transfer of learning among motor patterns with different relative timing. *Journal of Experimental Psychology: Human Perception and Performance,* 14: 241–252.

Heuer, H., Schmidt, R.A., Ghodsian, D. (1995) Generalized motor programs for rapid bimanual tasks: A two-level multiplicative-rate model. *Biological Cybernetics,* 73: 343–356.

Jagacinski, R.J., Marshburn, E., Klapp, S.T., Jones, M.R. (1988) Tests of parallel versus integrated structure in polyrhythmic tapping. *Journal of Motor Behavior,* 20: 416–442.

James, W. (1890) *The Principles of Psychology.* Vol. 1. New York: Holt.

Keele, S.W. (1968) Movement control in skilled motor performance. *Psychological Bulletin,* 70: 387–403.

Keele, S.W. (1986) Motor control. In *Handbook of Perception and Performance.* L. Kaufman, J. Thomas, K. Boff (Eds.), New York: Wiley.

Kelso, J.A.S., Southard, D.L., Goodman, D. (1979) On the nature of human inter-limb coordination. *Science,* 203: 1029–1031.

Klapp, S.T. (1979) Doing two things at once: The role of temporal compatibility. *Memory & Cognition,* 7: 375–381.

Klapp, S.T. (1981) Temporal compatibility in dual motor tasks. II. Simultaneous articulation and hand movements. *Memory & Cognition,* 9: 398–401.

Klapp, S.T., Hill, M., Tyler, J., Martin, Z., Jagacinski, R., Jones, M. (1985) On marching to two different drummers: Perceptual aspects of the difficulties. *Journal of Experimental Psychology: Human Perception and Performance,* 11: 814–828.

Konzem, P.B. (1987) *Extended Practice and Patterns of Bimanual Interference.* Los Angeles: University of Southern California (Unpublished doctoral dissertation).

Lashley, K.S. (1917) The accuracy of movement in the absence of excitation from the moving organ. *The American Journal of Physiology,* 43: 169–194.

Marteniuk, R.G., MacKenzie, C.L. (1980) A preliminary theory of two-hand coordinated control. In *Tutorials in Motor Behavior.* G.E. Stelmach, J. Requin (Eds.), Amsterdam: North-Holland, 185–197.

Meyer, D.E., Abrams, R.A., Kornblum, S., Wright, C.E., Smith, J.E.K. (1988) Optimality in human motor performance: Ideal control of rapid aimed movements. *Psychological Review,* 95: 340–370.

Meyer, D.E., Smith, J.E.K., Wright, C.E. (1982) Models for the speed and accuracy of aimed movements. *Psychological Review,* 89: 449–482.

Norman, R.W., Komi, P.V. (1979) Electromechanical delay in skeletal muscle under normal movement conditions. *Acta Physiologica Scandinavica,* 106: 241–248.

Peters, M. (1977) Simultaneous performance of two motor activities: The factor of timing. *Neuropsychologica,* 15: 461–465.

Pew, R.W. (1974) Human perceptual-motor performance. In *Human Information Processing: Tutorials in Performance and Cognition.* B.H. Kantowitz (Ed.), Hillsdale, NJ: Erlbaum.

Quinn, J.T., Sherwood, D.E. (1983) Time requirements of changes in program and parameter variables in rapid ongoing movements. *Journal of Motor Behavior,* 15: 163–178.

Rosenbaum, D.A. (1985) Motor programming: A review and scheduling theory. In *Motor Behavior: Programming, Control, and Acquisition.* H. Heuer, U. Kleinbeck, K.-H. Schmidt (Eds.), Berlin: Springer Verlag, 1–33.

Rosenbaum, D.A. (1991) *Human Motor Control.* New York: Academic Press.

Schmidt, R.A. (1975) A schema theory of discrete motor skill learning. *Psychological Review,* 82: 225–260.

Schmidt, R.A. (1976) Control processes in motor skills. *Exercise and Sport Sciences Reviews,* 4: 229–261.

Schmidt, R.A. (1985a) Identifying units of motor behavior. *The Behavioral and Brain Sciences,* 8: 163–164.

Schmidt, R.A. (1985b) The search for invariance in skilled movement behavior. *Research Quarterly for Exercise and Sport*, 56: 188–200.

Schmidt, R.A. (1988a) Motor and action perspectives on motor behavior. In *Complex Movement Behavior: "The" Motor-Action Controversy*. O. Meijer, K. Roth (Eds.), Amsterdam: North-Holland, 3–44.

Schmidt, R.A. (1988b) *Motor Control and Learning: A Behavioral Emphasis*. 2d ed. Champaign, IL: Human Kinetics.

Schmidt, R.A., Sherwood, D.E., Zelaznik, H.N., Leikind, B.J. (1985) Speed-accuracy trade-offs in motor behavior: Theories of impulse variability. In *Motor Behavior: Programming, Control, and Acquisition*, H. Heuer, U. Kleinbeck, K.-H. Schmidt (Eds.), Berlin: Springer-Verlag, 79–123.

Schmidt, R.A., Young, D.E. (1991) Methodology for motor learning: A paradigm for kinematic feedback. *Journal of Motor Behavior*, 23: 13–24.

Schmidt, R.A., Zelaznik, H.N., Hawkins, B., Frank, J.S., Quinn, J.T. (1979) Motor-output variability: A theory for the accuracy of rapid motor acts. *Psychological Review*, 86: 415–451.

Schneider, D.M., Schmidt, R.A. (1995) Units of action in motor control: Role of response complexity and target speed. *Human Performance*, 8: 27–49.

Schöner, G., Kelso, J.A.S. (1988) Dynamic pattern generation in behavioral and neural systems. *Science*, 239: 1515–1520.

Shapiro, D.C. (1977) A preliminary attempt to determine the duration of a motor program. In *Psychology of Motor Behavior and Sport*. Vol. 1. D.M. Landers, R.W. Christina (Eds.), Champaign, IL: Human Kinetics.

Shapiro, D.C. (1978) *The Learning of Generalized Motor Programs*. Los Angeles: University of Southern California (Ph.D. dissertation).

Smith, J.L. (1978) Sensorimotor integration during motor programming. In *Information Processing in Motor Control and Learning*. G.E. Stelmach (Ed.), New York: Academic Press, 95–115.

Summers, J.J. (1977) The relationship between the sequencing and timing components of a skill. *Journal of Motor Behavior*, 9: 49–59.

Summers, J.J., Kennedy, T.M. (1991) Strategies in the production of a 5:3 polyrhythm. University of Victoria, Australia (Submitted).

Summers, J.J., Todd, J.A., Kim, Y.H. (1993) The influence of perceptual and motor factors on bimanual coordination in a polyrhythmic tapping task. *Psychological Research*, 55: 107–115.

Swinnen, S.P., Walter, C.B. (1991) Toward a movement dynamics perspective on dual-task performance. *Human Factors*, 3: 367–387.

Swinnen, S.P., Walter, C.B., Lee, T.D., Serrien, D.J. (1993) Acquiring bimanual skills: Contrasting forms of information feedback for interlimb decoupling. *Journal of Experimental Psychology: Learning, Memory, and Cognition*, 19: 1328–1334.

Swinnen, S.P., Walter, C.B., Shapiro, D.C. (1988) The coordination of limb movements with different kinematic patterns. *Brain and Cognition*, 8: 326–347.

Swinnen, S.P., Young, D.E., Walter, C.B., Serrien, D.J. (1991) Control of asymmetrical bimanual movements. *Experimental Brain Research*, 85: 163–173.

Terzuolo, C.A., Viviani, P. (1979) The central representation of learning motor programs. In *Posture and Movement*. R.E. Talbot, D.R. Humphrey (Eds.), New York: Raven, 113–121.

Terzuolo, C.A., Viviani, P. (1980) Determinants and characteristics of motor patterns used for typing. *Neuroscience*, 5: 1085–1103.

Turvey, M.T. (1990) Coordination. *American Psychologist*, 45: 938–953.

Vorberg, D., Wing, A. (1996) Modeling variability and dependence in timing. In *Handbook of Perception and: Motor Skills*. Vol. 2. H. Heuer, S.W. Keele (Eds.), London: Academic Press, 181–262.

Wallace, S.A. (Ed.) (1989) *Perspectives on the Coordination of Movement*. Amsterdam: North-Holland.

Walter, C.B., Swinnen, S.P. (1992) Adaptive tuning of interlimb attraction to facilitate bimanual decoupling. *Journal of Motor Behavior*, 24: 95–104.

Weiss, P. (1941) Self differentiation of the basic patterns of coordination. *Comparative Psychology Monographs*, 17 (4).

Wing, A.M., Kristofferson, A.B. (1973) The timing of interresponse intervals. *Perception and Psychophysics*, 13: 455–460.

Woodworth, R.S. (1899) The accuracy of voluntary movement. *Psychological Review*, 3.

Woodworth, R.S. (1903) *Le mouvement.* Paris: Doin.

Wulf, G., Schmidt, R.A. (1996) Average KR degrades parameter learning. *Journal of Motor Behavior.*

Wulf, G., Schmidt, R.A., Deubel, H. (1993) Reduced feedback frequency enhances generalized motor programming learning but not parameterization learning. *Journal of Experimental Psychology: Learning, Memory, and Cognition,* 19: 1–18.

Young, D.E., Schmidt, R.A. (1990) Units of motor behavior: Modifications with practice and feedback. In *Attention and Performance XIII.* M. Jeannerod (Ed.), Hillsdale, NJ: Erlbaum, 763–795.

Young, D.E., Schmidt, R.A. (1991) Motor programs as units of movement control. In *Making Them Move: Mechanics, Control, and Animation of Articulated Figures.* N.I. Balder, B.A. Barsky, D. Seltzer (Eds.), New York: Morgan Kaufmann Publishers, Inc., 129–155.

Young, D.E., Schmidt, R.A. (1992) Augmented kinematic feedback for motor learning. *Journal of Motor Behavior,* 24: 261–273.

Zelaznik, H.N., Schmidt, R.A., Gielen, S.C.A.M. (1986) Kinematic properties of rapid aim hand movements. *Journal of Motor Behavior,* 18: 353–372.

16

How Are Explosive Movements Controlled?

*A.J. "Knoek" van Soest and
Gerrit Jan van Ingen Schenau*
Faculty of Human Movement Sciences, Free University,
Amsterdam, The Netherlands

Motor behavior of human beings is generally aimed at achieving environment-related goals. Given a goal, the task for the motor-control system is ultimately to generate a movement that will achieve this goal. Bernstein (1967, 1996) pointed out that one of the central problems in understanding motor control is redundancy, commonly referred to as the degrees of freedom problem. As pointed out by Latash (1996), one is confronted with a genuine degrees of freedom problem when the number of variables (degrees of freedom) to be controlled as a function of time is larger than the number of variables specified by the task's goal. When a genuine degrees of freedom problem exists, this gives rise to choice for the control system. In principle, choice is an infinite number of controlled variable combinations to achieve the task's goal. Tasks do exist, however, where the goal is to optimize some aspect of the performance, for example, a maximum-height jump or the 100 m dash. In such tasks, the environment-related goal, in theory, is sufficient to specify fully how the degrees of freedom are to be coordinated. Only one coordination pattern will result in an optimal performance with respect to the environment-related goal. For these tasks, it is not necessary to assume the existence of internal optimization criteria/constraints/coordination rules in order to arrive at a unique coordination pattern. Quite possibly, of course, internal constraints are taken into account, in which case the outcome results from the interplay of external and internal criteria. (If and how the system identifies this optimal coordination is another problem.) Since this chapter is largely concerned with this type of task, the degrees of freedom problem is not a main issue here.

Another aspect of the motor-control problem identified as problematic by Bernstein (1967, 1996) may be referred to as the controllability problem. From an engineering perspective, the optimal situation would be where a one-to-one relation exists between central commands to actuators (muscles) and the force generated by these muscles. Bernstein pointed out that due to a number of characteristics of the musculoskeletal system, this relation is not unambiguous at all. The most important of the complicating characteristics are muscle fiber characteristics, tendon elasticity, interaction between body segments, and interaction with the mechanical environment.

Muscle Fiber Characteristics

The force exerted by a sarcomere, the basic building block of the contractile element, does not depend solely on its neural activation. Even in the simplest model, the dependency of force on sarcomere length and velocity must be taken into account. This dependency itself is not a static property but rather depends instantaneously on factors such as activation and temperature (Petrofsky and Phillips 1980; Rack and Westbury 1969). In addition as a result of preceding contractions, both improvement (potentiation) or reduction (fatigue) of the muscle's force-generating capacity may occur (history effects). In short, the force produced by a muscle does not depend solely on its neural activation but also on its length, velocity, and contraction history, to name the most important variables. Thus, a one-to-one correspondence between neural activation and muscle force does not exist. From an engineering perspective, muscle is not an ideal actuator.

Tendon Elasticity

Muscle consists not only of the contractile machinery in the form of sarcomeres connected in series and in parallel but of a contractile element connected in series to a component. Since the time of A.V. Hill (1938), this component has been referred to as the series elastic component. The characteristics of this component largely reflect the properties of connective tissue, in particular tendons and tendon plates. Tendon strain is typically 3%–5% of tendon slack length at maximal isometric muscle force; tendon viscosity is negligible. Irrespective of its mechanical significance, e.g., as a power amplifier and an energy buffer, tendon elasticity has serious consequences for motor control. Clearly, the length and contractile velocity of a muscle-tendon complex can be calculated from the angle and angular velocity of the joint(s) spanned by that muscle. Tendon is a purely passive structure with a length that depends on the force exerted on it. Thus, no one-to-one correspondence occurs between muscle fiber length and velocity on the one hand, and joint angle and angular velocity on the other hand. Consequently, the control system cannot correct for the length and velocity dependence of muscle force based on knowledge of the desired joint kinematics, something that might be suggested from an engineering perspective. As a corollary, the discrepancy between fiber length and muscle-tendon complex length makes feedback-based control over joint angles problematic, because muscle spindles yield information only concerning length and velocity of muscle fibers. As rightly observed by Bernstein even if one could exert full control over the length of muscle fibers, this still would not imply perfect control over joint angles.

Interaction Between Body Segments

As noted by Bernstein (1996, essay 6), the movements of different segments influence each other through joint reaction forces. As a consequence, the effect of a particular muscle is not limited to moving the joint(s) that is (are) crossed by that muscle. Zajac and colleagues (e.g. Zajac 1993) have provided convincing examples of the way in which muscles crossing one joint affect the movement in other joints in a way that depends on both position and kinematic constraints. Even more, biarticular muscles may cause joint displacements opposite to the direction of the torque generated by that muscle. For example in cycling, knee extension is partly achieved by the hamstrings, a muscle group anatomically classified as a knee flexor. Clearly, this feature magnifies the problem of motor control. When attempting to correct the movement at one joint by adapting the activation of muscles spanning that joint, new disturbances will occur at other joints.

Interaction with the Mechanical Environment

Mechanically, the environment influences movements through forces that act from the environment on the body. Examples are gravitational, frictional, and (e.g., ground) reaction forces. As a consequence if kinematically identical movements are to be produced in different force fields, muscle forces will have to be adapted as well. Similarly, when starting a movement, the situation is unlikely to be exactly the same every time, even in stereotyped movements. Small deviations in initial position may lead to an unsuccessful movement if the muscle forces are not attuned to the external forces. In conclusion, the muscle forces needed to produce a certain joint movement depend strongly on external conditions.

Uncertainty is the Rule!

One might think that as long as a movement unfolds according to plan, the complicating factors mentioned are not that important. After all, the central nervous system must be capable of generating adequate muscle activation patterns. Unfortunately, it is unlikely that one ever completes a movement about which everything was certain at the start. In sawing, the resistance offered by the wood will vary continuously; in picking up a cup of coffee, the weight is not exactly known; in turning a key, the amount of force required is unknown; in pole vaulting, wind often contributes a disturbing effect; in cycling, the friction between tire and road varies with weather conditions; after a meal, gravitational forces are larger; and so forth.

These are some of the factors that led Bernstein to conclude that an unequivocal relation cannot exist between open-loop central commands and any parameter of the outcome, be it kinematic or kinetic. Rather, muscle forces as well as joint angle traces emerge from the confluence of a large number of factors. To what extent may feedback be used to correct for the resulting discrepancy between central commands and outcome? In the absence of firm data, opinions about the contribution of low-level mechanisms have shifted back and forth historically. At one point, reflexes were considered too weak to be significant (Tardieu et al. 1968). At another point, reflexes were the core idea in influential models of motor control (Merton 1953).

Feedback-Based Motor Control?

The possibility of basing motor control on feedback presupposes the presence of sensors. In that respect, humans are well equipped. First of all, people are able to sense important aspects of their relationship to the environment (vision, hearing, touch). Furthermore, people have sensors for the state of the musculoskeletal system. They can sense joint angle, muscle force, and muscle fiber length and velocity, although the sensitivity with which these variables can be sensed varies. Is the presence of sensors sufficient to conclude that feedback-based motor control is feasible? In the authors' and Bernstein's view, this is not necessarily true.

First, the mechanical interaction between body segments complicates feedback-based control. As discussed above, the effect of muscle forces is not limited to the joint spanned by the muscle. If feedback-based control is to compensate for these interactions, the activation of each individual muscle will have to be influenced by feedback from numerous other muscles. Even more problematic, the gain of these off-diagonal terms should be not only task dependent but position dependent as well.

The major complication in making extensive use of feedback, however, is related to feedback latencies. The central nervous system goes to great trouble to reduce these latencies, e.g., myelinization. Nevertheless, in fast movements, the gain of delayed feedback loops must be low if instability is to be prevented (Hogan 1990;

MacMahon 1984). Exactly how large are feedback loop latencies? The conduction delay of monosynaptic feedback loops for human lower extremity muscles is between 30 and 50 ms (Gottlieb and Agarwal 1979). That is, after a perturbation, the first adaptations in EMG are seen after about 40 ms. For upper extremity muscles, this delay is slightly shorter. An aspect often ignored, however, is that realizing a skeletal acceleration in reaction to a mechanical perturbation takes significantly more time. This is because the dynamics of excitation-contraction coupling and of intrinsic muscle dynamics introduce a phase shift between muscle activation and muscle force. When expressed in units of time instead of in units of phase angle, the resulting delay between change in EMG and change in muscle force is sometimes referred to as the electromechanical delay (EMD). For contraction frequencies in the order of 2 Hz, EMD values have been reported to be about 90 ms for human leg muscles (Inman et al. 1952; Vos et al. 1991). In the latter study, this value was determined by cross-correlating EMG and joint torque data for a task where subjects had to perform rhythmic isometric contractions of the quadriceps at a frequency of 2 Hz. Thus, the delay between mechanical perturbation and first adaptation of skeletal acceleration is about 130 ms. The duration of many movements of the extremities is small in comparison. For example, Muhammad Ali's left jab was reported by Schmidt (1988) to be completed within 40 ms. Somewhat slower, the push off for a human vertical jump is completed in about 300 ms.

In comparison with such movement times, this delay between command and resulting acceleration is quite substantial and would give rise to considerable instability. Motor control, on the basis of such delayed feedback, would be comparable to a chess game where one has to base the nth move on the opponent's position after $n-3$ moves. With Bernstein, the authors conclude that for fast movements of the extremities, the feedback delay is too large to adjust an ongoing movement based on sensory feedback as part of the organization of undisturbed movements. This severely restricts the applicability of closed-loop control theories. It supports Bernstein's statements (1996, essay VII) that in actions like throwing and jumping, all sensory corrections, "Should be introduced based on anticipation when the movement has not yet started." This is largely consistent with the view of Bekoff et al. (1987) and Hasan and Stuart (1988) about afferent information specifying initial conditions that modulate pattern-generating circuitry. Relevant to Feldman's lambda model to be discussed later, this also seems true for muscle spindle information since the activation of gamma-motoneurons appears to be aimed at optimizing the information flow from these sensors (Kuffler et al. 1984; Loeb et al. 1985).

If one accepts that neural feedback cannot play an important role while executing fast movements and keeps in mind the controllability problem, one would expect to find a high variability in the kinematic aspects of fast movements. This would particularly occur in tasks like vertical jumping where the skeleton is analogous to an inverted, multiple pendulum. This expectation, however, is not supported by experimental data about fast movements such as vertical jumping and sprint running. Either the above analysis is flawed and feedback does play an important role or a nonneural stabilizing mechanism operates to counteract disturbances occurring during movement. With regard to the latter possibility, the viscoelastic properties resulting from muscle's contractile properties are an obvious candidate. These zero-lag peripheral feedback systems (van Soest and Bobbert 1993) or, alternatively, preflexes (Brown and Loeb 1996) counteract disturbances without any time delay. However, the gain is admittedly lower than that resulting from neural feedback loops. In fact, these properties are a central aspect of the alpha version of the equilibrium-point theory. Thus, the first question to be addressed in this chapter is whether these preflexes are potent enough to guarantee robust execution of explosive movements such as vertical jumping. In particular, this chapter considers robustness against variations in the state (e.g., position perturbations), parameters (e.g., mass perturbations), and input (e.g., noise in the open-loop specification of muscle activation) using a modeling/simulation approach. Next, how this open-loop control may be generated and to what extent existing theories of motor control are applicable to the type of task discussed here will be considered.

Description of Task and Model

The task studied is a maximum-height squat jump, i.e., a jump initiated from a static squatting position where a countermovement preceding the push off is not allowed. Both in the experiments and in the simulation model, the contribution of arm movements was excluded. In the experiments, subjects were instructed to keep their hands on their back; in the model, the upper body was represented as a single, rigid segment.

For an extensive description of the mathematical model used in these simulations of human, vertical, squat jumping, the reader is referred to van Soest and Bobbert (1993). The model consists of a part describing the skeletal structure and a part describing the behavior of the muscles. Both will be briefly described here (see also figure 16.1).

The skeletal submodel is two-dimensional and consists of four rigid segments representing feet, lower legs, upper legs, and head-arms-trunk (HAT). These segments are connected in frictionless hinge joints representing hip, knee, and ankle joints. At the distal end of the foot segment, the skeletal model is connected to the rigid ground by a fourth hinge joint, which can be considered a representation of the metatarsophalangeal joint. Possible ground contact at the heel is assumed to be elastic. Acceleration-determining forces are gravitational forces, the force at the heel in case of ground contact, and the muscle forces, represented as net joint moments.

The skeleton is actuated by six "muscles," representing the major muscle groups that contribute to extension of the lower extremity, i.e., gluteal muscles, hamstrings, vasti, rectus femoris, soleus, and gastrocnemius. A Hill-type muscle model represents these muscles. It consists of a contractile element and a series elastic element. Behavior of the elastic elements is governed by nonlinear force-length relationships. Behavior of the contractile element is more complex. Contractile element

Figure 16.1   Schematic representation of the model used in simulations of vertical jumping. The two-dimensional skeletal model consists of feet, lower legs, upper legs, and upper body. It is actuated by six muscle groups: gluteal muscles, hamstrings, vasti, rectus femoris, soleus, and gastrocnemius, all represented by Hill-type muscle models. Reprinted from Bobber and Van Soest 1994.

contraction velocity depends on active state, contractile element length, and force. Force is directly related to the length of the series elastic element. This length can be calculated at any instant from the position of the skeleton and contractile element length (which are used as state variables) since muscle-tendon complex length is directly related to skeletal position. Active state is related to muscle stimulation STIM, the independent neural input of the model, by first-order dynamics as described by Hatze (1981). STIM ranges between zero and one. It is a one-dimensional representation of the effects of recruitment and firing rate of alpha-motoneurons.

Wherever possible, parameter values for the muscles were derived based on morphometric data. Most importantly, contractile element optimal lengths were based on sarcomere numbers; muscle moment arms were based on muscle-length versus joint angle measurements; relative values for maximal isometric forces were based on muscle cross-sectional areas. All these measurements were made on the same group of cadavers. Sarcomere numbers and moment arms were scaled in proportion to segment lengths. Absolute values of muscle forces were set in such a way that realistic maximal isometric moments were obtained. Hill's force-velocity parameters a/F_{max} and $b/L_{ce(opt)}$ were set to 0.41 and 5.2, respectively, for all muscles. The stretch of the series elastic element at maximal isometric force was set at

0.04 times its slack length for all muscles. A detailed discussion of the muscle model structure and the strategy followed to estimate parameter values has been presented elsewhere (van Soest et al. 1994; van Soest and Bobbert 1993).

In total, the model is mathematically described by a set of 20 coupled, nonlinear, first-order, ordinary differential equations. Given the initial state and independent control signals (i.e., STIM) as a function of time, the resulting movement can be calculated by numerical integration. A variable-order, variable–step size, Adams-Bashford predictor, Adams-Moulton corrector integration algorithm was used (Shampine and Gordon 1975).

The optimal jump is the jump resulting in maximum height reached by the body's center of mass. It is to be compared with experimental data and used in reference purposes when applying perturbations. Therefore, optimization of the STIM(t) pattern must be handled first. In this study, the following restrictions were imposed on STIM. First, the initial STIM level was set in such a way that the static squatting start position was maintained. Second, STIM was allowed to take on either this initial value or the maximal value of 1.0. Third, STIM was allowed to switch to its maximal value just once and thereafter remained maximal until takeoff. Under these restrictions, STIM(t) of each of the six muscle groups is described by a single parameter: the instant at which STIM switches from initial value to maximal value. This reduces the optimization problem to finding the combination of six switching times that result in maximal jump height. Although this is still a computationally demanding problem, it can be solved using standard algorithms. NAG subroutine E04UCF, a sequential quadratic programming algorithm, was used (NAG Fortran Library Manual Mark 13, Numeric Algorithms Group Ltd., Oxford, U.K.). In what follows, the optimal STIM pattern pertaining to the initial position derived from experimental data will be referred to as the standard STIM pattern. Similarly, the resulting movement will be referred to as the standard movement.

In an attempt to show that the model captures the essential features of the real system, the movement resulting from the optimal STIM pattern was compared with experimental data obtained from well-trained volleyball players. The assumption here is that well-trained subjects have optimized their control so as to achieve maximum height. A high degree of similarity was found to exist between experimental and simulation data regarding jump height, kinematics, and kinetics (see figure 16.2). Furthermore, the optimal STIM pattern was similar to EMG patterns in its basic characteristics. (See van Soest et al. 1994 for more details.)

Perturbation Studies

As stated, we will now analyze the robustness against perturbations to state, parameters, and inputs, using the model as described above.

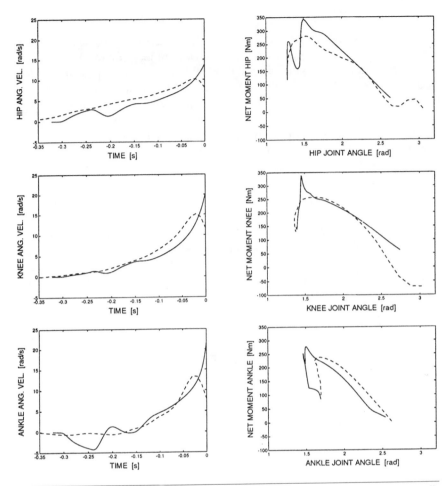

Figure 16.2   Comparison of simulation results with experimental data obtained from well-trained volleyball players. Left column: joint angular velocities versus time for hip, knee, and ankle joints. Time is represented relative to takeoff. Right column: net joint moments versus joint angles for hip, knee, and ankle joints. Zero angle indicates full flexion. Solid curves: simulation results; dashed curves: (calculated from) experimental data *(n = 6)*. Reprinted from Elsevier Science Ltd 1995.

The Effect of State Perturbations

In order to evaluate whether the system's behavior as modeled is robust against perturbations of the musculoskeletal system state, forward dynamic simulations were performed, the system was perturbed, and the resulting behavior was compared with that of the unperturbed system. More specifically, in order to assess if, indeed, muscle's force-length-velocity relation is responsible for this robustness, the authors compared the effect of the same perturbations on models with two different inputs. First, the optimal STIM(*t*) is the open-loop control signal, representing the neural input to the muscles that yields maximum jump height. In this case, muscle dynamics are included. The second kind of control signal is MOM(*t*), representing open-loop net joint moments. In this case, muscle dynamics are excluded. The open-loop MOM(*t*) are, in fact, the net joint moments that result from the optimal STIM(*t*) if no perturbations are applied. Thus, in the absence of perturbations, both types of input result in identical movements. See figure 16.3 for a schematic representation of the calculations performed in cases of STIM and MOM control.

The effect of preflexes on the system's sensitivity to perturbations is investigated by applying the same perturbations to both the STIM- and the MOM-controlled models. Resulting movements are compared with the unperturbed movement, yielding deviations. These deviations indicate the reduction in the effect of perturbations that results from preflexes. As an example, this chapter will show results concerning a perturbation as small as 0.01 radians of the initial foot segment position. For a more systematic investigation of the effects of position perturbations, the reader is referred to van Soest and Bobbert (1993).

An impression of the movement resulting from applying the optimal STIM pattern to the reference initial state is shown in figure 16.4 (top). Height reached by

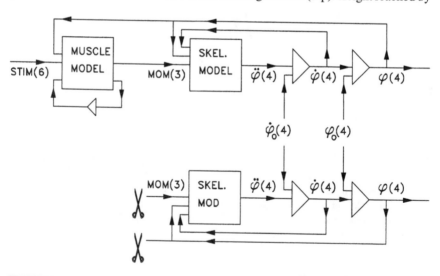

Figure 16.3 Schematic representation of the simulations in which the effects of perturbations are compared for STIM to MOM control. Numbers in parentheses indicate the dimension of the corresponding vectors. Triangles indicate integration over time. In case of STIM control (top), the joint moments depend on muscle dynamics and thus on skeletal position. In case of MOM control (bottom), net joint moments are directly prescribed as a function of time. These net joint moments were recorded from a simulation where STIM control was applied to the unperturbed system.
Reprinted from Bobber and Van Soest 1993.

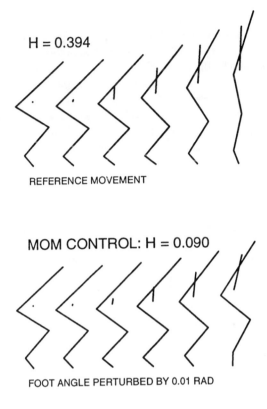

H = 0.394

REFERENCE MOVEMENT

MOM CONTROL: H = 0.090

FOOT ANGLE PERTURBED BY 0.01 RAD

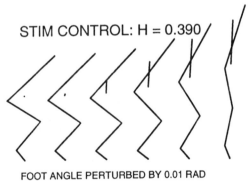

STIM CONTROL: H = 0.390

FOOT ANGLE PERTURBED BY 0.01 RAD

Figure 16.4 The effect of a position perturbation. (Top) Stick figure representation of the unperturbed, optimal push off movement. Left: starting position; right: position at takeoff. Individual figures are linearly spaced in time. The vector, which originates from the body's center of mass, represents velocity of the body's center of mass. (Middle) The same as shown at top, but the initial foot angle was perturbed by 0.01 radians and MOM control was applied. (Bottom) The same as in the middle except that STIM control was applied.

Reprinted from Bobber and Van Soest 1993.

the body's center of mass equals 1.48 m, which amounts to 0.394 m relative to upright standing. Figure 16.4 (middle) shows what happens if the initial foot angle is perturbed by 0.01 radians in the case of MOM control (muscle dynamics excluded). Comparison with the reference movement leads to the conclusion that the movement is dramatically affected by such a perturbation. As a consequence of the disintegration of the movement, jump height is seriously affected. Relative to upright standing, jump height is reduced to 0.09 m. Thus, a perturbation of foot angle by as little as 0.01 radians results in a loss of jump height of as much as 77% when muscle dynamics are excluded. For this task, the movement resulting from open-loop control in the absence of muscle properties is extremely sensitive to perturbations of the system's state. This, by the way, is largely due to the skeletal system in this task mechanically being analogous to an inverted pendulum and, thus, inherently unstable.

To what extent do preflexes reduce this sensitivity? Take a look at the effect of the same perturbation to the STIM-controlled model (figure 16.4, bottom). When this figure is compared with figure 16.4 top and middle, which show the unperturbed and the MOM-controlled perturbed movement, the conclusion is straightforward. The zero-lag peripheral feedback system formed by muscle's force-length-velocity relation is essential to assure robustness against a small perturbation of initial position. As shown in van Soest and Bobbert (1993), this conclusion is not specific for the perturbation used here as an example.

The Effect of Parameter Perturbations

Some parameters of the effector system change so slowly that it seems feasible to keep the open-loop control in tune with the values of these parameters. Examples include changes in muscle strength due to training/disuse and long-term changes of segment mass due to deposition or removal of adipose tissue. On the other hand, system parameters can change so quickly that it is hard to imagine the open-loop STIM pattern being adapted for every such change. For example, consider playing two volleyball matches interrupted by a meal. Obviously, this meal can be considered a perturbation of the upper body's mass. Is it now necessary to account for that change by adapting the open-loop STIM pattern? Instead, are preflexes potent enough to guarantee robust behavior following such a perturbation?

In order to assess the sensitivity to such a mass perturbation, the authors use an approach identical to that described above. Apply mass perturbations to both the STIM- and the MOM-controlled models, and compare the resulting movements against the unperturbed movement. Results shown here concern a perturbation of trunk mass similar to that caused by having lunch. The authors added 1 kg to the mass of the trunk segment. The resulting movement for MOM-controlled and STIM-controlled models is shown in figure 16.5. Everything else being equal, this addition of mass would be expected to reduce jump height by slightly less than 0.01 m. This occurs under STIM control. In fact, the work produced by the muscles is slightly enhanced in the case of added mass. This positive effect is offset by a

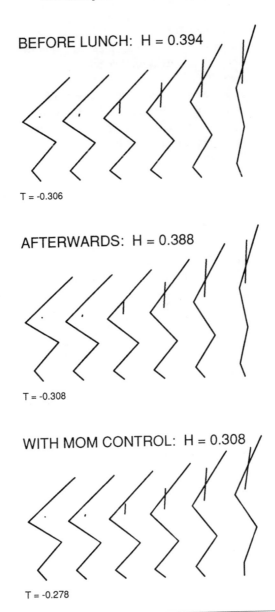

Figure 16.5 The effect of a mass perturbation. (Top) Stick figure representation of the unperturbed, optimal push off movement. Left: starting position; right: position at takeoff. Individual figures are linearly spaced in time. The vector, which originates from the body's center of mass, represents velocity of the body's center of mass. Time is expressed relative to takeoff. Jump height H is calculated relative to upright standing. (Middle) The same as shown at top, but 1 kg was added to the center of mass of the upper body and STIM control was applied. (Bottom) The same as in the middle except that MOM control was applied.

slightly larger rotational energy of the body at takeoff. Are preflexes important here? Figure 16.5 (bottom) shows the movement resulting from MOM control after having lunch. Here, jump height relative to upright standing is reduced to 0.308, a loss of well over 20%! This is reflected by both a loss in work produced by the muscles and a decreased fraction of that work contributed to jump height. The conclusion drawn from this example is, once again, muscle's force-length-velocity relation is highly effective in counteracting parameter perturbations of realistic size, in this case perturbations of body mass. If muscle were an ideal actuator in engineering terms, having lunch would lead to a major motor-control problem.

The Effect of Input Noise

A third type of perturbation present at all times is due to noise in the central nervous system. Due to fluctuations in hormonal levels, activity levels in related areas, cell death, and so forth, the actual firing pattern of alpha-motoneurons will have a stochastic component even if the highest level command is assumed to be unaffected by such factors. Hatze (1996) has recently suggested that due to the low-pass characteristics of the effector system, input noise does not affect the resulting movement. In order to test this assertion for the task considered here, systematic perturbations were applied to the optimal six-dimensional switching time vector. Certainly, in this case, where the perturbations are applied to the optimal control where the derivative of jump height with respect to STIM perturbations is zero by definition, one might expect only marginal effects provided that perturbations are sufficiently small. Results are shown concerning perturbations where the switching times of four muscles were perturbed by plus or minus 5 ms. Figure 16.6 shows the best and worst cases in terms of jump height taken from the 240 perturbed jumps. In perturbation 1, affecting behavior minimally, soleus and hamstrings were activated 0.005 s earlier than standard, and activation of gastrocnemius and vasti was delayed by the same amount. This led to a push off movement that differed only slightly from standard. In perturbation 2, the worst case, soleus and gastrocnemius activation was delayed by 0.005 s whereas gluteal muscles and hamstrings were activated 0.005 s earlier. This resulted in a serious disorganization of the movement, in particular, plantarflexion of the ankle was virtually absent now. This led to a reduced height of the body's center of mass at takeoff as well as to a reduced vertical velocity.

   From these figures, the authors concluded that, indeed, perturbations do exist that have a negligible effect on performance. On the other hand even though the optimal STIM was being slightly perturbed, perturbations in some directions dramatically affected performance. In this task, input noise may seriously affect the resulting movement, even if the reference control was optimal. Apparently, the optimum in control space is quite flat in some directions but extremely steep in others. In any case, these results certainly show that the room for errors in the open-loop control pattern is severely restricted for this task.

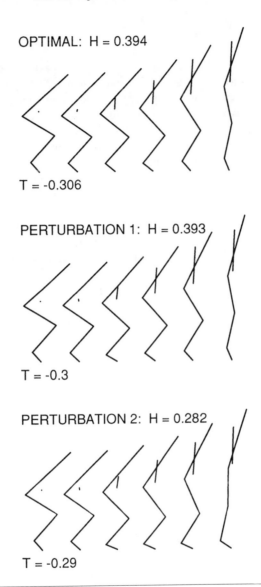

Figure 16.6 The effect of perturbing the switching times of muscles. Out of six muscles, the switching time of four was perturbed by ±0.005 s. (Top) Stick figure representation of the unperturbed, optimal push off movement. Left: starting position; right: position at takeoff. Individual figures are linearly spaced in time. The vector, which originates from the body's center of mass, represents velocity of the body's center of mass. Time is expressed relative to takeoff. Jump height H is calculated relative to upright standing. (Middle) The same as shown at top, but this figure pertains to the perturbation of switching times (see above) with the smallest effect on jump height. (Bottom) The same as at top, but this figure pertains to the perturbation of switching times (see above) with the largest effect on jump height.

Conclusion from Perturbation Studies

Can open-loop control of human vertical jumping, as an example of an explosive movement, result in robust behavior after state, parameter, and input variations? This question was addressed by using a modeling and simulation approach. Perturbations to state and parameters do not jeopardize successful execution of the task. This is due to the stabilizing effect of preflexes, i.e., the zero-lag peripheral feedback system formed by muscle's force-length-velocity relation.

Within the context of the model used, this stabilizing effect of preflexes arises from the Hill-type force-length-velocity relation. Thus, in the authors' model, the short-range stiffness and viscosity related to the dynamics of attached cross-bridges is not included. It is well-established that cross-bridge stiffness is much higher than the stiffness resulting from the force-length relationship (Kirsch 1996). As a result, restoring forces that counteract disturbances small enough to stay within the mentioned short range will be higher than those obtained from the present model.

The conclusion must not be made that neural feedback does not contribute at all. All that the simulation results show is that without neural feedback and short-range stiffness, the task can still be successfully performed under perturbations of a size to be expected in reality.

From the effects of input perturbations, the authors conclude that sensitivity to input variation is highly direction dependent. In general, the open-loop control must be well tuned to the system characteristics if successful performance is to be obtained. This then leads to the questions of which variables are open-loop controlled, how the values of these variables as a function of time are formed in real time, and how the ability to form these signals is acquired. These questions will be addressed here by considering the applicability of a number of existing theories of motor control.

On the Applicability of Equilibrium-Point Theories to the Control of Explosive Movements

The introduction reviewed a number of arguments forwarded by Bernstein that convincingly show that none of the observable characteristics of movements such as position, velocity, or force can be unequivocally specified by the motor-control system. Rather, movement emerges from the confluence of a large number of factors. That being said, the question is what variables in the motor-control systems plan explosive movements? The simplest proposition would be that the motor-control system plans directly in terms of the variable that forms the forward connection between the nervous system and the musculoskeletal system—activation of alpha-motoneurons.

STIM Control

In fact, a number of hypotheses are based on the assumption that movement is planned in terms of alpha-motoneuron activation, a signal referred to as STIM in the description of simulation results. This refers not only to motor program proposals (Gottlieb 1993) but also to the alpha version of the equilibrium-point theory (Bizzi et al. 1992). The central point in the latter hypothesis is that any fixed pattern of alpha-motoneuron activation defines a unique equilibrium position for the limb. This equilibrium position results from the instantaneous springlike properties of muscle itself, i.e., from the mechanism referred to above as preflexes in combination with the elastic behavior of passive structures. The primordial idea is that positioning movements might be realized by considering just the kinematics of the task. This would occur, in particular, by setting alpha-motoneuron activity of the limb's muscles in such a way that it represents an equilibrium point at the desired final position. This idea is highly attractive because it does not require any inverse calculations. That is, it results in convergence toward the desired position irrespective of initial position or perturbations applied during the movement. Unfortunately, reality is far more complex. The following lists just a few of the complications. The movement quality in deafferented animals was substantially reduced. Peripheral stiffness is such that severe co-contraction would be needed to achieve appropriate joint stiffness, which is not observed empirically. In deafferented animals, the equilibrium point was shown not to be moved instantaneously to the desired end point but rather was planned to move from the initial position to the desired final position. In other words, the theoretically possible independence of initial position, in the authors' view the most attractive feature of the hypothesis, was not observed. For a comprehensive overview of the problematic areas, see the commentaries to Bizzi et al. (1992). In spite of these problems, the alpha model is important because it focuses on stability and particularly on the contribution of peripheral mechanisms to stability.

To what extent is the alpha model applicable to explosive movements such as vertical jumping? Clearly, any pattern of alpha-motoneuron activity can, by a change of variables, be represented as an alpha equilibrium-point trajectory. Thus, from the optimal STIM pattern derived above, it is possible to calculate the alpha equilibrium-point trajectory. This is unlikely to make much sense, however, since the system is not aiming at achieving a stable static position by the end of the push off phase. Instead, it roughly aims at maximizing the vertical velocity of the body's center of mass at takeoff. Most of the time, the alpha equilibrium-point trajectory would correspond to an unrealizable hyperextended position, which, if clamped, would result in equilibrium at full extension of the joints. Hyperextension would be prevented only by passive structures. Only during the very last part of the push off, where deceleration of joint extension is observed experimentally (but not in simulation—see figure 16.2), will the corresponding alpha equilibrium point be within the physiological range of motion of the joint. This, however, has been argued to be specifically effected by the biarticular gastrocnemius muscle. Through proper timing of the biarticular gastrocnemius, the integrity of the knee joint structures is preserved. This muscle decelerates the knee angular velocity, using the absorbed rotational energy of the leg segments to support plantar flexion (Ingen Schenau 1989). Braking the knee joint extension using monoarticular muscles would be energetically inefficient. It is difficult to imagine how this action, which is uniquely related to the biarticularity of the gastrocnemius muscle, might emerge based on an organization where the specification of alpha-motoneuron activation does not essentially differ between monoarticular and biarticular muscles.

Note, by the way, that the alpha equilibrium-point trajectory does not specify the timing of the activation of any of the individual muscles involved. The same equilibrium point may result from different combinations of muscle activations (one aspect of the redundancy problem). In other words, it is not clear how this equilibrium-point trajectory would represent the intricate timing of individual muscles required to perform successfully (see above). In any case, the alpha equilibrium-point trajectory needed is not going to be a function of the task's kinematics. In particular, dynamic parameters of the musculoskeletal system have been shown to bear directly on the alpha-motoneuron activity pattern required. For example, in a modeling/simulation study by Bobbert and van Soest, it was found that if muscle strength increased while keeping the neural input identical, jump height actually decreased! If one assumes that alpha-motoneuron activation is the controlled variable in explosive movements, controlling it directly is preferable to controlling it in terms of the associated alpha equilibrium-point trajectory.

Lambda Control

An important objection to direct specification of muscle activation forming the essence of motor control is that after deafferentation, the animals perform much more poorly than before. Apparently, proprioceptive information, hierarchically superimposed on the preflex mechanism, does add to the quality of motor behavior. Stated differently, it is impossible actually to control alpha-motoneuron activity because low-level feedback loops have an impact on alpha-motoneuron activity (Feldman and Levin 1995). This finding lends support to the lambda version of the equilibrium-point hypothesis (Feldman and Levin 1995; Latash 1993). Here, the main controlled variable is assumed to be the threshold for the stretch reflex, referred to as lambda. It is thought to be implemented through appropriate activation of alpha- and gamma-motoneuron pools. Given this threshold, proprioceptive input from muscle spindles, in particular, determines the activation of alpha-motoneurons. At the organizational level of joints, control is conceptualized as R (for reciprocal) and C (for coactivation) commands instead of lambdas for agonist and antagonist muscles. In the absence of external forces, R and C code for the joint equilibrium position and the joint stiffness, respectively. In the case of positioning tasks, the R command shifts gradually and monotonously from the starting position to the desired final position, reaching that position at about the time where velocity peaks. As in the alpha model, for positioning tasks, specification of lambda over time can take place just on the basis of kinematic information about the initial and the desired final position of the effector. It is suggested, in addition, that the movement velocity can be influenced through the C command.

From a hierarchical point of view, the lambda model is richer than the alpha model. The lambda model places a neural feedback loop on top of the preflex mechanism. Unfortunately, in the theoretical work about the lambda model, the contribution of preflexes is often neglected so that the view emerges of two competing hypotheses rather than one being hierarchically on top of the other.

A number of problems have been mentioned relating to lambda control of positioning tasks. The first is specific to the lambda model, i.e., does not apply to the alpha model. From a control engineering perspective, control based on time-delayed feedback of position cannot result in stable behavior for intended movements that contain higher frequency components, as occur during high-speed positioning movements. In addition, measurements of joint stiffness, although both technically and conceptually involved (see Latash and Zatsiorsky 1993), have indicated that stiffness values appear to decrease during movement (Bennett et al. 1992), whereas the opposite would have to be observed in order to explain fast movements. One way out of this problem is to assume that higher-speed movements are achieved by tampering with the position reference signal, i.e., lambda. In the Latash version of the lambda model (Latash and Gottlieb 1991), it is indeed concluded that the R command changes in a nonmonotonous way from initial to desired position. Optimal tampering with lambda, however, requires solution of the full inverse dynamics problem, which was just what the theory was aimed to dispense with. Intermediate approaches may be conceivable where through parameterization of lambda(t) on the basis of task parameters and external forces, an approximation to the inverse dynamics problem is obtained.

Another way out of this problem has been suggested by McIntyre and Bizzi (1993). These authors argue that if lambda is to be a simple function of the desired kinematics and if stiffness is to be realistic, the only way to achieve fast positioning is to add a velocity reference signal. This proposal certainly makes sense from a control theoretical point of view. However, it again complicates the original elegance of the lambda concept. Moreover, it should be noted that the model of Latash and Gottlieb (1991) and the model of McIntyre and Bizzi (1993) both contain feedback loops in which the electromechanical delay indicated above is not incorporated. This results in an underestimation of the problems following from delays and phase shifts between sensing and acting.

A second problem that applies to both the alpha and the lambda hypothesis is that the way in which external forces and inertial effects are to be accounted for in specifying alpha or lambda is hardly touched upon. Given that intrinsic stiffness generally provides insufficient resistance against forces such as gravity and frictional forces (Bennett et al. 1992), these forces must be taken into account in planning lambda trajectories. In recent publications about the lambda theory (e.g., Feldman and Levin 1995), the important issue of how to do this is not discussed at all.

An extreme example of the effect of external forces has been discussed by van Ingen Schenau et al. (1992) (see figure 16.7). If the subject shown in this figure is to displace to the left the object located on a horizontal table, he or she will have to exert force on that object in the direction indicated. Inverse dynamic analysis shows that this force can be realized only by a combination of a net flexing torque in the elbow and a net horizontal adduction torque in the shoulder. However, if the object starts to move in the required direction, an elbow extension is observed. Thus, the direction of joint movement is opposite to the direction of the required net elbow torque. If lambda followed a path that takes it from the actual position of the end effector to the desired end position, a joint torque would result that was not only quantitatively off but of the wrong sign! In these so-called contact-control tasks, biarticular muscles may have distinctly different roles than monoarticular muscles (Ingen Schenau et al. 1995) that cannot be attributed to these muscles in current versions of equilibrium-point models. Based on examples such as the above, the authors are convinced that even in positioning tasks, kinetic aspects must be considered in planning lambda. Unfortunately, considering kinetics inevitably requires that substantial computations are performed in the process of generating lambda. Dealing adequately with external forces reintroduces the computational problems that the lambda hypothesis aimed to dispense with.

The extent to which the lambda model, which was developed in the context of single-joint positioning tasks, is applicable to the control of explosive movements will now be explored. In a reaction to the commentaries about his target article,

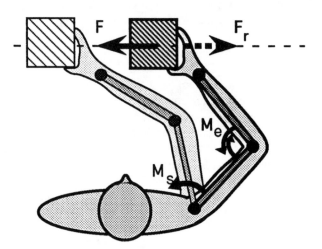

Figure 16.7   Example of a task where a conflict arises between elbow joint torque and change in elbow joint angle. Imagine that the subject is seated at a horizontal table and has to move a heavy object to the left. Inverse dynamic analysis shows that in the position drawn due to the reaction force F_r acting on the hand, the subject must generate a flexing net joint torque M_e at the elbow. However, kinematic analysis shows that in fact, the elbow is extending. See the text for a discussion of how this affects equilibrium-point theories.

Feldman and Levin (1995) argued that for example, the throwing action can be conceived as, "A fast transformation of the equilibrium body configuration, including the endpoint of the arm" (Feldman and Levin 1995, page 797). In the authors' view, it is true that both before and after any explosive action, static positions can be discerned. It seems equally true that controlling these static positions may be described in terms of lambda. It is absolutely false, however, to assume that the shift in lambda required to produce the explosive action is related to the lambda's pertaining to the two equilibria mentioned. As an example, consider the case where the two equilibria, i.e., the static positions before and after the action, coincide, which is perfectly possible. In that case, the explosive action in-between cannot follow from taking the system from equilibrium 1 to equilibrium 2, because these equilibria are identical! In other words even though control of the explosive movement may well be described in terms of a lambda trajectory, the trajectory producing the explosive movement cannot be related to the kinematics of that movement. This leads to the question of whether it is conceptually helpful to describe the supraspinal inputs on motoneuron pools in terms of lambda trajectories.

In addition, the argument presented above in the context of the alpha model applies to the lambda model as well. The timing of individual muscles required for maximum-height jumping critically depends on parameters of the musculoskeletal system. Stated differently, in order to perform optimally, the neural input of the individual muscles must be fine tuned to musculoskeletal parameters and external forces. In tasks with an external optimization criterion, such as maximum-height jumping, adherence to the hypothesis of anatomical correspondence as proposed by Feldman and Levin (1995) would profoundly limit optimal achievement. Since well-trained subjects appear quite able to produce muscle activation patterns that are close to optimal, the reduction in the number of degrees of freedom following from this hypothesis of anatomical correspondence must definitely be removable in well-trained subjects. Once again, the supraspinal inputs to motoneuron pools of individual muscles may be described in terms of lambda trajectories. The ability to generate these neural inputs, however, requires a representation of the dynamics of the effector system in relation to the environment and specification of the kinematic goal (jump height) of the task.

That being said, the authors do want to stress that the stabilizing effect of the physiological mechanisms at the heart of the alpha and lambda models, i.e., mass-spring behavior resulting from both intrinsic muscle properties and peripheral neural feedback loops, is unquestioned. The relative importance of these is task dependent. High-velocity tasks depend more on quite crude, instantaneous mechanisms. Slower tasks depend more on neural feedback loops that may be modulated.

From the above analysis, the conclusion is drawn that even though possible, it is not very helpful to conceptualize the control of explosive movements in terms of alpha equilibrium-point or lambda trajectories. This is particularly so for the inputs on motoneuron pools. In the authors' view, it is more helpful to describe these supraspinal inputs as a combination of two components. First, an open loop component accounts for effector system properties and mechanical interaction with the environment. Second, a feedback control component defines set points and gains for low-level neural feedback loops' carrying position, velocity, and force information. In the particular case of explosive movements, the feedback component is bound to be of minor importance due to the problems related to delays. That is not to say that the supraspinal part of the nervous system necessarily contains a full model of the musculoskeletal system in its relation to the environment. The system may aim at control that works, i.e., the simplest representation of the system that can yield successful performance.

In this context, the chapter will summarize in passing a study about vertical jumping where the authors questioned whether the control system is required to adapt the open-loop STIM pattern to large variations in the initial squatted position. From that analysis it was concluded that the optimal STIM pattern for any starting position yields successful performance for only small variations of that starting position. However, it was also found that one single STIM pattern can be found that yields close to optimal behavior over a large range of starting positions. This *control that works* was not optimal for any particular starting position. Even more interestingly, this STIM pattern that works resulted in successful performance for new starting positions as well. Using such strategies, the nervous system can reduce the complexity of the problem of generating STIM at an acceptable reduction of performance (van Soest et al. 1994). It is the expert's prerogative to take more variables into account in specifying STIM, resulting in closer to optimal performance for any situation encountered. The question remains, nevertheless, how exactly such a STIM pattern is generated and learned.

On the Primacy of Top-Level Planning in Terms of Kinematics

In the simulation studies described earlier in this chapter, the musculoskeletal system was studied. Consequently, alpha-motoneuron activation was the input and the resulting movement the output. From the subsequent discussion, particularly the lambda model, the authors concluded that inclusion of low-level spinal feedback loops in the subsystem studied is helpful. However, in the particular case of explosive movements, these feedback loops are expected to be of limited importance. Including these loops in the model would move the input to the level of supraspinal inputs on motoneuron pools. The question then is how these inputs are actually generated by the supraspinal system. More particularly, if top-level planning takes place in terms of extrinsic kinematics, an idea originally advanced by Bernstein, is it applicable to the control of explosive movements?

In particular, the study of positioning tasks has provided substantial support for the idea that top-level planning occurs in terms of extrinsic coordinates of the end effector. First of all, the movement is at its simplest in terms of end point coordinates. When intending to move the hand to an object, the hand movement in Cartesian coordinates (rather than movement at the joints, for example) is nearly along a straight line and has an almost symmetrical, bell-shaped velocity profile. Furthermore, over a wide range of conditions, these characteristics of end effector kinematics remain invariant, suggesting that end effector kinematics are primarily controlled (Atkeson and Hollerbach 1985; Shadmehr and Mussa-Ivaldi 1994; Soechting 1989). Additional support for the notion that top-level planning occurs in extrinsic end effector coordinates comes from neurophysiology. From the recordings of single cortical cells during positioning in monkeys made by Georgopoulos (Georgopoulous 1988; Georgopoulos et al. 1986), it may be concluded that at the cortical level, a population vector exists that specifies the direction of end effector movement.

Together, these findings have led many to take the position that top-level planning in terms of end effector kinematics is indeed a general feature of motor control. One might argue, however, that these findings are to a certain extent specific for the type of task studied. That the goal in these tasks is specified in terms of end effector kinematics may make that top-level control take place in these terms. It is recently becoming clear that there is indeed some ground for this line of reasoning. For example, Wolpert (1996) pointed out that in planning to move from one point to another while avoiding an obstacle, the observed end point kinematics reflect the dynamic characteristics of the arm. In particular, subjects appear to pass the obstacle in a position where the arm is most stable to perturbations. This stability is measured independently in the form of mobility ellipses. Apparently, adding obstacle avoidance to the task goal of a two-dimensional, point-to-point movement is sufficient to make the control system take kinetic factors into account in top-level planning.

While in Wolpert's case, top-level consideration of dynamic characteristics of the arm appears to be added to the still present kinematic requirements, explosive movements may be at the other end of a continuum. In explosive movements, dynamic factors, particularly properties of the musculoskeletal system and mechanical coupling to the environment, are the prime determinants of the required neural control. For example in a modeling/simulation study dealing with the relation between muscle strength, control of muscle activation, and optimal jump height, an increase in muscle strength actually results in a decrease in jump height unless the muscle activation pattern is fine tuned to the changed properties of the musculoskeletal system (Bobbert and van Soest 1994). In fact in vertical jumping, no direct relation exists between the goal (jump as high as possible) and any kinematic parameter of the push off. Rather, the optimal kinematics can be obtained only by actually applying the optimal inputs to a forward model of the musculoskeletal system. Thus, planning the kinematics here would presuppose that the control system is able to produce inputs to alpha- and gamma-motoneuron pools that result in that movement. Clearly, if the kinematics actually follow from the ability to generate the appropriate inputs to motoneuron pools, it makes little sense to consider the kinematics as the primarily controlled variable. That is, the authors would rather view these inputs to motoneuron pools as the controlled variables themselves.

Assuming that supraspinal control of explosive movements such as vertical jumping is directly involved in the assembly of a system able to generate inputs to motoneuron pools instead of deriving these in some way from desired kinematics, how do people learn to do this? At one extreme, one might imagine the learning process to be analogous to the optimization process for the presented mathematical model. With very little constraints and no structure whatsoever, the control space is explored using a trial and error approach in an attempt to localize the point in that space that corresponds to maximal achievement. Although this is admittedly speculative, the results of Sanes and Donoghue (1993) may be interpreted so as to imply that within the primary motor cortex, new combinations of muscle activation patterns can be formed and tried out within minutes. This ability may well be correlated with what is usually referred to as the cognitive phase in motor learning (Schmidt 1988). Due to strong dependency on feedback, however, this type of control cannot be used to generate fast movements. In fact, it may be impossible to generate an actual jump under such a regime, a possible explanation of why jumping is acquired relatively late in ontogenetic development.

Alternatively, acquisition of top-level control may have to do with coordination and tuning of lower-level circuits that implement relations (synergies) between the activation of several muscles by themselves. Some evidence shows that during the learning of new skills, phylogenetically younger stories of the brain teach the older structures (such as the brain stem and the cerebellum) to realize pattern-generating circuits. Kennedy (1990) provided evidence that an important aspect of learning is that cortico-spinal control is replaced by rubrospinal control, especially with respect to timing, velocity, and force of the movement. Interestingly, the output of the red nucleus also controls the coupling of various sources of unspecified (visual, proprioceptive, cutaneous) information to the output of Purkinje cells in the cerebellum. This occurs via their rubro-olivary projections and the olivary control of the climbing fibers of these Purkinje cells. Given the massive inputs from various areas in the cerebral hemispheres onto the red nucleus neurons involved in the rubro-olivary projections, this system may implement a form of supervised learning where the younger neocortex acts as a teacher for the older parts of the brain (Stein 1986, 1992). Ultimately, this may give rise to a lower-level system of circuits that generates inputs onto motoneuron pools as a function of musculoskeletal properties, the current sensory input, and the desired movement intensity (e.g., jump height) specified by higher centers. Clearly, the above is nothing more than a possible framework indicating how skill acquisition might work. It is interesting to note, though, that this framework is largely consistent with the cognitive, associative, and automatized phases of learning (Schmidt 1988).

Conclusion

Bernstein viewed the motor-control system as a hierarchical structure of increasingly more adaptive systems where complexities are taken care of by subsystems as low as possible in the hierarchy. In agreement with this, the authors propose that motor control in general depends on a number of processes on top of each other. At the very lowest level, the chapter identified muscle's force-length-velocity relation as the zero-lag peripheral feedback mechanism that counteracts small perturbations to state and parameters that occur during movement. On top of that, spinal reflex loops are present that form feedback loops of more intricate nature at the cost, however, of time delays. On top of these systems, a supraspinal system of neural circuits is assumed to generate the inputs onto the motoneuron pools as a function of the properties of the musculoskeletal system, the external forces, sensory input, and the desired movement intensity. The relative importance of these different levels was argued to be task dependent. Slow movements depend more on neural feedback operating at different levels. Explosive movements depend primarily on muscle properties for their stability. Through modeling/simulation, it was shown that in explosive movements, the lowest level indeed suffices to stabilize against small state and parameter perturbations. On theoretical grounds, it was argued that feedback during movement is unlikely to be effective during these movements due to time delays. Thus, in case of explosive movements, the role of sensory input at the highest level is limited to signaling of the initial state. Furthermore, it was shown that small errors in the supraspinal inputs can dramatically affect explosive movements. Thus, these supraspinal inputs must be finely tuned to the properties of the musculo-skeletal system and the external forces.

References

Atkeson, C.G., Hollerbach, J.M. (1985) Kinematic features of unrestrained vertical arm movements. *Journal of Neuroscience,* 5: 2318–2330.

Bekoff, A., Nusbaum, M.P., Sabichi, A.L., Clifford, M. (1987) Neural control of limb coordination I. *Journal of Neuroscience,* 7: 2320–2330.

Bennett, D.J., Hollerbach, J.M., Xu, Y., Hunter, I.W. (1992) Time varying stiffness of human elbow joint during cyclic voluntary movement. *Exp. Brain Res.,* 88: 433–442.

Bernstein, N.A. (1967) *The Coordination and Regulation of Movements.* London, Pergamon Press.

Bernstein, N.A. (1996) On Dexterity and its Development. In *Dexterity and its Development.* M.L. Latash and M.T. Turvey (Eds.), Mahwah, NJ: LEA Publishers, 3-244.

Bizzi, E., Hogan, N., Mussa-Ivaldi, F.A., Giszter, S. (1992) Does the nervous system use equilibrium point control to guide single and multiple joint movements? *Beh. Brain Sci.,* 15: 603–613.

Bobbert, M.F., Soest, A.J. van (1994) Effects of muscle strengthening on vertical jump height: A simulation study. *Med. Sci. Sports Exerc.,* 26: 1012–1020.

Brown, I.E., Loeb, G.E. (1996). "Preflexes"—programmable, high-gain, zero-delay intrinsic responses of perturbed musculoskeletal systems. In *Proceedings of the 9th EFC Conference on Biomechanics and Neural Control of Movement.* P.E. Crago, J.F. Winters (Eds.), Mt. Sterling, Ohio: Engineering Foundation, 10-11.

Feldman, A.G., Levin, M.F. (1995) The origin and use of positional frames of reference in motor control. *Beh. Brain Sci.,* 18: 723–806.

Georgopoulos, A.P. (1988) Neural integration of movement: role of motor cortex in reaching. *FASEB Journal,* 2: 2849–2857.

Georgopoulos, A.P., Schwartz, A.B., Kettner, R.E. (1986) Neural population coding of movement direction. *Science,* 233: 1416–1419.

Gottlieb, G.L. (1993) A computational model of the simplest motor program. *Journal of Motor Behavior,* 25: 153–161.

Gottlieb, G.L., Agarwal, G.C. (1979) Response to sudden torques about ankle in man: Myotatic reflex. *Journal of Neurophysiology,* 42: 91–106.

Hasan, Z., Stuart, D.G. (1988) Animal solutions to problems of movement control. *Ann. Rev. Neurosci.* 11: 199–223.

Hatze, H. (1981) *Myocybernetic Control Models of Skeletal Muscle.* Pretoria: University of South Africa.

Hatze, H. (1996) Modeling aspects relating to a large scale model of the human neuromusculoskeletal system and its simulation responses to neural control perturbations. In *Proceedings of the 9th EFC Conference on Biomechanics and Neural Control of Movement.* P.E. Crago, J.F. Winters (Eds.), Mt. Sterling, Ohio: Engineering Foundation, 28.

Hill, A.V. (1938) The heat of shortening and the dynamic constants of muscle. *Proc. Roy. Soc., 126B:* 136–195.

Hogan, N. (1990) Mechanical impedance of single- and multi-articular systems. In *Multiple Muscle Systems.* J.M. Winters, S.L.Y. Woo (Eds.), New York: Springer-Verlag, 149–164.

Ingen Schenau, G.J. van (1989) From rotation to translation: Constraints on multi-joint movements and the unique action of bi-articular muscles. *Human Movement Sci.,* 8: 301–337.

Ingen Schenau, G.J. van, Beek, P.J., Bootsma, R.J. (1992) Is position information alone sufficient for the control of external forces? *Beh. Brain Sci.,* 15: 804–805.

Ingen Schenau, G.J. van, Soest, A.J. van, Gabreels, F.J.M., Horstink, M.W.I.M. (1995) The control of multi-joint movements relies on detailed internal representations. *Human Movement Sci.,* 14: 511–538.

Inman, V.T., Ralston, H.J., Saunders, J.B., Feinstein, B., Wright, E.W. (1952) Relation of human electromyogram to muscular tension. *Electroencephalography and Clininical Neurophysiology,* 4: 187–194.

Kennedy, P.R. (1990) Corticospinal, rubrospinal and rubro-olivary projections: A unifying hypothesis. *Trends in Neuroscience,* 13: 474–479.

Kirsch, R.F. (1996) Implementing posture and movement control strategies through the peripheral motor system: Intrinsic muscle properties and segmental reflexes. In *Proceedings of the 9th EFC Conference on Biomechanics and Neural Control of Movement.* P.E. Crago, J.R. Winters (Eds.), Mt. Sterling, Ohio: Engineering Foundation, 20-21.

Kuffler, S.W., Nicholls, J.G., Martin, A.R. (1984) *From Neuron to Brain.* Sunderland MA: Sinauer Associates.

Latash, M.L. (1993) *Control of Human Movement.* Champaign, IL: Human Kinetics.

Latash, M.L. (1996) The Bernstein problem: how does the central nervous system make its choices? In *Dexterity and its Development.* M.L. Latash and M.T. Turvey (Eds.), Mahwah, NJ: LEA Publishers, 277-304.

Latash, M.L., Gottlieb, G.L. (1991) An equilibrium point model for fast single-joint movement II. Similarity of single-joint isometric and isotonic descending commands. *Journal of Motor Behavior,* 23: 179–191.

Latash, M.L., Zatsiorsky, V.M. (1993) Joint stiffness: myth or reality? *Human Movement Sci,* 12: 653–692.

Loeb, G.E., Hoffer, J.A., Marks, W.B. (1985) Activity of spindle afferents from cat anterior thigh muscles III. Effects of external stimuli. *Journal of Neurophysiology,* 54: 578–591.

MacMahon, T.A. (1984) *Muscles, Reflexes and Locomotion.* Princeton, NJ: Princeton University Press, 148–161.

McIntyre, J., Bizzi, E. (1993) Servo hypotheses for the biological control of movement. *Journal of Motor Beh,* 25: 193–202.

Merton, P.A. (1953) Speculations on the servo-control of movement. In *Spinal Cord, CIBA Foundation Symposium 1952.* Wolstenholme, G.E.W. (Eds.), Churchill, 247–255.

Petrofsky, J.S., Phillips, C.A. (1980) The influence of temperature, initial length and electrical activity on the force-velocity relationship of the medial gastrocnemius muscle of the cat. *Journal of Biomechanics,* 14: 297–306.

Rack, P.M.H., Westbury, D.R. (1969) The effects of length and stimulus rate on tension in the isometric cat soleus muscle. *Journal of Physiology,* 204: 443–460.

Sanes, J.N., Donoghue, J.P. (1993) Organization and adaptability of muscle representations in performance in primary motor cortex. In *Control of Arm Movement in Space: Neurophysiological and Computational Approaches. Experimental Brain Series.* Vol. 22. R. Caminiti, R. et al. (Eds.), Berlin: Springer-Verlag, 103–127.

Schmidt, R.A. (1988) *Motor Control and Learning: A Behavioral Emphasis*. Champaign, IL: Human Kinetics.

Shadmehr, R., Mussa-Ivaldi, F.A. (1994) Adaptive representation of dynamics during learning of a motor task. *Journal of Neuroscience*, 14: 3208–3224.

Shampine, L.F., Gordon, M.K. (1975) *Computer Solution of Ordinary Differential Equations. The Initial Value Problem*. San Francisco: W.H. Freeman & Co.

Soechting, J.F. (1989) Elements of coordinated arm movements in three dimensional space. In *Perspectives on the Coordination of Movement*. S.A. Wallace (Ed.), Amsterdam: Elsevier Science, 47–83.

Soest, A.J. van, Bobbert, M.F. (1993) The contribution of muscle properties in the control of explosive movements. *Biol. Cybern.*, 69: 195–204.

Soest, A.J. van, Bobbert, M.F., Ingen Schenau, G.J. van (1994) A control strategy for the execution of explosive movements from varying starting positions. *Journal of Neurophysiology*, 71:1390–1402.

Stein, J.F. (1986) Role of the cerebellum in the visual guidance of movement. *Nature*, 323: 217–221.

Stein, J.F. (1992) The role of the cerebellum in calibrating feedforward control. *Beh. Brain Sci.*, 15: 798–799.

Tardieu, C., Tabary, J.C., Tardieu, G. (1968) Étude mechanique et electromyographique des réponses a differentes perturbations au maintien postural. *J Physiol (Paris)*, 60: 243–259.

Vos, E.J., Harlaar, J., Ingen Schenau, G.J. van (1991) Electromechanical delay during knee extensor contractions. *Med. Sci. Sports Exerc.*, 23: 1187–1193.

Wolpert, D.M. (1996) Untitled abstract. In *Proceedings of the 9th EFC Conference on Biomechanics and Neural Control of Movement*. P.E. Crago, J.F. Winters (Eds.), Mt. Sterling, Ohio: Engineering Foundation, 22-23.

Zajac, F.E. (1993) Muscle coordination of movement: A perspective. *Journal of Biomechanics*, 26: 109–124.

Index

Contributors

Alexey Alexandrov
Institute of Higher Nervous Activity and
　Neurophysiology
Academy of Sciences of Russia
Moscow, Russia

Randall Beer
Rehabilitation Institute of Chicago
Chicago, IL, U.S.A

David Burke
Prince of Wales Medical Research Institute
Sydney, Australia

Paul J. Cordo
Robert S. Dow Neurological Sciences Institute
Portland, OR, U.S.A.

Jules Dewald
Rehabilitation Institute of Chicago
Chicago, IL, U.S.A.

Josef M. Feigenberg
Gilo, Jerusalem, Israel

Anatol G. Feldman
Research Center
Rehabilitation Institute of Montreal
University of Montreal
Montreal, Canada

Simon C. Gandevia
Prince of Wales Medical Research Institute
Sydney, Australia

Dina Ghodsian
Department of Psychology
University of California
Los Angeles, CA, U.S.A.

Stan Gielen
Department of Medical Physics
　and Biophysics
University of Nijmegen
Nijmegen, The Netherlands

Joseph Given
Rehabilitation Institute of Chicago
Chicago, IL, U.S.A.

Victor S. Gurfinkel
Institute for Information Transmission
　Problems
Russian Academy of Science
Moscow, Russia

John-Paul Hales
Prince of Wales Medical Research Institute
Sydney, Australia

Herbert Heuer
Abteilung Arbeitsphysiologie
Institut für Arbeitsphysiologie an der
　Universität Dortmund
Dortmund, Germany

J.A. Scott Kelso
Program in Complex Systems
　and Brain Sciences
　Center for Complex Systems
Florida Atlantic University
Boca Raton, FL, U.S.A.

Nam-Gyoon Kim
Center for the Ecological Study of Perception
　and Action
University of Connecticut
Storrs, CT, U.S.A.

Lev P. Latash
Chicago, IL, U.S.A.

Mark L. Latash
Pennsylvania State University
University Park, PA, U.S.A.

Jean Massion
Laboratoire de Neurobiologie et Mouvements
CNRS
Marseille, France

Jennifer C. McDonagh
Department of Physiology
University of Arizona College of Medicine
Tucson, AZ, U.S.A.

W. Zev Rymer
Rehabilitation Institute of Chicago
Chicago, IL, U.S.A.

Richard A. Schmidt
Human Performance Group
Failure Analysis Associates, Inc., and
Department of Psychology
University of California
Los Angeles, CA, U.S.A.

Douglas G. Stuart
Department of Physiology
University of Arizona College of Medicine
Tucson, AZ, U.S.A.

Esther Thelen
Department of Psychology
Indiana University
Bloomington, IN, U.S.A.

M.T. Turvey
Center for the Ecological Study of Perception
    and Action
University of Connecticut
Storrs, CT, U.S.A.

Bauke van Bolhuis
Department of Medical Physics
    and Biophysics
University of Nijmegen
Nijmegen, The Netherlands

Gerrit Jan van Ingen Schenau
Faculty of Human Movement Sciences
Free University
Amsterdam, The Netherlands

A.J. "Knoek" van Soest
Faculty of Human Movement Sciences
Free University
Amsterdam, The Netherlands

Sylvie Vernazza
Laboratoire de Neurobiologie et Mouvements
CNRS
Marseille, France

Erik Vrijenhoek
Department of Medical Physics
    and Biophysics
University of Nijmegen
Nijmegen, The Netherlands

Mario Wiesendanger
Laboratory of Motor Systems
Department of Neurology
University of Berne
Berne, Switzerland

Douglas E. Young
Human Performance Group
Failure Analysis Associates, Inc., and
Department of Kinesiology
California State University
Long Beach, CA, U.S.A.

About the Editor

Mark L. Latash, PhD, is an associate professor of kinesiology at Penn State University. Since the 1970s, he has worked extensively in the areas of normal and disordered motor control. His work has included animal studies, human experiments, modeling, and clinical studies.

Latash chaired the organizing committee of the international conference, "Bernstein's Traditions in Motor Control," which took place at Penn State in August of 1996. Chapters of *Progress in Motor Control, Volume 1* were written by invited speakers at the conference.

The author of *Control of Human Movement* (Human Kinetics, 1993), Latash also translated Bernstein's classic, *On Dexterity and its Development* (Erlbaum), in 1996.

Latash earned a master's degree in physics of living systems from the Moscow Physico-Technical Institute in 1976 and a PhD in physiology from Rush University in 1989. He is a member of the Society for Neuroscience and the American Society of Biomechanics.

Latash lives in State College, Pennsylvania. His leisure activities include spending time with friends, playing guitar and singing, and reading.

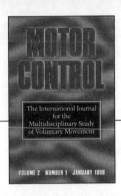

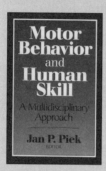
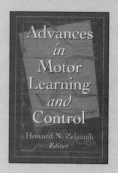